Urban China in Transition

Studies in Urban and Social Change

Published by Blackwell in association with the *International Journal of Urban and Regional Research*.

The Blackwell Studies in Urban and Social Change aim to advance debates and empirical analyses stimulated by changes in the fortunes of cities and regions across the world. Topics range from monographs on single places to large-scale comparisons across East and West, North and South. The series is explicitly interdisciplinary; the editors judge books by their contribution to intellectual solutions rather than according to disciplinary origin.

Published

Urban China in Transition
John R. Logan (ed.)

Getting Into Local Power: The Politics of Ethnic Minorities in British and French Cities
Romain Garbaye

Cities of Europe
Yuri Kazepov (ed.)

Cities, War, and Terrorism
Stephen Graham (ed.)

Cities and Visitors: Regulating Tourists, Markets, and City Space
Lily M. Hoffman, Susan S. Fainstein, and Dennis R. Judd (eds)

Understanding the City: Contemporary and Future Perspectives
John Eade and Christopher Mele (eds)

The New Chinese City: Globalization and Market Reform
John R. Logan (ed.)

Cinema and the City: Film and Urban Societies in a Global Context
Mark Shiel and Tony Fitzmaurice (eds)

The Social Control of Cities? A Comparative Perspective
Sophie Body-Gendrot

Globalizing Cities: A New Spatial Order?
Peter Marcuse and Ronald van Kempen (eds)

Contemporary Urban Japan: A Sociology of Consumption
John Clammer

Capital Culture: Gender at Work in the City
Linda McDowell

Cities After Socialism: Urban and Regional Change and Conflict in Post-Socialist Societies
Gregory Andrusz, Michael Harloe and Ivan Szelenyi (eds)

The People's Home? Social Rented Housing in Europe and America
Michael Harloe

Post-Fordism
Ash Amin (ed.)

Free Markets and Food Riots
John Walton and David Seddon

Fragmented Societies
Enzo Mingione

Urban Poverty and the Underclass: A Reader
Enzo Mingione

Forthcoming

Eurostars and Eurocities: Free Movement and Mobility in an Integrating Europe
Adrian Favell

Cities and Regions in a Global Era
Alan Harding (ed.)

Urban South Africa
Alan Mabin

Urban Social Movements and the State
Margit Mayer

Social Capital Formation in Immigrant Neighborhoods
Min Zhou

Urban China in Transition

Edited by
John R. Logan

BLACKWELL PUBLISHING
350 Main Street, Malden, MA 02148-5020, USA
9600 Garsington Road, Oxford OX4 2DQ, UK
550 Swanston Street, Carlton, Victoria 3053, Australia

First published 2008 by Blackwell Publishing Ltd

1 2008

Library of Congress Cataloging-in-Publication Data

Urban China in transition / edited by John R. Logan.
p. cm. – (Studies in urban and social change)
Includes bibliographical references and index.
ISBN 978-1-4051-6145-9 (hardcover : alk. paper) – ISBN 978-1-4051-6146-6 (pbk. : alk. paper)
1. Urbanization–China. 2. Rural-urban migration–China. 3. Sociology, Urban–China. I. Logan, John.
HT384.C6U73 2007
307.760951—dc22

2007017679

A catalogue record for this title is available from the British Library.

Set in 10.5/12 pt Baskerville
by SPi Publisher Services, Pondicherry, India.

The publisher's policy is to use permanent paper from mills that operate a sustainable forestry policy, and which has been manufactured from pulp processed using acid-free and elementary chlorine-free practices. Furthermore, the publisher ensures that the text paper and cover board used have met acceptable environmental accreditation standards.

For further information on
Blackwell Publishing, visit our website:
www.blackwellpublishing.com

Contents

Notes on the Contributors

Simon Appleton is Associate Professor of Economics at Nottingham University. He has worked mainly on issues of poverty, labor markets, health, and education in Africa and China. He is currently working with an international team of researchers to analyze health and migration in China.

Yanjie Bian is a Professor of Sociology at the University of Minnesota. He specializes in structural sociology, with a special interest in social stratification and mobility, economic sociology, social networks, and contemporary Chinese societies in East Asia.

Guoxuan Cai is Director of the Institute of Sociology, Guangzhou Academy of Social Sciences, China.

Chun-Ming Chen is Associate Professor of Public Policy and Management at Shih Hsin University, Taipei, Taiwan. His fields of research are democratic administration and public consultation, civil–military relations. He has conducted both nation-wide opinion surveys concerning *lizhan's* function and position and a citizen consensus conference on tax reform in Taiwan.

Yiu Por (Vincent) Chen is an Assistant Professor at Public Services Graduate Program, DePaul University. His research interests are labor migration, urbanization, the political economy of development, institutional analysis, and spatial inequality. His research includes the examination of rural labor mobility in China, higher educational labor mobility from China to other places, the urbanization process and rural-urban inequality in China.

Susan S. Fainstein is Professor in the urban planning program at the Harvard University Graduate School of Design. Her teaching and research have focused on comparative urban public policy, planning theory, and urban redevelopment.

C. Cindy Fan is Professor of Geography and Professor and Chair of Asian American Studies at University of California, Los Angeles. Her research interests center on the regional and social dimensions of transitional economies, focusing on regional policy, inequality, labor migration, marriage migration, gender, and the urban system in post-Mao China. She is currently working on a collaborative project with Chinese scholars on rural-urban migration and urban-rural return migration.

Theodore P. Gerber is Professor of Sociology and Director of the Center for Russia, East Europe, and Central Asia at the University of Wisconsin-Madison. His research examines socioeconomic stratification, labor markets, demographic processes, public opinion, institutional change, HIV/AIDS, and science in contemporary Russia. He has pulished widely and has edited several volumes. He has developed (in whole or in part) 16 large-sample surveys in Russia and Ukraine since 1998, and has also conducted numerous focus groups and in-depth interviews.

Youqin Huang is an Assistant Professor in the Department of Geography and Planning at the Center for Social and Demographic Analysis, University at Albany, SUNY.

Graeme Hugo is Federation Fellow, Professor of the Department of Geographical and Environmental Studies and Director of the National Centre for Social Applications of Geographic Information Systems at the University of Adelaide. His research interests are in population issues in Australia and South East Asia, especially migration.

Susanne Karstedt is Professor of Criminology at Keele University, United Kingdom. She has widely published on comparative criminology, cross-cultural criminology and crime in post-communist countries, and has conducted cross-national and cross-cultural research on corruption and crimes of the respectable, democracy and violence, and indicators of punishment.

Zai Liang is Professor of Sociology at State University of New York at Albany and Co-director of Urban China Research Network. His current research projects include residential mobility, migration and rural transformation, and migration and minority opportunities in China.

Jianhong Liu is Professor of Sociology at Rhode Island College. He is Vice Chairman of Academic Committee and Co-director of Center for Drug Crime and Public Policy at Southwestern University of Political Science and Law, China. He is currently conducting a survey on Drugs and Community in China funded by Southwestern University of Political Science and Law.

John R. Logan is Professor of Sociology and Director of the initiative on Spatial Structures in the Social Sciences at Brown University. His books include *Urban Fortunes: The Political Economy of Place* (California 1987) and *The New Chinese City: Globalization and Market Reform* (Blackwell 2002). He also founded the Urban China Research Network, supported by the Andrew W. Mellon Foundation.

Setha M. Low is Professor of Environmental Psychology, Anthropology and Women's Studies at the Graduate Center of the City University of New York and Director of the Public Space Research Group. She has completed ethnographies of plazas in Latin America and Western Europe, gated communities in New York, Texas and Mexico, and urban and hertiage parks in the United States.

Hanchao Lu is Professor of History at the Georgia Institute of Technology, where he has served as the Director of Graduate Studies in the School of History, Technology, and Society since 2004. He served as the president of the Chinese Historians in the United States (CHUS) and has been an honorary Senior Research Professor at the Shanghai Academy of Social Sciences since 1995. He is the co-editor of the refereed journal, *Chinese Historical Review* and the editor of a 12-volume series, *The Culture and Customs of Asia*. Among his recent publications are two award-wining books, *Beyond the Neon Lights: Everyday Shanghai in the Early Twentieth* Century (California, 1999/2004) and *Street Criers: A Cultural History of Chinese Beggars* (Stanford, 2005).

Hy Van Luong is Professor of Anthropology at the University of Toronto. He has conducted regular comparative fieldwork in different parts of Vietnam, and has published extensively on discourse, gender, social capital, and political economy in Vietnam.

Steven F. Messner is Distinguished Teaching Professor of Sociology, University at Albany, SUNY. His research focuses on social organization and crime, the spatial patterning of crime, and crime and social control in China.

Kaoru Nabeshima is a consultant for the Development Economics Research Group at The World Bank and has written on development issues such as education, technology transfer, and privatization of state-owned firms. His research interests lie in innovation capabilities of firms and industrialization in East Asia.

Benjamin L. Read is Assistant Professor of Political Science at the University of Iowa. His research focuses on grassroots organizations in China and Taiwan, particularly those with institutionalized links to government.

Joanna Regulska is Professor and Chair of Women's and Gender Studies, and Professor of Geography. She also serves as the Director of the International Programs, School of Arts and Sciences, Rutgers University. Her work focuses on women's agency, political activism, grassroots mobilization, and construction of women's political spaces in central and east Europe, and Caucasus.

Emily Rosenbaum is Professor of Sociology at Fordham University in New York City. She has written extensively in the areas of urban demography, immigration, and inequality. Her most recent publications include *The Housing Divide: How Generations of Immigrants Fare in New York's Housing Market* (NYU Press, 2007), co-authored with Samantha Friedman.

Jennifer Rudolph is Associate Professor of History at the Worcester Polytechnic Institute. She is one of the founding members of the Urban China Research Network. Among her recent publications is *Negotiated Power in Late Imperial China: The Zongli Yamen and the Politics of Reform* (2007). She is currently working on a book project that examines the impact of colonialisms in Taiwan and Taipei.

Christopher J. Smith is Professor of Geography and Planning at the University Albany (SUNY), and is also jointly appointed in the department of East Asian Studies. He is an urban geographer whose research interests are in the area of urban social problems, focusing on the implications of modernization and economic development in contemporary China. His most recent work has involved a consideration of the relationship between migration and the spread of HIV/AIDS in China.

Lina Song is Reader at the School of Sociology and Social Policy in Nottingham University. She has conducted studies of income inequality, migration, labour markets, the urban-rural divide and intra-household allocation in China. She is currently researching social discontent in China and working with an international team of researchers to analyze health and migration in China.

Michael J. White is Professor of Sociology at Brown University, where he is also Director of the Population Studies and Training Center. He was a member of the US National Academy of Sciences Panel on Urban Population Dynamics, and he has been a fellow of the Woodrow Wilson International Center for Scholars. He has served on the Board of Directors of the Population Association of America, and was recently chair of the NICHD Population Studies Committee. Currently, he serves as president of the Association of Population Centers, a US-based consortium of research centers engaged in demographic research and training. He serves on the scientific advisory committee of the INDEPTH network of demographic surveillance systems in developing countries. He works broadly on topics of population distribution and urban sociology. His other current work on China includes a study of temporary migration patterns.

Fulong Wu is Professor of East Asian Planning and Development and the Director of the Urban China Research Centre at the School of City and Regional Planning of Cardiff University. His recent research is in urban poverty and transition. He is co-author (with Jiang Xu and Anthony Gar-On Yeh) of *Urban Development in Post-Reform China: State, Market, and Space* (Routledge, 2007).

Weiping Wu is Associate Professor of Urban Studies, Geography and Planning at Virginia Commonwealth University, and a frequent consultant to the World Bank. She conducts research and publishes widely in the areas of urban economic geography, local innovation and university-industry linkage, migrant housing and settlement, and China's urban development.

Shahid Yusuf is Economic Adviser, Development Economics Research Group at The World Bank and is the team leader for the World Bank-Japan project on East Asia's Future Economy. He has written extensively on development issues. His research interests lie in the area of urban development, innovation systems, and industrialization in East Asia.

Min Zhou is Professor of Sociology and Asian American Studies at the University of California, Los Angeles.

Yixing Zhou is Professor and former Chair of the Department of Geography at Peking University. He is a council member of the Chinese Association of Geographers and a board member of China Planning Association, Regional Planning Research Committee, China Urban Economics Association, and China Urban Economics Society. He is on the editorial boards of several journals and he is the author of over a hundred publications, including *Urban Geography* and *Spatial Concentration and Diffusion of China's Costal Urban Regions*.

Series Editors' Preface

The Blackwell *Studies in Urban and Social Change* series aims to advance theoretical debates and empirical analyses stimulated by changes in the fortunes of cities and regions across the world. Among topics taken up in past volumes and welcomed for future submissions are:

- Connections between economic restructuring and urban change
- Urban divisions, difference, and diversity
- Convergence and divergence among regions of east and west, north, and south
- Urban and environmental movements
- International migration and capital flows
- Trends in urban political economy
- Patterns of urban-based consumption

The series is explicitly interdisciplinary; the editors judge books by their contribution to intellectual solutions rather than according to disciplinary origin. Proposals may be submitted to members of the series Editorial Committee:

Acknowledgments

This book has a history, and to acknowledge properly the people who contributed to it requires that I say something about its origins.

I would like to thank the University at Albany for its support of the Lewis Mumford Center for Comparative Urban and Regional Research during the years that I served as Director (1999–2004). The Mumford Center has proved to be an effective mechanism to promote urban research.

Next I wish to recognize the substantial financial support provided by the Andrew W. Mellon Foundation's Urban Studies and Demography Program (2000–2005). Mellon's funding enabled our Albany team to establish the Urban China Research Network with the purpose of stimulating research on China, particularly through grants to young scholars and annual conferences to showcase their work.

A supplementary grant from the Mellon Foundation in 2002 was intended to bring several of the most eminent China scholars together with specialists on other countries through a series of three workshops. After the first workshop we decided to focus on a book project with two main features. First, it would provide the most recent evidence on the directions of urban change as seen through the eyes of a China specialist and a non-specialist. Second, it would seek to interpret changes within a broader theoretical framework that could be useful in studies of other countries. I would like to thank my Albany colleagues who helped to sharpen and carry out that plan: Chris Smith and Youqin Huang (Geography and Planning), Steve Messner and Zai Liang

(Sociology), and Jennifer Rudolph (History). Each of these colleagues has also contributed as co-author of a chapter in this volume.

A final acknowledgment is due to the other scholars who had a hand in this process, including colleagues around the world who provided reviews of the early drafts of chapters and contributors who acceded to my persistent prodding to do better, to give more attention to the theories, and to give clearer explanations while reducing their word count. I give a special thanks to Susan Fainstein, whose experience and wisdom add much to our jointly authored introductory chapter.

John R. Logan

Introduction: Urban China in Comparative Perspective

John R. Logan and Susan S. Fainstein

We may still, at the beginning of the twenty-first century, think of China as a developing country. Almost certainly, however, this perspective will soon seem anachronistic. China is becoming a world power, and its major cities are becoming world cities. This historic and rapid transformation makes it hard for us to interpret the changes that we see today. To what do we compare China? Is China like any other country in our experience? Through what lens will the many transformations going on in the country make the most sense?

Narrowing our focus to urban China does little to resolve the issue of appropriate comparisons. We approach urban change as having many dimensions, and the contributors to this book deal with topics as dissimilar as rural-urban migration and the position of women, urban crime and suburbanization, neighborhood inequality and economic competitiveness. The chapters compare Chinese cities variously to US urban areas, to formerly socialist communities of Eastern Europe, and to megalopolises of the developing world. Each of these comparisons offers insights but also poses different questions about the trajectory that we foresee for Chinese urban development.

This is the specific contribution of the essays in this volume. Underlying the diversity of "urban" topics and international cases, all the contributors are grappling with the same theoretical question. How should we understand what is happening in China today? This is a hard question and a kind of question that specialists on any part of the world tend to avoid. The more you know the details of a country, its history, culture, and politics, the greater the temptation to reject all general theories. This has been a particular problem for social science research on China. John Friedman's recent overview of "China's urban transition" (2005) reflects a common reluctance on the part of China scholars to generalize. China, he argues, cannot be fitted neatly into the narrative of any grand theory, whether that be the narrative of modernization or globalization, urbanization or national integration – certainly not yet (because the future is so rapidly being made) and perhaps never (because China is not just another country, but a civilization that deserves to be understood on its own terms).

We take the opposite tack. If no single theory can explain everything, we look for ways to combine theories. China is important not only for itself but because it has lessons for the rest of the world. It offers both positive and negative examples of how economic expansion can be promoted in a late-developing country, how new forms of competition and partnership with multinational corporations and world powers can be negotiated, how the state can keep a strong hand in a market economy in a post-socialist era. Our purpose is not only to share the latest insights into how China is evolving, but to make sense of those shifts so that China can become a key case in twenty-first-century thinking about global change.

Here is our theoretical repertoire. From a modernization perspective, China is either a unique example of successful development or a case study in the costs of too rapid change. From a world system perspective it demonstrates the imperative of national autonomy, the impact of global economic and cultural connections, or the persistence of colonial heritage. The more specific concept of the developmentalist state turns our attention to the elite's capacity to direct and control change. The analogy to other cases of post-socialist transition raises the question of whether there was a socialist urban form and how the state's centrality in urban development differs in these cases from its role in a market system.

The mix of theories on which we draw throughout this volume varies from chapter to chapter. One clear common thread is that we all rely on more than one line of thought. Another is the surprising usefulness of modernization theory to many of us. Only recently, after all, China was one model of traditional society, and questions about how it found a path to sustained development ("take-off" in the classic vocabulary) and how it copes with the strains of rapid social change show up in several chapters. But if we grapple with the key questions of modernization theory, our answers are informed by an acute awareness of how state action has directed, created the conditions for, or intervened in changes in the society. The same party that created a twentieth-century state apparatus has overseen more than five decades of social engineering, economic planning, and reorganization of the mechanisms of control. Whatever our theoretical inclination, the state is a central player in every account.

Modernization

One theoretical starting point is the neoclassical development literature. In the optimistic years of the 1950s and 1960s, many observers expected that the historical experience of the Western industrial countries provided a model for the future of the developing world. "Modernization" was understood as a natural process of diffusion of capital, technology, and social organization (Slater 1986). Modern societies were understood to be urban, because cities served as the catalyst for the transformation of agricultural into industrial societies.

One aspect of modernization is the emergence of a market economy, including adoption of modern practices that are now widespread among corporations in industrialized countries, as discussed by Yusuf and Nabeshima (Chapter 1). China's large and midsized state enterprises provide a valuable perspective on this process, because of their continuing salience in the Chinese economy and their attempt to straddle two different systems. On the one hand, these enterprises are only reluctantly relinquishing the attributes of the socialist period in which these firms played an important role in providing steady employment and social services (as well as maintaining social control) for a large share of the urban workforce. Their privileged access to financing and close ties to government agencies historically protected them from competition. On the other hand, these enterprises are acquiring many of the characteristics of capitalist firms and are creating new forms of partnerships with local governments as a result of the push to reform and competition both from the domestic private and collective sectors and from foreign firms. Yusuf and Nabeshima are optimistic that the stronger trend is toward modernity.

Another aspect of modernization is the relationship between urbanization and demographic transition. Modernization theorists expected urbanization to proceed as part and parcel of industrial development, so that the growth of population would be absorbed into the urban economy. As the population became more mobile, it would also become more educated and capable, and this in turn would create the conditions under which natural growth rates would fall to a manageable level. In much of the Third World this premise turned out to be wrong (Kasarda and Crenshaw 1991; Preston 1979). Urbanization quickly outpaced industrial development in many countries (Hoselitz 1962). Too rapid rural-urban migration threatened to saturate city labor markets, truncate opportunity structures in rural areas, overburden public services, and ultimately retard economic growth. Compared to the West European experience, mortality declined earlier and more quickly than fertility, and urban populations consequently swelled from both rural-urban migration and natural increase (Beier 1976). Under those conditions, cities could not meet the demand for infrastructure, housing, social services, and employment.

China differs from most Third World countries in its relatively low level and modest pace of urbanization. Chen and Parish (1996) estimate that by the mid-1990s China was still only about 30 percent urban (not counting large numbers of rural residents who lived within the formal boundaries of a city or town). At the same time the country's economic expansion in the post-1978 reform period has been spectacular. The urban share of the population has risen more slowly than non-agricultural job growth since the 1970s. Hence, through the prism of modernization theory urban China might be seen to represent an ideal model of change, the exceptional case that has avoided the cultural lags and distortions that disrupted progress elsewhere.

Contrary to modernization theory, the creation of a market economy has been not the result of contact with the more advanced industrial societies but of an explicit policy decision by the state. The current low level of urbanization also results from governmental controls on both migration and population growth rather than from a naturally occurring demographic transition (see Liang, Van Luong, and Chen, Chapter 9). On the other hand, Chinese cities certainly incorporate a constellation of changes that reflect the "passing of traditional society" as Lerner (1958) understood it – industrial growth, urbanization, emergence of an independent middle class, rising levels of education and awareness of the external world, shifting gender and family patterns. Rapid change also raises concerns about imbalance across these dimensions and social discontinuity. For example Friedmann (2005) regards Chinese society as uneasily poised between the two poles of anarchy and a reimposition of totalitarian rule leading to stasis. Analyses in this volume on urban crime (Chapter 12) and on AIDS and sexually transmitted diseases (Chapter 13) give special attention to the issue of maintaining social control in a rapidly changing country.

Dependency and World System Theory

Theories of dependent development and the world system provide another prism through which to analyze urban change in China. Theorists in this tradition, which arose largely as a backlash to modernization theory, do not see over-urbanization (and associated phenomena of primate cities, squatter settlements, underemployment, and informal economies) as an aberration from a "normal" course of development. Rather, they adopt a more critical view of the impacts of industrial countries on Third World societies, many of which were once colonies. From this perspective uneven and imbalanced growth is endemic to capitalism. Relations between countries are inherently exploitive, organized to mine the resources (initially, literally from mines and plantations) of agricultural societies for the benefit of the metropole. In André Gunder Frank's (1969) expressive terminology, contact with the West stimulated a process of the development of underdevelopment in the Third World. Impoverishment of the countryside caused by the introduction of more productive, export-oriented agriculture pushed surplus labor off the land into cities, where migrants could survive only at marginal subsistence levels; foreign investment and foreign trade were encapsulated within enclaves that did not reinforce autonomous economic growth. Hence, stronger connections with the world economy, rather than stimulating development, generated excessive urban growth and retarded national economic improvement (Bradshaw 1987, Evans and Timberlake 1980).

A dependency perspective leads to a focus on the economic and political relationships between China and the rest of the world. These have undergone profound shifts over time. In the nineteenth century a weakened Qing Dynasty was forced to share sovereignty over most major cities with foreign powers in the aftermath of the First Opium War with Britain (1839–42). Many cities, though not Beijing itself, were designated as treaty ports and opened to foreign trade and investment on terms dictated from the outside. They became in a sense colonial cities, with a duality typical of such places, "not a single unified city, but, in fact, two quite different cities, physically juxtaposed but architecturally and socially distinct" (Abu-Lughod 1965, see also King 1976). Once established, the juxtaposition of foreign enclaves with preindustrial zones in port cities became an enduring feature of their built environment. In their contribution to this volume, Rudolph and Lu (Chapter 7) assess how this colonial history has been incorporated into the contemporary identities of Chinese cities.

With the reunification of the country under the Communist Party after World War II, China maintained a defensive posture toward the West and had tense relations with the Soviet Union. There were many internal struggles within the Party over economic development and urbanization policies after 1949 (Mingione 1977). During the period of the Cultural Revolution, policy was aimed at creating self-reliant domestic development. The government pursued an anti-urban agenda, seeking to decentralize production and reduce inter-regional and rural-urban inequalities, with a large share of the cost financed by transfers from historically more productive centers like Shanghai. Chen and Parish (1996) identify this as the "distinctive" period of China's urban and regional form, when state power was used aggressively to restrict city growth. Rural-urban migration was regulated, and undocumented migrants did not have access to urban housing or food rations. In addition, as in other socialist countries at this time, the state discouraged consumption, investing little in housing and maintaining an unusually small service sector (Murray and Szelenyi 1984).

Most contributors to this volume have little positive to say about the uses of state policy in the socialist period. Messner, Liu, and Karstedt (Chapter 12) argue that state controls create the risk of an institutional imbalance that promotes pervasive corruption, even if the state succeeds in moderating other forms of crime. The *hukou* (registration) system that was used to control migration is now seen as long past its usefulness, a remnant of the old regime that needlessly obstructs migrants' integration into the city in a period when all agree that a growing urban labor force is needed.

Other major initiatives of this period – communal agriculture, forced relocation of urban youth to the countryside, and experiments in creating primary industries in rural areas – have been judged in retrospect to be failures. But from a dependency theory perspective the creation of an effective, autonomous central state bureaucracy with the capacity to carry out national policies at the local level distinguishes China from most developing countries. In Wallerstein's (1976) analysis of the capitalist world system, development of an efficient national state was the critical achievement in those countries that became the dominant core, including England and France. Here a modernizing bourgeoisie was able to mobilize state power in support of commercial and industrial development. Weak states in the periphery, such as much of Eastern Europe, ended up as exporters of raw materials and foodstuffs. Cardoso and Faletto (1979) make a similar distinction between weaker and stronger states in Latin America. The least developed Central American economies had the same peripheral role (mining and plantation agriculture), and a similar political dependence on foreign powers. More dynamic processes could be found in countries like Argentina, Brazil, and Chile, where a more developed civil society provided better prospects for relatively progressive and independent national policies.

Theories of the Developmentalist State

The pessimism of the dependency theorists was challenged by the success of the newly industrializing countries (NICs) of East Asia and to some degree by Latin American achievements (Evans 1995, Woo-Cummings 1999). Haggard (1990), in his investigation of the Asian NICs, attributes their rapid rise to land reform resulting in increased agricultural productivity, a relatively egalitarian distribution of income, a strong state able to suppress internal discord, and investment in education. Amsden (2001) stresses similar factors, and like Haggard argues that governmental priority to the development of export industries was key to growth. She notes the effect of Japan's economic success on those countries that followed in its path: "Inductive role models may influence economic policy-making more than (or as much as) deductive abstract theories. The influence of a role model was striking in the case of East Asia's highly successful export promotion policies … based on Japan's past trade regime" (Amsden 2001, p. 290).

Once it reversed its earlier course of separating itself from the global economy, China followed a similar model and can be described as a developmentalist state after the 1978 reforms. For most contributors to this volume, this political breakthrough is the key to understanding the present, although most also express concerns about lingering policies associated with the old order.

During the reform period the national government has undertaken a series of structural changes that eventually enabled the reincorporation of China's economy into the capitalist world system. As in the successful NICs, its economic strategy depended on agricultural surpluses that enabled urban development and investment in higher education to produce an educated elite that could both staff the government offices directing the economy and provide technical skills to industry. Although the modernization theorists had envisioned unimpeded market processes as the engine of growth, the NICs relied on state designation of export industries worthy of investment to foster economic development. The utility of export-oriented production as the means for economic expansion in China and elsewhere resulted from changing opportunities within the international economy. Western firms have increasingly moved to outsourcing and subcontracting to reduce their costs of production, thus creating new markets for industrial production in formerly peripheral areas. The developmentalist states, while far more active than modernization theory had anticipated, did succeed, as we commented earlier, in achieving one of the expectations of modernization theory – that cities would be capable of absorbing rural outmigration.

Most important, China has been able to rely on sources of capital unavailable to other developing countries:

> Overseas Chinese business networks are … the main intermediaries between global capital, including overseas Chinese capital, and China's markets and producing/exporting sites. … *China's multiple link to the global economy is local, that is, it is performed through the connection between overseas Chinese business and local provincial government in China,* the *sui generis* capitalist class … [of] "bureaucratic entrepreneurs." (Castells 2000, p. 317, emphasis in original; see also McGee 1986)

China's efforts thus have been facilitated by the presence of a Chinese investment community, which is deeply rooted in Hong Kong, Singapore, and Taiwan. Hong Kong in particular has played an important role in nearby South China. As its own wage structure became uncompetitive in export-oriented manufacturing, Hong Kong investors increasingly made use of familial and political connections in Guangdong Province to transfer production there. In this way even small village-centered enterprises became tied to global commodity chains.

Property markets in China have also captured substantial capital investment from abroad, partly searching for speculative profits and partly in recognition of the long-term prospects for appreciation, especially in the more cosmopolitan coastal cities. China was a safe haven for investments during and after the Southeast Asian economic crisis in the late 1990s. And in an unusual twist on the typical Third World situation, China's export earnings and attractiveness to foreign direct investment have made it a capital exporter. Even while maintaining a high rate of spending on infrastructural projects and supporting a rapid improvement in living standards, China is now a major investor in capital markets in the United States, lending money that allows the US to continue to run large deficits in its balance of payments. In turn, this strategy permits American firms to purchase Chinese goods.

Despite China's evident economic success, some of the social changes that we now observe in China resemble typical processes of dependent development (Dos Santos 1970) – excessive investment in port cities, increasing regional and rural-urban inequalities, marginalization of a large sector of the urban population, and repressive labor policies. Yet from a dependency perspective, China is an example of a country that is succeeding in reframing its relationship to the world system. This calls attention to the exercise of state power and to the geopolitical circumstances in which this was possible. Political issues arise in important ways at the local level. One question is the capacity of local government officials to deal with global investors (see Yusuf and Nabeshima, Chapter 1). Another is the capacity of the state to maintain a high degree of internal political control. Read and Chen (Chapter 14) show that the state's efforts on this score reach deeply into urban neighborhoods. Others (Messner et al., Chapter 12, and Smith and Hugo, Chapter 13) point out that rapid change nevertheless undermines the social order, a typical expectation of modernization theorists.

Transition from Socialism

Modernization theory, dependency/world system theory, and the theory of the developmental state have broad applicability throughout the Third World. China, however, differs from most countries of the "global South" in having moved from a thoroughgoing socialist system to one of quasi-capitalism. Consequently two more theoretical frames can be applied to changes there, those of "socialist urbanism" and "market transition."

A relatively small body of literature analyzed neighborhood inequalities in the "socialist city" (Andrusz, Harloe, and Szelényi 1996; French and Hamilton 1979) and its subsequent transformation (Dangschat and Blasius 1987; Daniell and Struyk 1994; Kovacs 1994; Ladanyi 1989; Musil 1993; Sykora 1999). China has moved more slowly toward dismantling socialist institutions than did the former Soviet bloc. It is still in a situation of "partial reform," although Nee (Nee and Matthews 1996) believes state sector enterprises (representing the centrally planned economy) are being rapidly displaced by the private sector (representing a free market economy). His theoretical imagery envisions these two sectors as remaining separate, one shrinking while the other grows; at the point sometime in the future that the economy becomes predominantly private, reform will have been completed.

This same logic shows up in other domains. Housing has increasingly become a market commodity, and many city residents have been able to purchase their apartments outright. Housing construction also is being marketized. Real estate development firms have been created with the mission of building for a profit; land that was formerly simply allocated for new projects is now bought and sold. The result, potentially, is a new process of urban development and housing allocation, the replacement of a socialist city by a market-driven form.

State influence over economic development nevertheless continues. Work units and other actors in the state sector increasingly must take market processes into account, but at the same time, ostensibly "private" transactions often have a strong public connection, as evidenced by the occasional punishment of high officials using their insider status for profit. This interpenetration, however, is not simply a sign of corruption. The fact is, in China it is the state, embodied in the Central Committee of the Communist Party, which decided to embark on the path of market reform:

> China's economic development and technological modernization, within the framework of the new global economy, were (are) pursued by the Chinese communist leadership both as an indispensable tool for national power, and as a new legitimacy principle of the Communist party. In this sense, Chinese communism in the early twenty-first century represents the historical merger of the developmental state and the revolutionary state. (Castells 2000, pp. 312–13)

Thus, choices made at every step of the way are marked by the state's continuing intervention. The persistence of Communist Party rule implies the continuation of an economy where private interests are interpenetrated with public agencies. As will be discussed below, this phenomenon is evidenced by local growth coalitions that in some respects mirror those found in the US and Western Europe but with a more prominent state role than found in those countries.

Older mechanisms of urban adaptation in China have ceased, but the new system is still in the process of formation. Although the system of land use and development has become increasingly market driven, the public sector continues to play a role more reserved to private entrepreneurs in capitalist societies. Furthermore, the great transformations of the political economy notwithstanding, positions of influence rooted in the earlier period of the Communist regime still maintain themselves and provide access to resources. A central question for research on the new Chinese city is, therefore, how new is it? To be more precise: what aspects of the current explosive urbanization should be interpreted as outgrowths of pre-existing processes of planning, control, and distribution – characteristic of socialist China – and what should be understood as the urban impacts of emerging market processes? The discussion of spatial inequality in this volume by White, Wu, and Chen (Chapter 5) emphasizes the effects not only of marketization but also of continuing policy intervention, as does the review of metropolitan expansion by Zhou and Logan (Chapter 6). Market forces have been only partially unleashed.

The same question arises in the case of the Eastern European countries that have undergone market transitions in this era. Enyedi (1992) argues that socialism made little difference in the outcomes of urban development, though specific policies may have affected the pace at which cities reached various stages in a global urbanization process. Dangschat and Blasius (1987) adopt the more widely accepted view that although the levels and patterns of sociospatial inequality are similar in Eastern Europe and Western Europe, they certainly stem from different processes. Musil (1993) further notes that the urban systems of Central Europe have undergone simultaneously a societal revolution associated with post-socialism and a technological revolution related to production, administration, communication, and information processing.

Comparison with urban patterns in Eastern Europe remind us that we must be cautious in our view of what was specifically "socialist" about the pre-transition situation in China. Ruoppila (2004) gives greater weight to the carryover of patterns of class segregation from the pre-socialist period, as well as to housing allocation policies during the socialist era that tended to favor elites and highly skilled workers. Szelényi argues that although there were distinctively socialist patterns, they did not necessarily result from the designs of socialist planners. Rather he attributes them broadly to "the consequences of the abolition of private property, of the monopoly of state ownership of the means of production, and of the redistributive, centrally planned character of the economic system" (1996, p. 287). These features included under-urbanization (relative to capitalist systems at the same industrial level), low levels of spatial differentiation, unusually low density in central areas, and few signs of socially marginal groups. Old neighborhoods were allowed to deteriorate, while new construction was focused in high-density blocs in peripheral zones.

There are, of course, great differences between the Chinese and other post-socialist experiences. Market reform in China was introduced by newly ascendant members of the old regime, rather than by an entirely new governing coalition, and it was implemented in the context of economic expansion rather than collapse. Szelényi (1996) believes that economic crisis in East European cities slowed down some potential effects of privatization; the Chinese experience – where the economy has expanded rather than contracted – offers a test of his view.

The Context of Urban Change

Having reviewed several alternative theoretical perspectives that all shed some light on urban change in contemporary China, we now turn to how they are applied to specific questions in the following chapters. We identify two dimensions of the urban transformation: 1) changing patterns of social differentiation and creation of new social classes; and 2) shifts in the way that cities and regions are built and urban space is restructured.

Patterns of social differentiation

The first trend is change in social differentiation. While it is possible to characterize the contrasts between the new rich, the recently laid-off workers and poor migrants as social polarization (Gu and Liu 2002), income inequality remains low in comparison to most market societies (Khan and Riskin 2001). Moreover, as a manifestation of the state's objective of building a "relatively comfortable society" and the associated growth of consumerism, China has a growing middle class and strong overall economic growth that trickles down to most of the population (Lu 2000).

Extensive research has been carried out by sociologists on China's social stratification (Bian 2002; Nee and Matthews 1996; Walder 1989). The theory of market transition and its modified version advocated by Nee (1989) claim that the emergence of a market undermines socialist redistributive economies, thus devaluing state and Party hierarchies. Specifically, the theory predicts that the market transition benefits 'direct producers,' as measured by the increasing return to 'human capital'. In other words, possessors of talent and educational qualifications will move into top places, while bureaucratic position and party membership will lose their value. On the other hand, Bian and Logan (1996) argue that movement toward an economy based on technological capability can maintain and even reinforce the advantage of the privileged, since political position offers the easiest route to acquisition of other forms of capital. Thus, although the correlation of educational attainment with economic return can be explained by human capital theory, it can also be attributed to privileged access to educational institutions by members and offspring of the state elite.

The contributors to this volume take somewhat different positions on these changes. Appleton and Song (Chapter 2) argue that the key process here is the modernization of the Chinese economy. Their analysis leaves aside the position of rural-urban migrants, but finds that among native urban workers those at the bottom of the income distribution are less poor now than before. In this respect, the official policy to "let some get rich first" seems not to have been at the expense of the urban poor. Rising inequality itself created pressures toward increasing poverty, as did massive layoffs and forced early retirements. But these were more than offset by income growth even among the poorest Chinese and by the fact that many unemployed persons live in households where other family members have earnings.

Bian and Gerber (Chapter 3) portray their chapter as a test of the explanatory power of institutional changes associated with market transition (found in both China and Russia) vs. economic trajectory (expansion in China, severe contraction in Russia). They highlight the classic modernization model of the Kuznets curve: at early stages of development economic growth is accompanied by increasing inequality, while at more advanced stages of development, growth is accompanied by decreasing inequality. Presuming that China began at a relatively low development level, the increasing inequality between higher and lower status occupations throughout the 1980s and 1990s is consistent with this prediction. Intersectoral and inter-regional inequality appears to increase steadily as the economy has grown. But inequality grew in Russia, also, despite economic decline. Therefore, they conclude that institutional factors linked to the transition from a socialist to a market economy offer the stronger explanation.

Attention has also been given to gender relations in urban China (a recent review is found in Hershatter 2004). One issue is participation in the labor market. As in many socialist countries, women have had high rates of labor force participation in China. The net wage disparity of about 20 percent between men and women has been surprisingly stable over the last several decades, despite some speculation that the growth of the private sector would disadvantage women (Shu and Bian 2003). But they have been disadvantaged in the public sector, because women have borne the brunt of layoffs (often in the form of forced early retirement) from state-owned and collective enterprises.

Fan and Regulska (Chapter 4) argue that neoclassical and other conventional theories are inadequate in accounting for the persistent gender inequality in the labor market. By comparing China and Poland, they offer a feminist perspective that focuses on patriarchal ideology and women's active roles in improving their situations. Fan and Regulska believe that in both China and Poland the socialist state acted in some ways to improve women's standing, especially in terms of labor force participation. But they also observe that the state's repositioning into a "developmentalist" role, a point echoed in several other chapters in this volume, led not only to withdrawal of protection but also to overt discrimination against women. The transition from socialism has created some new opportunities for women while also requiring a new feminist mobilization to compensate for the retreat of the state from equality-based policies.

A special aspect of social stratification in China is the division of migrants from permanently registered urban households. The *hukou* system of household registration constrains migrants' access to the urban labor market and to various services (Chan 1996; Cheng and Selden 1994; Solinger 1999). Possessors of a local *hukou* receive preference in job allocation, while temporary migrants must rely on personal, family, and ethnic networks to find employment (Feng, Zuo, and Ruan 2002). City governments often adopt policies restricting the jobs available to migrants. A small proportion of these migrants may attain high occupational status, but they remain far more likely than permanent residents to remain in low-level service and manual laborer occupations (Feng et al. 2002). Although formal restrictions on migrants' rights are gradually being eliminated, most migrants continue to experience discrimination and prejudice leading to lower social status and lower incomes (Feng et al. 2002; Gu and Liu 2002).

Three chapters in this volume focus on the issues of urban growth and migration and show the importance of state regulation, rather than simply general processes of development, in contributing to the formation of social exclusion. They point to the usefulness of theories of the developmental state for understanding urban change in China, but also show that patterns of development under state leadership differ from country to country.

Liang, Van Luong, and Chen (Chapter 9) seek to identify the factors underlying urban growth and to evaluate the differing effects of natural increase and migration. This is a classic issue in demographic theories of Third World urbanization. As we noted above, the model of demographic transition posits that at an early point cities would grow mainly through rural-urban migration, and both modernization and dependency theorists have expressed concern at the apparent "overurbanization" of developing countries. Liang and his collaborators note that both China and Vietnam have urbanized relatively slowly. In these countries state policy has had a decisive impact on urbanization patterns. First, strong policies of fertility control have sharply limited the natural growth of the population, especially in cities which is where these policies were more forcefully enforced. Second, though controls on rural-urban migration historically dampened growth from migration, the volume of movement has skyrocketed in recent years. The state's interest in social control appears to be giving way to concerns about the availability of cheap labor.

Zhou and Cai (Chapter 10) examine how residential patterns reinforce social marginalization and, more specifically, how varied housing modes constrain (or ease) the adjustment of the diverse migrant population to urban life. Based on a 2003 qualitative study of migrant workers in an industrial development district in Guangzhou, they find that migrant workers are often disparaged and residentially segregated. They also find that state and local development policies concerning migrant workers have been primarily intended to keep them productive at work while preventing claims for social services and welfare benefits. Because of the treatment of migrants, Chinese social stratification resembles that predicted by the theorists of dependent development – due not to the influence of the global system but to contradictory state policies on labor mobility.

Focusing on the concept of citizenship, Wu and Rosenbaum (Chapter 11) examine how the treatment of migrants as non-citizens plays itself out in the housing system. The "local–non-local" divide resulting from the *hukou* system discourages Chinese migrants from seeking secure tenure, and thus differentiates their experience from that of migrants in other developing countries. Like their counterparts elsewhere, however, Chinese migrants invest little in housing improvement in order to minimize their costs while in the city and thus maximize the amount of money they can bring home. Like Zhou and Cai, Wu and Rosenbaum also compare the status of Chinese migrants to illegal immigrants in the US and find similar processes of exclusion. Their investigation of the micro-processes involved in the search for housing elaborates the ways in which social isolation develops. Because their focus is mainly on legal status and public policy, they find that the theories of developmental state and post-socialist transition are most useful to their analysis.

Differentiation by place of origin is another factor in the position of migrants. Honig (1990, p. 273) argues that:

> the analyses of Chinese officials and Western scholars alike, focusing on class and ethnicity, have overlooked a form of inequality that is perhaps most basic to China's largest urban centre, Shanghai, namely that based on native-place identification. Throughout the twentieth century, social inequality, discriminatory practices and popular prejudice in Shanghai have been largely based on or correlate to a distinction between people of different local origins.

This pattern of differentiation has persisted through socialist transformation and market resurgence and has arisen again in the treatment of new migrants.

Rudolph and Lu (Chapter 7) show that place identities evolved in similar ways in Taipei and Shanghai. This chapter points to the applicability of all the theoretical frameworks discussed above: modernization, dependency, developmental state, and the transition from socialism. The authors begin with the two cities' common history of dependent development within a colonial or semi-colonial situation, as both cities displayed the typical dual city form of colonial domination. Overlaid on top of this legacy after independence was a nationalist state. Then both cities also adapted to immigration – of migrants from the countryside in Shanghai, of mainland Chinese in Taipei. Both cities have strong global connections, creating the possibility of a place identity that has both local and international components. Of course, place identities in both cities are largely constructed and reinforced through governmental action, since both national and municipal authorities have profound interests in the future of these places.

City building and urban development

In the period of market reform, commercial land development has occurred on a large scale in urban China, including major initiatives like the development zones of the 1980s (that created the new city of Shenzhen) and more recently creation of a new international financial and trade center in Shangai's Pudong Development Area. In these projects, and in myriad smaller and more local projects, powerful national and/or local authorities open the doors to international property capital. At the same time, speculative investment from public enterprises from around the country is channeled toward these zones, diverting capital from what managers perceive as less profitable uses in their own enterprises. The emerging Chinese urban growth coalition (Zhou and Logan 1996, Zhu 1999) also includes recently formed real estate development companies, which are typically affiliates of larger state enterprises that provide the capital and the political connections through which they operate. Like urban growth machines in the West, local governments generally seek a development strategy that can stimulate growth and expand its revenue base.

A wide range of players participates in the development business in China. Government itself, at both national and local levels, is the biggest speculator as it takes upon itself the task of altering the course of growth of a whole city or city district. On a smaller scale, multiple and shifting growth coalitions have emerged. Developers may acquire land from any of several different public agencies at a politically negotiable price. State banks provide construction loans to those builders with enough authority to demand them. As a reward for their participation, growth coalition members may obtain access to apartments at lower prices than are paid by foreign firms or overseas Chinese. The agency that provides the land, the bank that offers the loan, the enterprise that organizes the project, or any other party to the transaction might capture below-market housing for members of its work unit.

The Chinese form of urban growth machine reveals the contradictions and complexities of a partially marketized system. The close relationship between municipal government and the development industry means that the industry has not become wholly competitive (Leaf 1995). All land remains state owned, and developers are dependent upon municipalities releasing land through a dual system of either long-term leases or direct administrative allocation. In principle this gives government very strong control over land use, but in practice the local state's weak planning capacity and hunger for revenue and foreign investment undermine this power (Wu 1998a; Zhang 2000). The local level of the Chinese "developmentalist state" is becoming more relevant to the country's development, but compared to national policy makers may operate with a shorter time perspective, greater emphasis on quick returns from its promotion of local growth and in many cases a weak administrative and revenue base.

White, Wu, and Chen (Chapter 5) address the results of this system for the spatial structure of urban China including inequalities across regions and within metropolitan areas. Like many of our contributors, they draw partly on modernization theory, presented here in the form of a theory of economic and demographic transition (Kuznets 1955). In early stages of development, urbanization is stimulated by migration from rural areas to urban nodes of opportunity. This is associated with growing inequality across regions, and at the same time it results in increased disparities (between migrants and natives) in cities. In a later stage of industrialization and urbanization, both forms of spatial inequality are expected to decline. But ultimately they argue that state policies have the greatest weight in urban development patterns.

A specific aspect of spatial inequality is the emergence of gated communities, discussed by Huang and Low (Chapter 8). The specific form that gating is taking (upper middle class apartment complexes or villas surrounded by walls and protected by private security personnel) is new to China. But Huang and Low point out that it is not simply a consequence of growing social inequality, and they argue that in the Chinese context it is not even exclusionary. Rather the wall around a housing compound is interpreted as an expression of solidarity and collective identity. And it is this function of gating that has led the state to promote it, because social organization is an ingredient of social control. Paradoxically, then, gated communities are an antidote to the breakdown of other traditional forms of organization in a rapidly changing society.

Zhou and Logan (Chapter 6) focus more specifically on development of the urban fringe around major cities. The typical Third World city characteristically has a dense core surrounded by sprawling squatter settlements interspersed with gated suburban enclaves (Gugler 1997). Although the periphery of Chinese cities similarly displays a broad range of wealth, it lacks the presence of shantytowns and has a greater (and growing) presence of middle- and working-class communities, often under the sponsorship of work units. One stream of suburban development is composed of higher status employees who receive favorable treatment from their employers. Even in the 1980s the "cadre" and "intellectual" zones around Guangzhou extended outward to the east by quite a substantial margin between 1984 and 1990 (Yeh, Xu, and Hu 1995). Many work units have also made it possible for ordinary workers to leave traditional work unit compounds for commodified suburban housing estates. At the same time, substantial numbers of people affected by redevelopment projects were resettled in suburban resettlement housing blocks. The result is a great diversity of housing types in the urban fringe.

This "suburban" trend is complicated by the emergence of large marginalized zones inhabited by the floating population. Migrant housing is concentrated in peripheral areas where private leasing of substandard buildings and unauthorized construction are common practices. The price of housing certainly plays a role, as the rent is more expensive inside the main urban areas than in peri-urban areas. Land use is subject to more stringent control inside the urban core, while the supply of cheap peripheral rental housing is made available through converting local farmers' houses into rental housing.

Dynamics of these migrant enclaves will be dependent on changes in the *hukou* system and other institutional aspects. Since 2001, the system has been adjusted through various experimental schemes. These changes will inevitably change the constraints on migrant residential relocation; however, the choices of migrants (e.g., preferring low cost housing so as to save money to remit back to families left behind in the villages; combining of production and residence through the network of place of origin) will not change overnight.

Appropriate Comparisons

The contributors to this volume seek to apply the varied theoretical prisms of modernization, dependency/world system, the developmental state, and market transition to changes in contemporary urban China. We do so in part by making specific comparisons to cities in the rest of the world. It is striking that the places selected for comparison are as diverse as the available theoretical approaches, and that each shares certain characteristics with Chinese urban areas. We have struggled with the question of which category of city best captures the Chinese experience – modern industrial, Third World, developmentalist, post-socialist?

A recent visitor to China immediately notices resemblances to US and European cities. In less than two decades China's settlements have transformed from relatively compact, low-rise centers to sprawling metropolises with high-rise cores surrounded by suburban subdivisions and mega-malls along the US model. Increasingly the small shop-houses that were the traditional locus of retail trade are being squeezed out by superstores along the model of the "category killers" and giant supermarkets devastating "mom and pop" stores in the US. In addition, as formerly independent cities begin to fall within the penumbra of major cities (e.g., as in the Suzhou-Shanghai corridor), resemblances increase to the megalopolises of Tokyo-Yokahama, New York-Washington, and the British Southeast. Sudjic (1992) refers to these as "100 mile cities." Although this is particularly the case for Beijing and Shanghai, many other metropolitan regions are also moving in this direction. The trend is being reinforced by the construction of major freeway links between cities and peripheral highways around them. At the same time, continued elite preference for the urban core results in a socio-spatial pattern similar to the European one, wherein, despite the presence of some upper-class suburbs, the inner city continues to have a highly heterogeneous social mix with significant and growing fashionable areas. Gentrification activities in the coastal cities also follow the established Western model, whereby architecturally interesting traditional dwellings are rehabilitated, and disused factories become converted to loft living.

At the same time, the growing physical similarity to the cities of the wealthy countries partially disguises the continued existence of Third World levels of poverty and substandard living conditions. The vitality of coastal metropolises has its counterpoint in deindustrializing interior cities and remote rural areas with very low standards of living. Chinese cities lack the enormous peripheral shanty towns prevalent in most of the developing world but nevertheless suffer in some districts from high levels of overcrowding, poor sanitation, unsafe construction, air and water pollution, and lack of open space. Despite the prominence of the state role in urban development, much construction proceeds without regulation. Informal markets operate on a much larger scale and for much longer hours than in Western cities. Traffic flows are frequently chaotic. Although automobile ownership is growing rapidly, most of the population relies on foot or bicycle transportation. As is common in other developing countries, poor neighborhoods are vulnerable to sudden removal to make way for state-sponsored projects, and residents find themselves displaced to locations far from the center and their places of work. This phenomenon, of course, has also occurred within developed countries, especially during the heyday of urban renewal. The extent to which it occurs in China and the political powerlessness of those affected, however, give it a Third World quality.

There is no single Third World model with regard to social differentiation or political democracy. Some cities in less-developed countries are divided by ethnic rivalries while others have fairly homogeneous populations. Some have active participation by women in the labor force and afford women a fair amount of autonomy, while in others women are strictly subordinate. Most but not all have high levels of income inequality, and few provide adequate levels of schooling and social protection to their citizens. Chinese cities, while displaying some level of conflict and discrimination based on ethnicity or place of origin, are not deeply divided except in relation to legal status. Among urban citizens there are substantial social entitlements, but the large numbers that do not possess formal citizenship rights are excluded from these benefits. The migrant population, however, does not suffer from the extreme levels of unemployment and underemployment prevalent throughout the Third World. Although China certainly has considerable underemployment, urban migrants generally do have jobs, albeit often the worst jobs available. In Chinese cities, as in Latin America and parts of Africa, a high percentage of women are economically active, but as everywhere they suffer from discrimination in wages and status.

Chinese urban development perhaps conforms most closely to the trajectory set by East and South Asian countries that have likewise displayed rapid economic development under the guidance of the developmental state. As was noted above, Japan set the template that others then followed. We can see many similarities in the growth of Tokyo, then Taipei, Hong Kong, Seoul, Bangkok, and Kuala Lumpur – and now Shanghai, Beijing, Shenzhen, and other large Chinese cities. All these cities reveal the marks of rapid conversion from relatively sleepy entrepôts to thriving industrial centers, resulting in the disappearance of traditional housing and commercial structures and their replacement by skyscrapers, large apartment blocks, and shopping malls. What gives them their particular flavor is a somewhat anarchic quality – enormous traffic jams, sharp juxtapositions of primitive forms of industry with ultra-modern technologies, extraordinary levels of street life, and shiny new trophy buildings mixed in with remnants of traditional building forms, The difference between Chinese cities and the other cities mentioned is in their socialist history and the continuing dominant role of the state. Although the state has played an important role in guiding development elsewhere in Asia, except in Singapore it does not have the authority possessed by the Chinese state, nor does it have the presence of an equivalent to the Chinese Communist Party.

The post-socialist cities of Russia and Eastern Europe do share a similar history, and to some degree a similar physical appearance. The parts of the city that predate the transition to privatization manifest the same endless blocks of shabby concrete apartment houses. Empty storefronts in the center, left vacant by pre-socialist merchants, have been occupied by restaurants, clubs, and flashy purveyors of stylish goods. Successful businesspersons frequently have made their fortunes as a consequence of their previous rank in the Communist hierarchy or, in contemporary China, because of connections with Party cadres. In China, Russia, and some of the Eastern European countries, the imposition of Communism broke down the extreme subordination of women that had characterized traditional culture, while the introduction of market reforms has forced women into an unequal competition within labor markets. None of the other post-socialist cities, however, has witnessed the kind of economic boom that has catapulted Chinese cities into their present form.

Hence, although we find all these commonalities with other cities, at the same time we find differences. In particular, there are three factors that make Chinese cities unique. First, as mentioned earlier, is the explosive character of growth. Second is the amount of new construction, of both infrastructure and buildings, in a very short time period. While a number of Asian cities have been host to major projects, no other country has seen in its cities the simultaneous construction of so many bridges, subway lines, expressways, technology parks, office developments, factories, and residential units as China. Third is the particular situation of the migrant population, which, although resembling that of undocumented immigrants in the US and Europe, is a distinctively Chinese phenomenon that differentiates the native urban population based on legal status.

Studying Chinese cities, as a consequence of both their unique and familiar characteristics, resembles examining a cubist painting. The writers of the chapters in this volume approach the object of their study from a variety of angles. The authors were teamed based on their expertise in the particular topic (e.g., public health or crime), but typically only one of the pair is a China expert. The purpose of these matches was to produce a broad range of comparison.

We seek to apply the varied theoretical prisms introduced in this introductory chapter – modernization, dependency/world system, the developmental state, and market transition – to specific aspects of change in contemporary urban China. No one model seems satisfactory. Modernization theory, despite its age and the attacks it has endured over the years, is surprisingly resilient, and some contributors rely mainly on its insights. But most look for a hybrid solution. As White, Wu and Chen (Chapter 5) put it: "We disagree with modernization theory in its determinism of economic stages, disagree with world system theory in its determinism of world politics, disagree with the developmental state approach in its rosy picture of state capacity, and disagree with post-socialist transition in its notion of 'socialist legacies.'"

Yet we also find ideas to agree with, and we evaluate these theories in part by making specific comparisons between Chinese cities and cities in the rest of the world. The wide range of theories has its counterpart in the variety of cities that the authors have chosen as the basis of comparison. They include cities in both the developed and less-developed world, cities that have made the transition from colonialism to independence and others that have moved from socialism to capitalism, and cities that have rapidly industrialized under the aegis of the developmental state. We have struggled over which type of city makes the most appropriate comparison to China's urban areas and have concluded that Chinese cities are both *sui generis* and share many aspects of cities elsewhere. It is, in fact, the combination of common characteristics with a spectrum of cities and certain unique Chinese factors that makes Chinese cities so difficult to categorize.

The chapters in this volume add to our understanding of the theoretical frameworks through which we approach the study of urban transition in China. They do so by providing a close-up look at the processes through which change is occurring and the social and spatial outcomes that have resulted so far. Given the speed of change in China, we expect urban transformation to continue apace, pushed by growing ties to the global economy, continuing privatization, highly active property developers, and an increasingly literate, self-conscious population.

REFERENCES

Abu-Lughod, J. (1965) "Tale of Two Cities: The Origins of Modern Cairo." *Comparative Studies in Society and History* 7: 429–57.

Amsden, A. (2001) *The Rise of "The Rest."* Oxford: Oxford University Press.

Andrusz, G.D., M. Harloe, and I. Szelényi (eds.). (1996) *Cities after Socialism: Urban and Regional Change and Conflict in Post-Socialist Societies.* Oxford: Blackwell.

Beier, G. (1976) "Can Third World Cities Cope?" *Population Bulletin* 31: 3–35.

Bian, Y. (2002) "Chinese Social Stratification and Social Mobility." *Annual Review of Sociology* 28: 91–116.

Bian, Y., and J.R. Logan. (1996) "Market Transition and the Persistence of Power: The Changing Stratification System in Urban China." *American Sociological Review* 61: 739–58.

Bradshaw, Y. (1987) "Urbanization and Underdevelopment." *American Sociological Review* 52: 224–39.

Cardoso, F.H., and E. Faletto. (1979) *Dependency and Development in Latin America.* Berkeley: University of California Press.

Castells, M. (2000) *End of Millenium.* Oxford: Blackwell.

Chan, K.W. (1996) "Post-Mao China: A Two-Class Urban Society in the Making." *International Journal of Urban and Regional Research* 20: 134–50.

Chen, X., and W. Parish. (1996) "Urbanization in China: Reassessing an Evolving Model." In J. Gugler (ed.), *The Urban Transformation of the Developing World.* New York: Oxford University Press: pp. 61–90.

Cheng, T., and M. Selden. (1994) "The Origins and Social Consequences of China's Hukou System." *The China Quarterly* 139: 644–68.

Dangschat, J., and J. Blasius. (1987) "Social and Spatial Disparities in Warsaw in 1978: An Application of Correspondence Analysis to a 'Socialist' City." *Urban Studies* 24: 173–91.

Daniell, J., and R. Struyk. (1994) "Housing Privatization in Moscow: Who Privatizes and Why." *International Journal of Urban and Regional Research* 18: 510–25.

Dos Santos, T. (1970) "The Structure of Dependence." *American Economic Review* 60: 231–6.

Enyedi, G. (1992) "Urbanisation in East Central Europe: Social Processes and Societal Responses in the State Socialist Systems." *Urban Studies* 29: 869–80.

Evans, P. (1995) *Embedded Autonomy: States and Industrial Transformation.* Princeton: Princeton University Press.

Evans, P.B., and M. Timberlake. (1980) "Dependence, Inequality, and the Growth of the Tertiary: A Comparative Analysis of Less Developed Countries." *American Sociological Review* 45: 531–52.

Feng, W., X. Zuo, and D. Ruan. (2002) "Rural Migrants in Shanghai: Living under the Shadow of Socialism." *International Migration Review* 36: 520–45.

Frank, A. Gunder. (1969) *Capitalism and Underdevelopment in Latin America.* New York: Monthly Review.

French, R.A., and F.E. Ian Hamilton (eds.). (1979) *The Socialist City.* Chichester: John Wiley and Sons.

Friedmann, J. (2005) *China's Urban Transition.* Minneapolis: University of Minneapolis Press.

Gu, Ch.L., and H.Y. Liu. (2002) "Social Polarization and Segregation in Beijing." In J.R. Logan (ed.), *The New Chinese City: Globalization and Market Reform.* Oxford: Blackwell: pp. 198–211.

Gugler, J. (ed.). (1997) *Cities in the Developing World.* New York: Oxford University Press.

Haggard, S. (1990) *Pathways from the Periphery.* Ithaca, NY: Cornell University Press.

Hershatter, G. (2004) "State of the Field: Women in China's Long Twentieth Century." *The Journal of Asian Studies* 63: 991–1065.

Honig, E. (1990) "Invisible Inequalities – The Status of Subei People in Contemporary Shanghai." *China Quarterly* 122: 273–92.

Hoselitz, B. (1962) *Sociological Aspects of Economic Growth.* Glencoe, IL: Free Press.

Kasarda, J., and E. Crenshaw. (1991) "Third World Urbanization" *Annual Review of Sociology* 17: 467–501.

Khan, A.R., and C. Riskin (eds.). (2001) *Inequality and Poverty in China in the Age of Globalization.* New York: Oxford University Press.

King, A. (1976) *Colonial Urban Development: Culture, Social Power and Environment.* London: Routledge and Kegan Paul.

Kovacs, Z. (1994) "A City at the Crossroads: Social and Economic Transformation in Budapest." *Urban Studies* 31: 1081–96.

Kuznets, S. (1955) "Economic Growth and Income Inequality." *American Economic Review* 65: 1–28.

Ladanyi, J. (1989) "Changing Patterns of Residential Segregation." *International Journal of Urban and Regional Research* 13: 555–70.

Leaf, M. (1995) "Inner City Redevelopment in China: Implications for the City of Beijing." *Cities* 12, 3: 149–62.

Lerner, D. (1958) *The Passing of Traditional Society: Modernizing the Middle East.* New York: Free Press.

Lu, H. (2000) "To Be Relatively Comfortable in an Egalitarian Society." In D.S. Davis (ed.), *The Consumer Revolution in Urban China.* Berkeley, CA: University of California Press: pp. 124–44.

McGee, T.G. (1986) "Circuits and Networks of Capital: The Internationalisation of the World Economy and National Urbanisation." In D. Drakakis-Smith (ed.), *Urbanisation in the Developing World.* London: Croom Helm: pp. 23–36.

Mingione, E. (1977). "Territorial Social Problems in Socialist China." In J. Abu-Lughod and R. Hay, Jr. (eds.), *Third World Urbanization.* New York: Methuen: pp. 370–84.

Murray, P., and I. Szelényi. (1984). "The City in the Transition to Socialism." *International Journal of Urban and Regional Research* 8: 90–108.

Musil, J. (1993) "Changing Urban Systems in Post-Communist Societies in Central Europe: Analysis and Prediction." *Urban Studies* 30: 899–905.

Nee, V. (1989) "A Theory of Market Transition: From Redistribution to Markets in State Socialism." *American Sociological Review* 54: 663–81.

Nee, V., and R. Matthews. (1996) "Market Transition and Social Transformation in Reforming State Socialism." *Annual Review of Sociology* 22: 401–35.

Preston, S. (1979) "Urban Growth in Developing Countries: A Demographic Reappraisal." *Population and Development Review* 5: 195–215.

Ruoppila, S. (2004) "Processes of Residential Differentiation in Socialist Cities: Literature Review on the Cases of Budapest, Prague, Tallinn and Warsaw." *European Journal of Spatial Development* 9: 1–24.

Shu, X., and Y. Bian. (2003) "Market Transition and Gender Gap in Earnings in Urban China." *Social Forces* 81: 1107–45.

Slater, D. (1986) "Capitalism and Urbanisation at the Periphery: Problems of Interpretation and Analysis with Reference to Latin America." In D. Drakakis-Smith (ed.), *Urbanisation in the Developing World.* London: Croom Helm: pp. 7–21.

Solinger, D.J. (1999) *Contesting Citizenship in Urban China: Peasant Migrants, the State, and the Logic of the Market.* Berkeley: University of California Press.

Sudjic, D. (1992) *The 100 Mile City.* London: A. Deutsch.

Sykora, L. (1999) "Processes of Socio-Spatial Differentiation in Post-Communist Prague." *Housing Studies* 14: 679–701.

Szelényi, I. (1996) "Cities under Socialism – and After." In G. Andrusz et al. (eds.), *Cities after Socialism.* Oxford: Blackwell: pp. 286–317.

Walder, A.G. (1989) "Social Change in Post-Revolution China." *Annual Review of Sociology* 15: 405–24.

Wallerstein, I. (1976) *The Modern World-System: Capitalist Agriculture and the Origins of the European World-Economy in the Sixteenth Century*. New York: Academic Press.

Woo-Cummings, M. (ed.). (1999) *The Developmental State*. Ithaca: Cornell University Press.

Wu, F. (1998) "The New Structure of Building Provision and the Transformation of the Urban Landscape in Metropolitan Guangzhou, China." *Urban Studies* 35, 2: 259–83.

Yeh, A.G.O., X.Q. Xu, and H.Y. Hu. (1995) "The Social Space of Guangzhou City, China." *Urban Geography* 16: 595–621.

Zhang, T. (2000) "Land Market Forces and Government's Role in Sprawl." *Cities* 17, 2: 123–35.

Zhou, M., and J.R. Logan. (1996) "Market Transition and the Commodification of Housing in Urban China." *International Journal of Urban and Regional Research* 20: 400–21.

Zhu, J. (1999) "Local Growth Coalition: The Context and Implications of China's Gradualist Urban Land Reforms." *International Journal of Urban and Regional Research* 23: 534–48.

Part I

Market Transition in Work Units and the Labor Market

1

Two Decades of Reform: The Changing Organization Dynamics of Chinese Industrial Firms

Shahid Yusuf and Kaoru Nabeshima

Introduction

Urban development under communism followed a path that diverged in significant respects from that taken by predominantly capitalist economies. This divergence which varied among countries and is now being narrowed can be explained by four factors.

First, communist regimes viewed urban development as being led by the manufacturing sector because the objective of planners was to maximize so-called "productive investment." Economic power and national security were linked with deepening manufacturing capability specifically of the metallurgical and engineering industries. Thus a majority of urban jobs were either directly or indirectly associated with manufacturing or with the administrative functions of different levels of government.

A second, and related, factor was the low priority assigned to the services sector which is a far more prolific source of urban jobs. By deemphasizing services industries, planners in communist countries constrained the principal driver of urbanization and suppressed a wide range of activities that impart dynamism to the economic and social life of cities.

Third, the narrow view of urban development influenced investment in urban housing and infrastructure. Because the scale of service industries and commercial activities remained modest, cities were dominated by administrative and industrial units with investments in urban infrastructure and housing kept to a minimum to cater to these needs. This made urban areas drab, utilitarian, and uninviting.

Fourth, and finally, a variety of regulations interfered with the free movement of people making it difficult, for instance, for rural residents to find jobs, housing and services in urban areas.

For these reasons, a development strategy focused primarily on industrialization resulted in a narrow and limited form of urbanization. This was more so in countries such as China than in some of the Eastern European economies, in part because the former was at an earlier stage of modernization and in part because in China there was a stronger ideological bias against urbanization and favoring heavy industry.

The reform of the socialist economy started a process of convergence towards the urban norms of market-based economies. China has converged faster because it started from a lower income level than others and its economic growth has been much faster. However, several elements of reform are contributing to the rate of change, in China as well as in other transition economies.

The move towards much greater openness and integration with the global economy has led to a rapid broadening of the industrial base and underscored the role of services. "Narrow" urban development has been succeeded by "normal" urban development with services increasing their shares and generating jobs.

While retaining its developmental role, the Chinese state has adopted a decentralized approach, devolving policy-making discretion to municipalities and dismantling some of the barriers to the flows of labor and capital within the country. This has stimulated the growth of the larger cities, which benefit from agglomeration economies. It has also created numerous other urban foci of growth. Furthermore, by pursuing enterprise reform, the state is providing existing state-owned enterprises (SOEs) and new firms with the incentives and opportunities to grow and to diversify. This is imparting greater impetus to urban development and generating demands that attract workers to urban centers. In response, and recognizing that development is inseparable from urbanization and that livable, well managed cities are key, the state is pouring resources into building urban housing and infrastructure. In China, this big push to urbanization is now interwoven with the overall development strategy. The massive construction activity attendant upon urbanization, coupled with the equally massive migration to urban areas, is adding additional fuel to economic growth.

China's transition has already lasted more than a quarter of a century and it is far from over. It has now permeated every corner of Chinese life and no organization is untouched. Yet remarkably for an economy where the ratio of trade to gross domestic product (GDP) approaches 65 percent, industrial organization in China – especially in the state sector – has been slow to shed many of the distinctive structural characteristics of the Maoist era state enterprises. Both SOEs as well as many of the non-state enterprises remain embedded in a web of formal and informal relationships qualitatively different from those of most other East Asian developmentalist states that link producers to the administrative or financial organs of the state. Similarly, while the evidence of change in the urban sector is everywhere and is impinging on organization, notable also is the tenacity of institutions and administrative practices that emerged in the pre-reform era when the Chinese economy was on a different trajectory. China's level of urbanization, which reached 43 percent in 2005, is still significantly below that of countries at comparable stages of development and of developed countries.[1] Many cities continue to enforce *hukou* (residence) based restrictions on permanent migration and have only slowly started to dismantle the plethora of regulatory controls over every form of economic activity that were the hallmark of a communist society.

Chinese policymakers recognize that the urban sector can achieve a "balanced" development that holds poverty and inequality in check only if industrial reforms provide a sufficient growth impetus and jobs for the swelling workforce.[2] The main prong of the industrial strategy in support of urban change is ownership reform that transforms SOEs into corporate entities with majority state ownership or places them wholly into private hands, in the process also bolstering the incentives for and the dynamism of the private sector. While the central government spearheads the ownership reform initiative in the majority of cases, the actual implementation is in the hands of municipal, county, and prefectural governments that must work with other actors influencing urban changes. However, the ultimate success of the reforms rests substantially with the firms themselves, in particular, their managers and directors who must steer these enterprises through an environment that is increasingly competitive locally and integrating rapidly with the wider national and international economies.

The purpose of this chapter is to situate industrial change within the context of urban development and to examine the interplay of the broad reform strategy with local implementation, and its actual practice by the reformed firms themselves.

Medium and large state-owned enterprises (MLSOEs) and organizational change

Since 1997, the authorities have moved aggressively to reform the ownership structure of MLSOEs and to privatize, divest, or close approximately 120,000 smaller SOEs. By introducing new governance and management practices, ownership changes are reinforcing pressures exerted by market competition on the state enterprise sector. Thus, the large economic and social roles of MLSOEs in the urban economy and their exposure to the global environment, to foreign direct investment (FDI), to new technologies and to reforms, make them a natural focus for research.

What makes these firms especially interesting is their ongoing effort to arrive at organizational forms and modes of operation that viably fuse elements of legacy systems with other elements borrowed from East Asian and Western corporate models. All this is occurring in an urban environment in which control exercised by municipal or other agencies of the state is still pervasive and instigates, guides, modulates or obstructs corporate experimentation as MLSOEs venture deeper into the market economy and are exposed to the pressures of globalization. Trends that have surfaced in China during the past decade and international experience both suggest that urban development will be increasingly concentrated in large agglomerations along the East coast.

Reforming China's Industry

China's industrial enterprises in 1980 were the production units in a centralized command economy. The organizational structure of the typical MLSOE was relatively flat. Production, input supplies, and interfirm transactions were regulated by the plan and implemented by the industrial bureaus. Managers enjoyed limited discretion – their job was to ensure that production targets were met – and because all wages were fixed, there were few rewards for initiative at any level in the enterprise. Goods produced were distributed through state-created channels as there were very few markets. Transactions among firms were settled through a mono-banking system[3] and the enterprise was permitted to hold only miniscule cash balances. In effect the purpose of enterprises was to manufacture and to care for their workers from the time they were recruited till the end of their lives.

Enterprise reforms sought to enhance the performance of SOEs by giving them some autonomy to produce for newly expanded markets and permitting the deployment of contract-based incentive schemes for enterprise managers and their workforces. By devolving fiscal responsibilities to sub-national entities, they also encouraged municipal, county, and township governments to act entrepreneurially and promote pre-existing commune and brigade enterprises and new 'non-state enterprises' that would generate revenues and provide jobs.[4] With sub-national, and particularly municipal authorities taking the lead, the incremental elaboration of enterprise and associated price and market reforms continued through the mid 1990s with some improvement in the productivity and profitability of SOEs. The non-state sector benefited even more. Its share of industrial output rose from 24 percent in 1980 to 66 percent in 1995 while that of the state sector fell to 34 percent by 1995.

From the standpoint of the MLSOEs, reforms in four areas were of vital significance.[5] First, the creation of markets and freeing of many hitherto controlled prices opened a vast range of opportunities for producers and consumers. By the early 1990s, most consumer goods were being distributed through market channels leaving only 89 industrial goods and transport prices set by the state (Lardy 2002).

In parallel with the freeing of markets, a second set of reforms progressively opened the economy to trade, starting initially with producers in four special economic zones (SEZs): Shenzen, Zhuhai, Shantou, and Xiamen which quickly attracted investment from Hong Kong.[6] Chinese enterprises were encouraged to export in order to enlarge the flow of urgently needed foreign exchange (Lardy 1992).

A third, and related, reform, which coincided with the establishing of the four SEZs and the efforts to develop export-oriented industries, was the proffering of incentives to foreign investors at first in export-oriented industries. By the mid 1990s, FDI was being welcomed into a much broader range of sectors including and in particular, urban commercial real estate and urban infrastructure. A trickle of FDI in the 1980s grew rapidly after 1993 and swelled to a flood in the late 1990s and in 2005 China ranked as the third largest recipient in the world after the UK and the US (UNCTAD 2006).

The gradual integration of the domestic market with the dismantling of many barriers to trade was a fourth reform initiative. This built upon the increasing scope given to market transactions as well as investments by central and sub-national governments in transport and distribution infrastructure.

Together, these four reforms opened new vistas for China's urban economies. FDI became a conduit, mainly by way of joint ventures, for capital, skills, and technology transfers. International trade, a more integrated domestic market, and the emergence of the non-state sector, exposed the SOEs to much needed competitive pressures, in some cases spurring them to adapt so as to make the most of opportunities now within their reach.[7] Other reforms, many introduced by the major coastal municipalities, further stimulated market development, openness, and access to FDI. Municipalities such as Shenzhen, Guangzhou, Dongguan, and Foshan led the way. The purpose was to give enterprises further autonomy and incentives to exploit market opportunities and in the process, to raise often deplorably low levels of efficiency and quality. Management contracts and later, leasing of businesses, bonus schemes for production in excess of targets, greater freedom to determine the product mix, some latitude to adjust the labor force, permission to borrow from banks and implement investment plans, and less oversight by supervisory agencies, were some of the steps taken to motivate the SOEs and make them more responsive to market opportunities.

These measures began devolving responsibilities to the SOE and pushing it in the direction of acquiring the organizational capabilities of a firm and away from the status of a simple production unit that also served as the key element in the urban social security system. From the early 1990s, the authorities began to form enterprise groups modeled on the lines of Japanese *keiretsu* so as to take advantage of scale economies and the benefits of lower transaction costs from intra-group trading and capital markets.[8] By yoking together profitable and struggling enterprises, central and sub-national governments frequently attempted to revive or sustain loss-making firms. One important consequence of the reforms was to reduce the scope of planning by state agencies and micro-level controls over enterprises were both reduced although the process was far from even. At the formal level, planned production targets were attenuated, without the command system being dismantled or the autonomy of enterprises being formally specified in a way that established legal rights. However, the flexibility introduced made it possible for provincial and county level officials – and enterprise managers – to take initiatives. A form of state corporatism began taking root with bureaucrat entrepreneurs essentially commandeering state-owned industrial assets and exploiting the new freedoms afforded to enterprises to redirect their operations (Gore 1998). In many instances, these enhanced the productivity and profitability of enterprises, but just as often, the autonomy for managers and the loosening of rules governing the activities of industrial bureaus led to asset stripping (as was commonplace in other transition economies) or plunged enterprises into deeper losses.[9]

The governance exercised by industrial bureaus or enterprise groups that took over some of the functions of the bureau was inadequate and needed to be supplemented, or displaced, by institutions that defined ownership rights and aligned incentives with the market. A privatization of state-owned industrial enterprises seemed the logical course to take but instead a typically Chinese partial reform was brokered. This reform privatized thousands of the small scale enterprises, and loss-making ones for which no buyers could be found were closed. However, MLSOEs accounting for most of the output of the state industrial sector were either left untouched or converted into limited liability shareholding firms in which the state continues to own the majority of shares and/or to exercise control rights. Thus there are now effectively two classes of state firms: those with some outside shareholders and others that remain pure SOEs. In theory, the former are closer to the threshold of the market system and are subject to governance from supervisory boards or boards of directors and hence more likely to be accountable to shareholders, to maximize shareholder value, and therefore, to take their cues from the market with respect to management, structure, strategy, and technology. In practice, there is a lot of variance with the average shareholding MLSOE not having evolved very far from an unreformed SOE but with some reformed MLSOEs (or collectives) such as CIMC, Lenovo, Changhong Electric, and Haier appearing to be blazing new trails and registering large gains in performance.[10]

Although the central government establishes the broad guidelines for the changes being introduced, much of the initiative and the responsibility for implementation rest with the urban centers. Before looking more closely at the operation of firms, it is important to take account of both the urban context in which reforms are unfolding and the contribution of the municipal governments to the business environment, the pace of enterprise reform, and to the forces guiding organizational change.

Coordinating Urban with Industrial Change

With urbanization accelerating, the role of China's cities in promoting industrial development is becoming even more critical. Cities provide the matrix of institutions and generate the localized knowledge spillovers that are valuable for businesses. In China, substantial political powers vested in municipal authorities also means that they can take the lead in trying to build what Feldman and Martin (2004) call "jurisdictional advantage" and nudge the business community towards a supportive consensus. Jurisdictional advantage accrues from the unique assets that a particular location bestows on a firm. Many cities, especially coastal ones, have adopted such a proactive stance because they recognize that they must acquire differentiated industrial capabilities in order to sustain competitive advantage. Reliance on low production costs must now be superseded by an approach that is focused more on routinizing productivity gains and innovation in one or more industrial subsectors (Woetzel 2003). This can be a long drawn process and early movers that exercise good strategic foresight can acquire a commanding lead over other urban centers. The municipality can propose broad objectives or a vision but it needs the support of the business sector to realize them. Cities better endowed with locational advantages, human capital, and financial institutions, have an easier time in mobilizing the business community with incentives and through investment in the physical infrastructure and in institutions, but ultimately, it is the vitality of the business sector that is decisive. Whether it is the formation of clusters or the steady creation of high value-adding jobs, it is the actions of a multitude of firms reacting to municipal policies and through their own efforts at enhancing competitiveness by investing in skills, research, local universities, or the urban infrastructure that determine the scale of urban agglomeration benefits.

In this race to widen the margin of competitive advantage, the larger cities such as Shanghai and Guangzhou start with a number of advantages: they have a wider spectrum of industries and such diversity promises larger urbanization economies; they have a broader revenue base and deeper banking and financial resources to support development; some of the ablest party cadres are appointed to manage these cities and through their networks they maintain close ties with the leading policymakers in Beijing, which gives them the latitude to pursue bolder reforms; and they are more attractive to foreign investors who are sensitive to the scale of agglomeration benefits and are seeking a vibrant urban environment that offers more varied lifestyle choices. These are not necessarily overwhelming advantages but when combined with geographical location and the benefits of an early start, they do favor the major cities in the Pearl River Delta region and the Yangtze Basin region. Shanghai and its surrounding cities not only have an edge in terms of scale, they also command a vast and increasingly affluent hinterland of close to 400 million people. But medium and smaller sized cities are competing fiercely and typically growing much faster than the larger cities because they are readier to absorb the influx of migrants from the countryside,[11] because they enjoy cost advantages, and because they have the leverage that comes from localization economies.

What are the policy and reform options available to China's cities as they scramble to industrialize and how do these impinge upon the performance of the business sector? For most cities with industrial aspirations, the tendency is to use fiscal incentives and bank financing to promote industrial development and to buttress this with investment in physical infrastructure. Where there is an abundant local workforce or a ready supply of migrant workers, such an investment push has frequently been sufficient to initiate an industrial spiral, especially in the medium and small cities and townships. However, for most of the larger cities, the stakes are higher. Generic incentive mechanisms are no longer sufficient.[12] A deepening of industrialization requires a much closer engagement between municipality and firms.

Cross-country and Chinese experience shows that industrial growth, productivity, and innovativeness as well as the formation of dynamic urban clusters is closely related to the ease with which new firms can enter an industry and on barriers to the exit of failing firms. Smaller firms are also responsible for the bulk of the employment generated in the urban economy. Hence, institutional conditions facilitating entry are among the most effective means for a municipality to induce industrial growth. By streamlining licensing and registration requirements, reducing regulatory impediments, improving access to finance, and providing basic services with the minimum of red tape, one of the essential conditions for a thriving urban market economy can be met. A related requirement is the ease of exit because if failing firms are enabled to continue producing and investing, they divert resources from others and depress the profitability of viable firms. While in some cities such as Shenzhen and Hangzhou, the exit barrier is low, this remains a major problem in other Chinese cities and is related to enterprise reform that we will return to below (Dollar et al. 2003).[13]

One of the main sources of agglomeration economies is the pool of skills available to firms in a metropolitan region. The volume, quality, and depth of skills hugely influences the productivity, and innovativeness of existing firms, the entry of new firms, and the overall flexibility of the urban economy, in particular its capacity to evolve in new directions to meet demand or acquire new clusters of activities. Cities like Beijing, Chongqing, Hangzhou, Guangzhou and Shanghai have an abundance of skills but firms in other cities cannot draw on a substantial local pool of skilled workers, resulting in sub par performance (Dollar et al. 2003). Most municipalities recognize that to advance to a higher stage of industrialization, they must augment human resources if they have the financing to do so. This is no easy matter especially for the small and medium-sized cities. Investing in universities and other training institutions is only a part of the process, a lot depends on how much the business sector is motivated to invest in formal and on-the-job training as well. There is, in addition, the problem of retaining workers and attracting workers from elsewhere. This depends on the institutions governing the supply and affordability of housing,[14] the supply of municipal services, especially schooling, and the quality of the urban environment.[15] In each of these areas, private providers have a role but in China, the dominant player is the municipal authority that spells out the ground rules and is directly or indirectly the source of a significant part of the financing. Cities like Shanghai, Beijing, and Hangzhou have a lead in this regard, but land use, financing,[16] environmental quality, and other issues relating to the integration of the national labor market continue to bedevil these and other cities whose future prospects depend upon the presence of a broad, constantly refreshed, and sound base of skills.[17]

Another advantage of agglomeration economies is the knowledge spillovers from the investment in research and development (R&D) and in technological extension services provided by firms. With the majority of China's research institutes still in the public sector and either reporting to the municipal government or affiliated with SOEs, the municipality exercises substantial say in the level of research conducted (Sun 2002). Coastal municipalities have taken a lead in attempting both to promote applied research and to push researchers to commercialize their findings by linking with business firms. Municipal bureaus such as in Shanghai have been especially aggressive in this regard, establishing stringent criteria to gauge the performance of universities and research institutes with respect to Science and Technology (S&T) performance. This heightened pressure is a departure from past practice and it runs the risk of diverting too much energy away from teaching and basic research. One can also question the likely short-term gains in the form of useable research findings from a sudden ratcheting up of the effort to produce commercializable findings.[18] Nevertheless, China's national and subnational governments are unequivocally committed to increasing the level and quality of R&D and are encouraging SOEs to view this as the means to long-term competitiveness (Sigurdson 2005).

Cities throughout China have wooed FDI as a means of mobilizing resources, modernizing industry, and increasing exports. More recently, FDI has come to be viewed as one of the primary vehicles for raising industrial productivity and transferring technology as well as often being the only source of financing available to medium and small-sized firms. Municipal governments are eager to attract FDI in order to set up joint ventures with local firms or wholly-owned foreign subsidiaries that will advance the technological frontiers and through spillovers and demonstration effects, contribute to business practices that will raise industrial productivity.[19] These cities are also using the opportunities presented by the large pool of knowledge workers coupled with other incentives to persuade MNCs to set up research laboratories in China. By 2005, nearly 750 foreign companies had invested in R&D centers in China and while few companies have considered transferring the core research function to China,[20] they are helping to build a capability that is shaping the strategies of firms such as Huawei, Ningbo Bird, and Wanxiang, and will filter into other firms as well. Once again, while the policies of firms are decisive, municipalities play a major instrumental role in attracting FDI and are active in determining its composition with an eye to maximizing the knowledge spillovers.

Building a municipal labor market that will be attractive for dynamic and well-managed firms requires close attention to the urban environment. This is also crucial for pulling in FDI into higher technology activities. The quality of the urban environment, the standard of services, and amenities provided and the level of housing facilities is a function of many factors. Chief among them is administrative capacity and governance, which remains variable throughout urban China and increasingly is at the forefront of the central government's attention. Several cities, especially in the East Coast region, enhanced transparency and eased transaction costs for businesses by introducing electronic websites and permitting electronic processing of certain transactions (see Table 1.1). These changes are mirrored by the business climate in these cities and the entry of firms both local and foreign.

The efforts to reduce transaction costs are being buttressed by investment in infrastructure and housing. In the highly charged competitive environment, the quality of services provided by public utilities, urban transport, and the abundance of housing can tip the economic balance in the favor of one city over another. As we noted above, the leading cities are able to provide an adequate level of infrastructure services for industry but in many instances, they still fall short of the standard achieved by the industrialized countries. The building of highly capital-intensive physical infrastructure is hampered by the state of the capital markets and in particular by the thinness of structured bond markets that would permit cities to raise funds more efficiently and at lower cost for long-term projects. The full and efficient usage of the infrastructure already constructed is also an issue in many cities because of deficiencies in interagency coordination, regulatory practices and the pricing of services. Thus, sewerage facilities in several cities remain unused because no adequate system is in place to finance their operation, and the disposal of solid

Table 1.1 Investment climate scores of 23 cities in China

City	*ICA score*
Hangzhou	A+
Guangzhou	A+
Shanghai	A+
Shenzhen	A+
Chongqing	A
Changchun	A
Jiangmen	A
Wenzhou	A
Tianjin	A–
Dalian	A–
Beijing	A–
Zhengzhou	A–
Wuhan	B+
Nanchang	B+
Xi'an	B+
Changsha	B+
Chengdu	B
Guiyang	B
Kunming	B
Nanning	B
Harbin	B–
Lanzhou	B–
Benxi	B–

Source: Dollar et al. (2003).

waste is becoming a worsening burden as cities run short of landfill sites. Similarly, the regulation of facilities in use remains uneven.

The lag in the development of institutions for long-term financing is almost unavoidable because these markets need time to build up the matrix of skills, depth, experience, rules, and credibility. The same can be said for local regulation. What the lag does though is to widen the advantages enjoyed by the major metropolitan governments. These are able to use their superior bargaining power to bid for the resources that are available in abundance and to invest in and operate urban infrastructure. Such investment has localized multiplier/accelerator effects, stimulates growth, and also pulls in other private investment. This is reflected in the performance of the cities shown in Figure 1.1, especially the large coastal cities.

Improving the business environment is one major strand of the urban economic strategy in China. Equally important is the role of municipal and county governments in modulating and implementing enterprise ownership

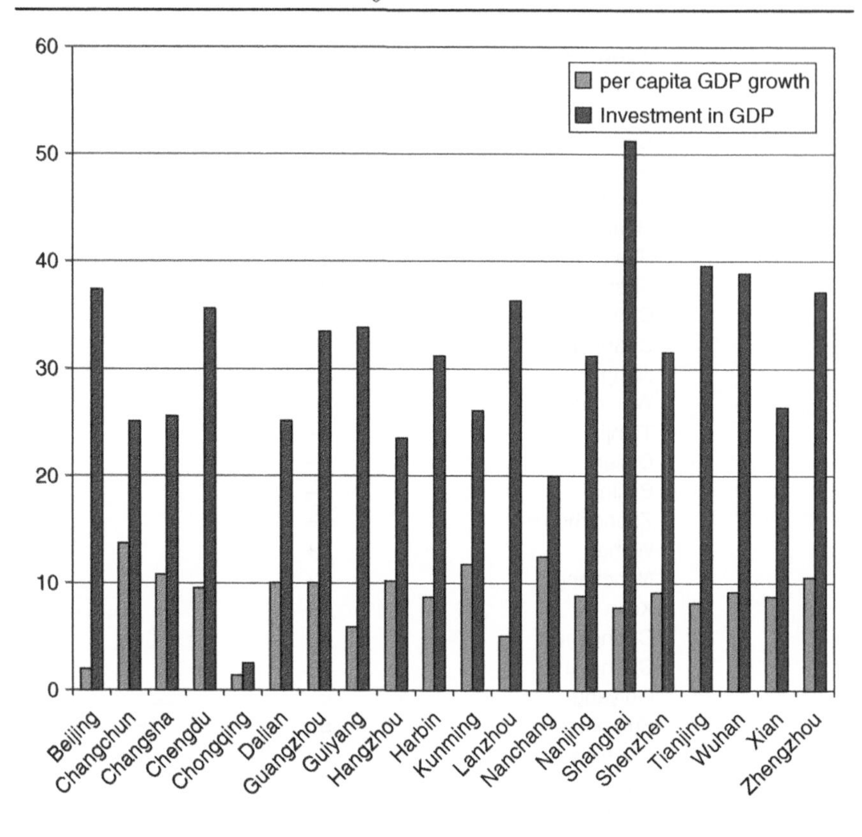

Source: Lin and Song (2002).

Figure 1.1 Per Capita GDP Growth and Investment Ratio of Selected Cities (1991–8)

reforms. Although the overarching guidelines have been determined by the central government, the microstructure of the reforms is in the hands of subnational governments. It is they that are setting the pace of privatization and it is the local industrial bureaus that determine how much autonomy is permitted to the corporatized state enterprises[21] and the governance structures being crafted to monitor their performance. CIMC (and also Huawei) are striking examples of how the autonomy extended by the Shenzhen municipality has permitted the company's management to achieve dramatic results through horizontal mergers, innovation, and an aggressive strategy to win local and foreign markets. TCL, ZTE, Haier, and Wanxiang are other examples of companies that have autonomy. But there are also numerous instances of municipal governments intervening to prop up failing enterprises, using their ownership clout to force companies to take over loss-making entities, and micromanaging companies under their control, requiring them to maintain jobs and to "contribute" to municipal finances.

In short, the policy discretion enjoyed by municipalities in China's decentralized economic system means that they have a large role in managing local industrial change but that still leaves a large and increasing responsibility for introducing organizational changes, devising strategy, and conducting business in the hands of enterprises.

From State-Owned Enterprises to Modern Firms

MLSOEs are spread across the spectrum performance-wise but are somewhat more narrowly grouped with respect to certain attributes relating to structure and organizational dynamics.

Governance. Ownership and governance are the defining characteristics of enterprise reform in China and these are likely to be critical to future success. Starting from a condition of state ownership, plan-based production, and control exercised through industrial ministries and bureaus, reform has created two classes of MLSOEs. One consists of wholly state-owned firms directed either by central ministries or more usually by provincial, municipal, or county bureaus. While the formal structure of governance is usually well-defined, the actual degree of accountability varies by industrial sector, jurisdiction, and bureaucratic entrepreneurship. A second category of MLSOEs consists of enterprises that have been corporatized into limited liability corporations (LLC) and shareholding limited liability corporations (LLSC). Smaller SOEs were converted into LLCs and larger SOEs were commonly converted into LLSCs under the company law passed in 1994. They are accountable, in principle, to the Board of Directors (BOD), a supervisory body, and a public agency responsible for the industry to which an enterprise belongs. The governance structure of the LLC is quite similar to a private company with a BOD providing oversight and day-to-day business conducted by a manager. The LLSC has similar governance structure to the LLC, but, the BOD for the LLSC has a stronger supervisory role than is the case for the LLC (Keister and Lu 2001). However, given the dominance of the government's share, government appointees on the BOD and supervisory agencies provide the framework of accountability. Our research shows that the formation of LLSCs does seem to have a large positive effect on performance, so that ownership reform is having predictable consequences with the change in governance contributing to the improvement.[22] Nevertheless, findings from around the world and the Eastern European countries suggest that fully privatized firms are the ones that are the most efficient manufacturers. The ongoing ownership reform has introduced some elements of modern corporate governance into the SOE sector. The next stage is full privatization and the transfer of the state's control rights in the majority of the MLSOEs so that firms can truly transform themselves, and acquire professional managers fully accountable to shareholders.

Structure. The vast majority of MLSOEs started out as single plant operations with a vertically integrated production system, little outsourcing and a comprehensive suite of internally provided services ranging from schooling to transport. Many still cling to this structure. The more dynamic firms, however, have acquired multiple plants – sometimes as a result of mergers forced upon them by the authorities. They have become allied with enterprise groups,[23] and are divesting themselves of non-core operations, in some instances, to other members of the group, and outsourcing some of the services previously provided in-house.[24] It is too early to tell how far these trends will continue but so far deverticalization, outsourcing of non-core functions, and greater focus on core specialization is the exception in even the leading Chinese firms.

Labor. Most MLSOEs now rely more on contract workers and have flexibility in reassigning workers among activities, but some tenured staff remain and many firms continue supporting their pensioners. Even with an increasing share of contract workers, political pressure can make it difficult for firms to lay off workers especially in the smaller cities where alternative employment options are few.

Innovation. In sum, the MLSOEs are undergoing a change in their ownership and broader governance structure. They have begun to modify management, strategy, and the internal organization of the firm in response to the reform efforts of national and municipal governments, market pressures, and globalization. However, inertia, continued oversight, and interference from supervisory agencies, and the soft budget constraints, undercut competitive pressures to streamline their activities and focus on core products.

Creating a World-Class Corporate System

Observers of Chinese MLSOEs have remarked that China still lacks world-class industrial companies (Nolan 2001). With the exception of Haier and possibly Huawei and Lenovo, no Chinese industrial firm or enterprise group has the size, management skills, and the mix of capabilities to operate on a global scale. This is not surprising, because the MLSOEs are having to telescope in a few decades managerial and organizational changes that occurred in the US over more than a century.[25] Undoubtedly, the ranks of the handful of leading firms will thicken and FDI will assist in the process, but much will depend on the future course of state ownership and the state's role in determining industrial organization and maintaining a soft budget constraint. International experience, although grounded in a past and in institutions very different from those of China, does offer some guidance with respect to government policy towards industry and urbanization. From among these offerings, several merit attention by the Chinese.

Other business models

The literature on urban economics and on urban geography show that if the diseconomies of urban size can be controlled then scale promises large productivity gains via agglomeration effects.[26] The average size of cities in China is below the optimum and the low Gini coefficient of 0.43, as against 0.65 for Brazil and Japan, indicates that there are a small number of very large urban areas. Thus, from the standpoint of economic advantage, there is potential for urban growth to be exploited. For instance, if the size of a city is 50 percent below the optimum size, then doubling the city size can increase the value-added per worker by 35 percent. But China's expanding cities will have to invest heavily in livability in order to continue attracting industry.

Livability depends on the quality of the physical infrastructure, social amenities, recreational facilities, and the environment. In many Chinese cities, these are barely adequate. In the face of continued migration and the expansion of motor vehicle use, living conditions could deteriorate in the absence of well-planned investment and effective regulation.[27]

Municipal governments must also work to enhance the competitiveness of the local economic environment In order to facilitate the entry and exit of firms. In this regard, their policies towards the SOEs will be important. It is becoming clear that industrial policies that rely on directed credit, subsidies, trade barriers, and government purchases of products to induce the development of targeted industries, are frequently a costly and ineffective approach to local industrialization. Moreover, many of the instruments used in support of such policies are now disallowed by the World Trade Organization (WTO). Government efforts to impose mergers and to create enterprise groups have not yielded robust results anywhere in East Asia. They can burden successful firms with unwanted baggage rather than creating larger and more dynamic entities. It is much better to let market forces do their job, and manage development with the help of a sound competition policy.

Privatization of industrial enterprises has been shown to raise productivity and lead to positive business outcomes in the vast majority of cases.[28] Furthermore, there is little evidence from other transition economies to suggest that a gradual and elaborately sequenced ownership reform has advantages over a swifter form of privatization[29] nor that a slower pace of reform resulted in the introduction of effective social safety nets.[30]

There can be no denying the contribution of professional management and strategy making to the competitiveness, growth, and profitability of firms. A small number of Chinese MLSOEs have the leadership but few have as yet equipped themselves with the other attributes. The absence of managerial depth also shows up in the structure of MLSOEs. While the trend worldwide is towards flatter hierarchies, large firms with national or international operations must have the organizational resources and managerial hierarchies to deal with complex and widely ramified operations. Privatization can contribute to the strengthening of management and subnational governments can provide an added push by ensuring that the institutions of corporate governance enforce the needed discipline.

Competitive and rapidly growing firms that invest in R&D encourage different forms of innovation. Municipal agencies can support this tendency by making it easier for firms to enter into joint research ventures and to contract with universities and research institutes. Thus far, MLSOEs have rarely adopted these practices. Competing on the basis of technology will become necessary once companies are fully exposed to market pressures and seek to expand overseas.

The leading MLSOEs have begun adopting stronger incentive regimes appropriate for the market environment and with the encouragement of municipal authorities, they are giving more attention to enhancing their human resources, but not enough is known about training practices, pay scales, promotion ladders, and bonus schemes to assess the adequacy of the systems now in place. Similarly, the current composition of the internal labor markets and hiring policies of firms are uncertain. However, the future competitiveness of firms will be increasingly tied to the upgrading of the workforce. This will require the concerted efforts of firms and of the municipalities, which have a big stake in raising the level of skills.

While the internal workings of a firm are certainly key, as MLSOEs deverticalize and outsource, their competitiveness also depends on the efficacy of their cooperative arrangements with partners, suppliers, providers of business services, and buyers. Whether one considers Japanese *keiretsu*, Korean *chaebol*, American auto companies or the leading manufacturers of electronics, garments and machinery, the best firms are ones that have a clear strategy and based on this, have created a support system that leverages the resources, technologies, and innovativeness of others. There is no single best practice but plenty of good models. The vertically integrated and insular MLSOE does not approximate any of these models, nor do the loosely knit enterprise groups created by governmental fiat. If international experience is a guide, building international production networks and alliances to conduct research or product development will be necessary for the growth of MLSOEs that eventually graduate into the ranks of successful private sector firms.

Finally, as the Chinese economy becomes more integrated domestically and closely linked with the global market, it will be harder to preserve or create sheltered domestic niches. Manufacturing firms across the world are finding that, little by little, local markets are becoming coextensive with global markets as trade barriers recede, transport charges fall, and the Internet enormously enhances the availability of information and cuts down transaction costs. In this kind of environment, survival is coming to depend upon preparedness and flexibility. Much more so than in the past, strategy and innovativeness matter. Also location matters. The ably managed and innovative firms will grow and agglomeration economies can be the springboard to their success.

Conclusion

In this chapter, we have closely examined two interrelated parts of the development strategy pursued by the Chinese state since the start of reforms in the late 1970s. One is the reform of the industrial system and specifically state-owned enterprises. The other is the urban policy context which modulates enterprise reform and the industrialization of Chinese cities.

Enterprise reform is still incomplete and even the most successful of the former SOEs have yet to achieve a structure, governance, and managerial capability comparable to leading firms elsewhere, especially multinational corporations (MNCs). However, the reforms introduced to date have transformed the small and medium SOEs and enormously stimulated the non-state sector. For 25 years, industrial growth has remained in the double digit range with no evidence of the momentum slackening.

Industrial change is being paced by rapid urbanization which is providing industry with the environment and workforce needed to sustain development. Municipal governments have taken advantage of administrative and fiscal decentralization to promote industrialization with larger cities taking the lead. The challenge now is to ensure that urban development is not at the cost of livability and that it is backed by needed fiscal and regulatory measures.

ACKNOWLEDGMENTS

The authors would like to thank, in particular, John Logan and Susan Fainstein for their helpful suggestions and Shiqing Xie for providing excellent research assistance. The findings, interpretations, and conclusions expressed in this study are entirely those of the authors and should not be attributed in any manner to the World Bank, to its affiliated organizations, or to members of its Board of Executive Directors or the countries they represent.

NOTES

1 China's modest rate of urbanization stems from past and remaining checks on migration and the low fertility rate in urban areas because of the successful enforcement of the one-child policy and the restriction on migration. For large cities such as Shanghai, the urban growth is almost entirely the result of the inflow of migrants (see Liang, Van Luong, and Chen 2005).

2 So far, the Chinese government has been able to reduce the number of absolute poor in the urban areas, an impressive achievement compared to the experience elsewhere. However, inequality within the urban area is rising (see Appleton and Song 2005).

3 This was finally dismantled in 1984 with the creation of the People's Bank of China and four major state owned banks.

4 For differing perspectives on local initiatives, see Duckett (1998), Gore (1998), and Oi (1999).

5 On reforms through the early 1990s see Naughton (1996).

6 See Howell (1993) and Wu (1999) on the creation of the early SEZs and the widening of trade channels.

7 Such competition also eroded profit margins and tax revenues transferred to various levels of governments. The result was a sharp drop in the tax/GDP ratio, now gradually being reversed through a broadening of the tax base.

8 Many of these groups were created out of enterprises affiliated with an industrial bureau. Many industrializing economies have sought to use the creation of enterprise groups as a means of overcoming institutional gaps or missing markets.

9 In many instances, mergers not only failed to turnaround loss makers, they also compromised the performance of their profitable partners.

10 The profit of CIMC increased by 120 percent for the first 6 months of 2005 compared to the same period the previous year (CIMC Semi-Annual Report 2005). Huawei's profit increased 63 percent from 2003 to 2004 and 10 percent from 2004 to 2005 (Huawei Technologies Annual Report 2005).

11 See Zhou and Cai (2005) for a discussion on migrants' lives in a new city.

12 A number of studies find that the correlation between the quantity of investment and economic growth is rather weak (Dollar et al. 2003; Lin and Song 2002). Dollar et al. (2003) find that the 23 Chinese cities they studied have built urban infrastructure to a point where shortages no longer constrain the performance of firms.

13 Many cities suffer from excess capacity of 20–30 percent or more because too few firms are exiting (Dollar et al. 2003).

14 On the issue of housing for migrants, see Wu and Rosenbaum (2005).

15 Crime is always high on the list of concerns for people in the larger urban centers. See Messner, Liu, and Karstedt (2005) for the recent trends in crime in urban China.

16 Many cities have raised a large volume of funding by repossessing and rezoning surrounding agricultural land from farmers and allocating it for commercial, residential, or industrial use. This provides a one-time fiscal bonus but is not a device for meeting longer-term fiscal requirements. Such encroachment also displaces farmers and runs the risk of exacerbating urban sprawl and eating into the limited supply of arable land.

17 Although Beijing, Guangzhou, and Shanghai do not face constraints in terms of supply of skills, performance of firms in these cities can still be greatly enhanced if labor markets are made more flexible (Dollar et al. 2003).

18 Without a corresponding increase in supply of experienced human capital to support R&D, a sudden infusion of funds or generous incentives to conduct R&D can lead to misallocation of scarce resources (Yusuf, Wang, and Nabeshima 2005).

19 The demonstration effects from FDI seem to be rather strong in coastal cities, judging by the rapid increase in the external design patent applications (Cheung and Lin 2004).

20 This reluctance arises from the current limits on research capacity in China and the perceived inadequacy of the protection for intellectual property.

21 Corporatized enterprises are those that have sold some of their equity to the non-state sector and have been organized as registered corporate entities with traded shares that are subject to the rules of corporate law.

22 Yusuf, Nabeshima, and Perkins (2005) find that former SOEs reformed into LLSCs perform the best, followed by joint venture firms. Although former SOEs reformed into LLCs performed better than the non-reformed SOEs, the positive effect stemming from the ownership reform was much smaller than that for LLSCs and joint ventures.

23 These '*jituan*' were formed after December 1991 following a State Council Directive.

24 In the case of First Auto Works in Changchun, the SOE is relocating some of its assembly operations to other parts of the country, including to Guangzhou.

25 Ferguson and Wascher (2004) note that before the Civil War in the US, most companies were sole proprietorships or partnerships, but this changed with advances in transport and communications and the growth of markets all of which favored larger hierarchical organizations. Further changes came in the wake of financial market developments, with multi-plant operations, and most recently, the advent of IT.

26 Alfred Marshall recognized the externalities associated with co-location of many firms in a city. The agglomeration of business activities often confer benefits to firms located in vicinity through access to specialized inputs, thicker labor markets, better information flow, and knowledge spillovers. See Ottaviano and Thisse (2004) for a recent review on this.

27 As migration increases, authorities will need to pay closer attention to the risks from infectious diseases such as HIV and sexually transmitted disease (See Chapter 13 this volume).

28 See the reviews by Djankov and Murrell (2002) and Megginson and Netter (2001).

29 Both Balcerowicz (2003) and Havrylyshyn (2004) note that fast reformers on balance registered somewhat better performance overall during the 1990s. Slow

reformers have often become bogged down, China included, with new reforms being opposed by strongly entrenched vested interests.

30 Traditionally SOEs have been responsible for providing housing, education, health care, and other public services. As SOEs are reformed, many of these functions are now becoming the responsibilities of individuals themselves or of various levels of government in China. So far, there has been no indication that a slower pace of SOE reform has made it possible to reform social policies that other chapters in this volume touch upon (Yusuf, Nabeshima, and Perkins 2005).

REFERENCES

Appleton, S., and L. Song. (2005) "The Myth of the 'New Urban Poverty'? Trends in Urban Poverty in China, 1988–2002." Presented at Urban China in Transition, New Orleans, Louisiana, January 15. Chapter 2 of this volume.

Balcerowicz, L. (2003) "Post-Communist Transition in a Comparative Perspective." Presented at Practitioners of Development Seminar Series, Washington, DC, November 18.

Cheung, K.-Y., and P. Lin. (2004) "Spillover Effects of FDI on Innovation in China: Evidence From the Provincial Data." *China Economic Review* 15, 1: 25–44.

Djankov, Si., and P, Murrell. (2002) "Enterprise Restructuring in Transition: A Quantitative Survey." *Journal of Economic Literature* 40, 3: 739–92.

Dollar, D., A. Shi, S. Wang, and L.C. Xu. (2003) *Improving City Competitiveness Through the Investment Climate: Ranking 23 Chinese Cities* Washington, DC: World Bank.

Duckett, J. (1998) *The Entrepreneurial State in China: Real Estate and Commerce Departments in Reform Era Tianjin.* London: Routledge.

Feldman, M., and R. Martin. (2004) "Jurisdictional Advantage." NBER Working Paper 10802. Cambridge, MA: National Bureau of Economic Research.

Ferguson, R.W. Jr., and W.L. Wascher. (2004) "Lessons from Past Productivity Booms." *Journal of Economic Perspectives* 18, 2: 3–28.

Gore, L.L.P. (1998) *Market Communism: The Institutional Foundation of China's Post-Mao Hyper-Growth.* Hong Kong: Oxford University Press.

Havrylyshyn, O. (2004) "Avoid Hubris but Acknowledge Successes: Lessons from the Postcommunist Transition." *Finance and Development* (September): 38–41.

Howell, J. (1993) *China Opens Its Doors.* Colorado: Lynne Rienner.

Keister, L.A., and J. Lu. (2001) "The Transformation Countries: The Status of Chinese State-Owned Enterprises at the Start of Millennium." *NBR Analysis* 12, 3: 5–31.

Lardy, N.R. (1992) *Foreign Trade and Economic Reform in 1978–1990.* Cambridge, UK: Cambridge University Press.

Lardy, N.R. (2002) *Integrating China into the Global Economy.* Washington, DC: Brookings Institute.

Liang, Z., H. Van Luong, and Y.P. Chen. (2005) "Urbanization in China in the 1990s: Patterns and Regional Variations." Presented at Urban China in Transition, New Orleans, Louisiana, January 15. Chapter 9 of this volume.

Lin, S., and S, Song. (2002) "Urban Economic Growth in China: Theory and Evidence." *Urban Studies* 39, 12: 2251–66.

Megginson, W.L., and J.M. Netter. (2001) "From State to Market: A Survey of Empirical Studies on Privatization." *Journal of Economic Literature* 39, 2: 321–89.
Messner, S.F., J. Liu, and S. Karstedt. (2005) "Economic Reform and Crime in Contemporary China: Paradoxes of a Planned Transition." Presented at Urban China in Transition, New Orleans, Louisiana, January 15. Chapter 12 of this volume.
Naughton, B. (1996) *Growing Out of the Plan.* New York: Cambridge University Press.
Nolan, P. (2001) *China and the Global Economy: National Champions, Industrial Policy, and the Big Business Revolution.* New York: Palgrave.
Oi, J.C. (1999) *Rural China Takes Off: Institutional Foundations of Economic Reform.* Berkeley, CA: University of California Press.
Ottaviano, G.I.P. and J.-F. Thisse. (2004) "Agglomeration and Economic Geography," in J.V. Henderson and Ja.-F. Thisse (eds.), *Handbook of Regional and Urban Economics.* North Holland: Elsevier: pp. 2563–608.
Sigurdson, J. (2005) *Technological Superpower China.* Northampton, MA: Edward Elgar Publishing.
Sun, Y. (2002) "China's National Innovation System in Transition." *Eurasian Geography and Economics* 43, 6: 476–92.
UNCTAD. (2006) *World Investment Report 2006.* New York: United Nations.
Woetzel, J.R. (2003) *Capitalist China: Strategies for a Revolutionized Economy.* Singapore: John Wiley & Sons (Asia).
Wu, W. (1999) *Pioneering Economic Reform in China's Special Economic Zones.* Aldershot, UK: Ashgate Publishing.
Wu, W., and E. Rosenbaum. (2005) "Migration and Housing: Comparing China With the United States." Presented at Urban China in Transition, New Orleans, Louisiana, January 15. Chapter 11 of this volume.
Yusuf, S., K. Nabeshima, and D.H. Perkins (2005) *Under New Ownership: Privatizing China's State-Owned Enterprises.* Stanford, CA: Stanford University Press.
Yusuf, S., S. Wang, and K. Nabeshima. (2005) "Fiscal Policies for Innovation." processed. Washington, DC: World Bank.
Zhou, M. and G. Cai. (2005) "Trapped in Neglected Corners of a Booming Metropolis: Patterns of Residence and Adaptation Among Migrant Workers in Guangzhou." Presented at Urban China in Transition, New Orleans, Louisiana, January 15. Chapter 10 of this volume.

2

The Myth of the "New Urban Poverty"? Trends in Urban Poverty in China, 1988–2002

Simon Appleton and Lina Song

Introduction

Until the second half of the 1990s, policy-makers and researchers regarded urban poverty in China as either non-existent or a very marginal problem. Government anti-poverty programs focused on rural areas and, in particular, on selected poor counties. As Khan (1998, p. 42) commented "China's official poverty reduction strategy is based on the assumption that poverty is a rural problem." Within academia, few studies focused on urban poverty, in contrast to the large literature on income distribution and inequality more generally. This neglect of urban poverty arose partly from using low poverty lines – such as a "dollar a day." Reflecting the great urban-rural divide in China (Knight and Song 1999), a significant proportion of the rural population fell below these poverty lines but only 1 percent of the urban population were classified poor. With urban poverty in China being defined so as to concern only a very small minority, it is scarcely surprising that the issue was marginalized by government and scholars alike. However, this attitude would be surprising to people in middle and high income countries, where poverty remains a relevant concept despite only very small proportions of citizens living on less than one dollar a day. Arguably, the main role of poverty lines is to focus attention on the living standards at the lower end of the income distribution. When set inappropriately low, they can have the opposite effect and help marginalize any discussion of the effect of policy reform on poorer urban residents. This is what happened in urban China prior to the late 1990s.

Things began to change in the second half of the 1990s as concern grew over what was seen as a "new urban poverty" caused partly by a wave of rural-urban migration and partly by mass unemployment following a program of retrenchment in state-owned enterprises (SOEs). These new forms of urban poverty differed from the old urban poverty which was often characterized as the "three withouts" – roughly corresponding to the disabled, the sick, and the orphaned (Wong 1998). By the turn of the century, opinion makers began to assert that urban poverty had risen during the 1990s, taking some of the shine off China's exceptionally high rates of economic growth. For example, *The Economist* (2001, page 39) declared "And in the cities, absolute poverty is increasing . . ." while the Chinese government magazine *Liaowang* (June 27, 2002) also argued that urban poverty had increased. Underpinning such commentary was a concern that rising urban poverty would lead to political unrest, jeopardizing the reforms that had enabled China's rapid economic growth (see Wu 2004). However, urban poverty in China continued to be measured using low poverty lines such as "a dollar a day" that, despite perceived adverse developments, still only categorized around 1 percent of the urban population as poor.

In this chapter, we use a range of higher poverty lines in order to consider more broadly how lower income urban households fared in the 1990s. In particular, we present new evidence on trends in urban poverty from 1988 to 2002 based on our analysis of the best available data set on income distribution in China. The data come from the surveys in 1988, 1995, 1999, and 2002 conducted as part of the Chinese Academy of Social Sciences (CASS) *Household Income Project*, which provides a much fuller accounting of household income than the biased measures used in official statistics (see Khan et al. 1993). An important limitation of our analysis is that the CASS surveys are based on the government's official sampling frame. This has the advantage of making our samples representative of all Chinese with urban registration (*hukou*). However, it excludes the "floating population" of rural-urban migrants who lack urban *hukous*. This omission is regrettable since rural-urban migration increased dramatically in the period and rural-urban migrants were no doubt poorer as a group than residents with urban *hukou*. Nonetheless, all large-scale statistical studies of urban poverty in China in the period are subject to the same limitation as a result of the government's failure to properly cover rural-urban migrants in its official statistics. For brevity, we will not continuously repeat this caveat and use the term "urban poverty" in this chapter to refer to poverty rates among those with urban *hukou*. While important equity issues arise when considering rural-urban migrants – notably in their lack of access to government services compared to urban residents – there is no real suggestion that migrants as a group have impoverished themselves by moving to the cities. If any thing, the presumption is that migration has provided a means by which they can escape poverty (Park, Du, and Wang 2004). What data we have on rural-urban migrants in 1999 show that unemployment rates among them are negligible

(Appleton et al. 2002). By contrast, the second half of the 1990s saw the emergence of mass unemployment among residents with urban *hukous*. Employees in loss-making SOEs found themselves laid-off and enduring long spells of unemployment. If one is to look for potential losers from China's reform process, our focus on the urban residents rather than the migrants seems appropriate.

Despite focusing on the urban residents, the potential losers from reform, we find that concern over adverse poverty trends in the 1990s appear misplaced. There has perhaps been an over-reaction to the previous neglect of urban poverty in China, with unwarranted pessimism about the living standards of poorer households. In particular, we challenge the assertion that poverty among registered urban residents rose in absolute terms. While this claim makes dramatic headlines, it is does not appear to be supported by the evidence. Undoubtedly there have been some losers from China's urban reforms, but the CASS surveys imply that this has not translated into a worsening of real income of those at the bottom of the income distribution. Indeed, while real incomes have grown less at the lower end of the distribution than higher up, they nonetheless have grown. As a consequence, whatever fixed point is taken as the poverty line, urban poverty appears to have fallen in the period. Although the existing literature on this question is mixed, we argue our result is consistent with official household statistics, despite the latter's biased measurement of household incomes. Consequently, while we do not deny that the concept of the "new urban poverty" identifies new forms of *relative* poverty and that migrants are likely to be poorer than registered urban residents, the idea that these trends have led to *absolute* impoverishment appears to be something of a myth.

The structure of the chapter is as follows. The next section frames the subject of this chapter within the wider international debate about growth, poverty, and inequality. This allows straightforward comparisons between our findings for China and those commonly observed elsewhere in the world. The small existing literature on trends in urban poverty in China is then reviewed. Our own findings on household income regarding trends in growth, inequality, and poverty are documented, and the chapter concludes by considering which of the four theoretical perspectives outlined in the introduction to this volume best explains our findings.

International Patterns of Poverty Reduction and Growth

The study of the poverty trends during China's rapid growth is of general interest because it informs the international debate on poverty reduction and growth. Although poverty reduction is widely agreed to the central overarching aim of economic policy in developing countries, opinions differ on how much emphasis to put on economic growth and on redistribution as the means to achieve this goal. One extreme would be to rely solely on economic growth to reduce poverty – sometimes characterized as the "trickle down" approach, although this term attributes an unwarranted degree of passivity to the role of poor in bettering their own lot. At the other extreme, some argue that it is essential for the government to undertake large-scale redistribution through progressive taxation and welfare policies. How growth is actually distributed clearly has some bearing on this debate. If the poor did not benefit from growth alone, then redistribution would be necessary for poverty reduction. Fishlow's (1972) work on Brazil was influential in this regard, with the finding that the poor benefited little during a period of rapid growth spurring the World Bank's advocacy of a strategy of "Redistribution with growth" (Chenery et al. 1974). However, almost as soon as this strategy was proposed, there was a renewed emphasis on the primacy of economic growth based in part on a reaction to the contrasting fortunes of East Asia and lagging regions such as Africa. Several fast-growing East Asian economies enjoyed tremendous success in reducing poverty after 1960 while absolute poverty rose in Africa due to slow economic growth after 1980.

The debate on poverty reduction and growth is particularly relevant to China, as the government position on this issue shifted dramatically as it switched from the planning to the reform period. In the era of planning, the Chinese state played a very strong redistributive role – for example, in urban areas, wage inequalities were compressed, unemployment avoided through implicit job guarantees and a variety of welfare benefits provided by work units. However, during the reform period, the Chinese government's approach to poverty reduction is closer to "trickle down" in emphasizing growth over redistribution – in the words of Deng Xiao-Ping, "let some get rich first." In the interests of promoting economic efficiency and hence growth, enterprises have been given more freedom in letting worker remuneration reflect productivity, excess workers in SOEs have been made unemployed and many state transfers have been removed. As Khan (1998) argues, post-reform, the Chinese government has tended to reject a "relief approach" to poverty reduction in favour of efforts to increase income generation. Indeed, Khan suggests that the reluctance of the Chinese government to adopt a redistributive approach to poverty reduction may partly be a backlash against the extreme egalitarianism of the planning period. Clearly, the efficacy of China's current strategy of emphasizing economic growth over redistribution depends partly on the extent to which growth has actually reduced poverty.

The weight of cross-country evidence implies that there is no systematic tendency for the distribution of income to change in either direction during growth. International experience since 1960 has refuted early theorizing that growth may initially worsen the distribution of income before later improving it (the "inverse U" hypothesis proposed by Kuznets 1955). It is true that middle income countries tend to have a more unequal distribution of income, but this result is driven by the inclusion in their ranks of most of Latin

America, which has a particularly uneven distribution of land. Analysis of different country experiences after 1960 implies that economic development does not have a systematic effect on inequality, if country-specific effects are controlled for (Fields 2001). This implies that it is the type of growth that is crucial in determining the extent to which the distribution of income (and hence relative poverty) changes.

Since there is no systematic tendency for inequality to change during growth, it follows that one should expect growth to tend to raise the incomes of the poor on a one-for-one basis. Again, cross-country evidence since 1960 supports this implication (Dollar and Kraay 2001). However, these generalizations are based on averages and individual country experiences vary widely, for example, from the growth with equity observed in high performing East Asian economies in 1960–90 to the very inegalitarian growth of Brazil observed by Fishlow (1972) in the 1960s. We cannot be sure whether income distribution in a particular country such as China will follow the typical pattern observed since 1960 or stray from it. Consequently, there is a case for poverty monitoring even in times of strong economic growth. Nonetheless, for growth to increase *absolute* poverty would be a rather unusual occurrence. For example, a close reading of Fishlow's paper does not support the idea that Brazil in the 1960s was an example of such immiserating growth, despite the very inegalitarian trends.

So far we have focused on how growth affects income distribution and poverty, but there has also been interest in the reverse causality. Traditionally, economists have argued that there is a trade-off between equity and efficiency. Improving incentives has been at the heart of China's pro-market reforms and arguably accounts for a substantial part of its growth. The conventional wisdom, therefore, is that China's high growth was enabled by a reduction in equality. However, growth theorists have recently begun to question this presumption – arguing that equality promotes growth. For example, it may reduce the need for redistributive policies or it may allow the poor to overcome credit constraints in order to make investments in human or physical capital. The orthodox view of an equity-efficiency trade-off would seem to have a large element of truth, although some have questioned its empirical basis. For example, recent studies have found that growth has been fastest in areas of China where income was most equally distributed (Benjamin, Brandt, and Giles 2004; Ravallion and Chen 2004).

The Literature on Trends in Urban Poverty

There are relatively few studies of urban poverty in China and these present seemingly contradictory conclusions on trends from 1988 to 2002. Table 2.1 compiles estimates of the headcount of the urban poor made by these studies.

Table 2.1 Compilation of estimates of poverty rates in urban China, 1988 onwards (percentages)

Poverty line anchor	*Khan (1998)*		*Khan (1996)*	*Meng, Gregory and Wang (2004)*		*Fang et al. (2002)*		*Chen and Wang (2002)*			*Ravallion and Chen (2004)*	*Appleton and Song (this chapter)*	
	2,100 cal.	*80% of 2,100 cal. line*	*2,150 cal.*	*2,100 cal. (upper) but varies)*	*2,100 cal. (lower) but varies)*	*$1/day*	*$1.5 /day*	*$1/day*	*$1.5/day*	*$2/day*	*2,100 cal.*	*$2/day*	*$3/day*
1988	6.70	2.20		2.63	1.50						2.07	7.33	36.36
1989			7.42	2.62	1.59						7.05		
1990			7.39	1.91	0.97			1.00	8.60	20.70	2.58		
1991			4.73	2.49	1.29						1.66		
1992				3.62	1.72	2.09	13.74	0.80	3.90	13.20	1.13		
1993				5.33	2.30			0.70	4.20	13.80	1.01		
1994			5.90	5.11	2.63	2.73	13.18	0.90	4.60	13.50	1.19		
1995	8.00	4.10		5.35	2.57	1.65	10.27	0.60	3.00	9.70	0.85	7.00	23.81
1996				4.94	2.28	1.69	8.41	0.50	2.60	9.30	0.61		
1997				5.28	2.48	2.00	9.21	0.50	2.70	9.10	0.70		
1998				4.83	1.85	2.06	8.86	1.00	3.40	9.00	1.16		
1999				4.21	1.70			0.50	2.20	6.80	0.57	3.66	12.39
2000				3.97	1.71						0.63		
2001											0.50		
2002											0.54	2.88	8.52
Data source	CASS	CASS	NBS grouped	NBS	NBS	Subset of NBS	Subset of NBS	NBS grouped	NBS grouped	NBS grouped	NBS grouped	CASS	CASS

(1) The most common source of data is the official NBS surveys, although typically researchers only have access to the grouped tabulations. The CASS surveys are used in the original analysis in this chapter and are discussed in the text.

(2) Poverty lines are typically absolute, either working as multiples of $1 a day (Purchasing Power Parity adjusted) or calculating the cost of obtaining a certain amount of calories per person per day. Meng et al. (2005) re-estimated their poverty line for each year of data.

Some studies report that poverty has increased in the 1990s, others that it has shown no trend and yet more that that it has fallen. In this section, we review these studies and attempt to adjudicate between them. The task of adjudication is made somewhat easier by the fact that all rely on one of two main sources of data – either the official National Bureau of Statistics (NBS) household survey results or the CASS Household Income Project surveys used in this chapter. We have argued above that the CASS data is preferable due to its fuller accounting of income but it is not clear that a difference in data accounts for the conflicting results on poverty trends. As previously discussed, both data sources cover only residents with urban registration *hukou* and so exclude most rural-urban migrants.

Poverty analysis using the CASS surveys has been restricted to a comparison of their results for 1988 and 1995 (Khan 1998; Khan and Riskin 2001). Anchoring the poverty line on the cost of obtaining 2,100 calories per capita per day, 6.7 percent of urban residents were estimated to be poor in 1988, rising to 8.0 percent in 1995. Taking an extreme poverty line, equal to 80 percent of the 2,100 calorie line, the rise was sharper – going from 2.2 percent to 4.1 percent. These results are the most comparable to those we present later in this chapter, as we take the same CASS data, but add in surveys for 1999 and 2002. We do not dispute Khan's analysis of the 1988 and 1995 survey, but show that the rise in poverty in that period was not sustained and indeed was clearly outweighed by the fall in poverty from 1995 to 2002. It seems likely that the CASS surveys imply less favorable estimates of poverty trends than the NBS figures. For example, Khan (1996) used tabulated NBS data to estimate that poverty fell from 7.42 percent in 1989 to 5.9 percent in 1994. The discrepancy probably arises from the fact that only the CASS surveys include in their income measure the various food and other subsidies to households that were gradually withdrawn between 1988 and 1995.

However, the picture of rising urban poverty given by Khan's comparison of 1988 and 1995 finds some support by the recent study by Meng, Gregory, and Wang (2005) using NBS data. A key feature of this study is that it estimates a different poverty line for each year (and indeed each region). Meng et al. report a rise in urban poverty from 1988 to 1995 using either upper or lower bound estimates of the poverty line. Thereafter, poverty trends are down but sufficiently slowly that even by 2000 the headcount remains higher than in 1988. Meng et al. argue that re-estimating the poverty line for each year is appropriate because it makes better allowance for the changing availability and price of foods, and for increased non-food requirements due to the withdrawal of various subsidies. These arguments have some merit, but the solution arguably causes greater biases than it corrects. In a period of economic growth such as urban China has enjoyed, people are likely to consume more expensive food, raising the cost of calories. As a result, the food component of the poverty lines used by Meng et al. will be rising over time. Moreover, non-food requirements are estimated as a mark-up based on the non-food share of the poor. By Engels' law, rising income leads to a rising non-food share and thus increases in estimated non-food requirements. Hence the non-food component of the poverty lines used by Meng et al. is also likely to be rising over time. Allowing poverty lines to rise in this way can lead to poverty appearing to rise despite increases in the real incomes of poorer urban residents.

The other studies using NBS data adopt poverty lines that are fixed in real terms but produce no consensus on poverty trends when using poverty lines that are set very low. Two studies find no strong trend. Fang, Zhang, and Fan (2002) used a subset of the NBS data – one representative city from each province. Chen and Wang (2001) used the full NBS data set but like most other studies were dependent on official tabulations (grouped data). When using low poverty lines such as a "dollar a day," both studies found find urban poverty to fluctuate in the 1990s with no clear trends. Nonetheless both studies show marked falls in urban poverty in the 1990s when it is more broadly defined. More positive conclusions are made in two other studies of NBS income data. Wang, Shi, and Zheng (2002) use interpolations from the officially tabulated grouped income and look for generalized Lorenz dominance from 1981 to 1999. They find that each year, with the exception of 1988, the generalized Lorenz curve of real income per capita dominates that for the previous year. This implies that social welfare rose year-on-year and hence poverty fell, regardless of what level the poverty line was set – so long as the line was constant in real terms over time. In perhaps the most authoritative study of NBS data, Ravallion and Chen (2004) devised a poverty line for urban China based on the cost of basic needs in collaboration with government statisticians. This is likely to be adopted as China's official poverty line. Comparing household real incomes with the line they derive, they estimate urban poverty rates to have been 2.07 percent in 1988, falling to 0.85 percent in 1995, 0.57 percent in 1999 and 0.54 percent in 2002.

How can we adjudicate between these seemingly conflicting results? Analysis of NBS data has reached mixed results when using low poverty lines – a dollar a day or thereabouts. The conflicting results of these studies using NBS data may be partly the outcome of using low poverty lines that define only a very small proportion of the urban population to be poor. Surveys may be less reliable in obtaining estimates for the very poorest – there are particular problems in sampling them and results may be very influenced by low outliers. These difficulties are likely to be particularly acute given that nearly all the studies use only official tabulated data with rather crude groupings. This requires interpolation that may be particularly difficult when dealing with only the extreme low end of the distribution. Somewhat higher poverty lines that are fixed over time reveal an improvement in the living standards of low income urban households during the 1990s. Arguably, the work on the CASS data is the most authoritative since these surveys provided the most comprehensive measurement of income. To the extent that the period 1988 to 1995 saw a large reduction in government transfers, not adjusting for this is likely to lead to too optimistic an assessment of poverty trends. However, until this chapter, analysis of poverty trends using the CASS data has been limited to a comparison of 1988 and 1995.

New Evidence on Trends in Poverty, Inequality and Growth

Household income growth and inequality

Real incomes have risen substantially over the 14-year interval covered by the surveys. The Household Income Project (HIP) data imply real income per capita for urban residents in 2002 is around twice as high as it was in 1988. There has also been a marked change in the structure of income between 1988 and subsequent years. In particular, subsidies and income in kind constitute a much smaller part of total household income after 1988. Ration coupons were abolished and housing subsidies shrank from 18 percent of total income in 1988 to a mere 2.8 percent in 2002. This is important, as some studies of income inequality in China focus more narrowly on cash wage earnings. Neglecting to account for subsidies and in-kind, which have been largely withdrawn after 1988, will tend to overstate the rise in income during the period. In our data, the share of cash earnings by working household members has risen from 43 percent of all household income in 1988 to 60 percent in 2002.

The more inclusive measure of income in the HIP surveys probably explains the discrepancy between the growth estimates from that data compared to those from the larger household surveys conducted by the NBS. NBS data imply higher growth during the period – 6.8 percent per annum compared to our estimate of 5.1 percent. HIP data record slower growth in the period 1988 to 1995, as subsidies were withdrawn, and also 1999 to 2002, when housing items contributed a smaller share to total income. By contrast, HIP data imply higher growth between 1995 and 2002 as housing rental values rose rapidly. In passing, it is interesting to note that China's real GDP per capita (nationally, not urban-only) is estimated to have grown by 7.4 percent between 1988 and 2002. The HIP data thus imply substantially less income growth than official data, although a sustained growth rate of 5 percent per capita is still very high by international standards.

The focus of this chapter is not on the average level of growth, but how growth has varied across the distribution of income and hence the impact on poverty and inequality. The HIP data show that income growth between 1988 and 2002 is greater, the higher up the income distribution one goes. We have already noted that average incomes virtually doubled in the period. However, for the lowest decile, the tenth percentile, real income per capita only increased by a half (49 percent higher). For the highest decile, the ninetieth percentile, incomes increased by 130 percent. As a consequence, growth rates for the highest decile averaged 6.0 percent per annum, more than twice the 2.8 percent growth experienced by the poorest decile.

The interval between the first survey in 1988 and the second in 1995 is largely what accounts for the unequal pattern of growth over the full period. Between 1988 and 1995, incomes of the poor grew slowly: the poorest decile saw only slow growth in income in this early period (0.8 percent per annum). By contrast, the top decile enjoyed very fast growth of 6.1 percent per annum. In the subsequent intervals between surveys, the pattern of growth across the deciles is much flatter and less marked. The fact that income grew less for poorer deciles than for more affluent ones implies an increase in inequality. By any of the conventional indices, inequality rises substantially from 1988 to 2002. For example, the Gini coefficient rises from 0.24 to 0.33. However, it is noticeable that most of the rise occurs between 1988 and 1995 – for example, the Gini coefficient is virtually unchanged between 1995 and 2002. Looking at the numbers in more detail, there is a modest rise in inequality between 1995 and 1999. This is more or less offset by a modest fall in inequality between 1999 and 2002. The slight fall in inequality between the last two surveys is perhaps surprising given the poorer deciles experienced growth below the median. However this is outweighed by the fact that the most affluent also enjoyed below average growth.

Poverty trends

The fact that incomes grew across the deciles implies that poverty fell, so long as a constant, reasonably broad poverty line is used. Figure 2.1 presents estimates of the percentage of urban residents who are poor for a continuum of poverty lines. The graph shows a dramatic fall in poverty during the full period from 1988 to 2002. This is true whatever poverty line is chosen, provided the poverty line is held constant in real terms (i.e. measures absolute, rather than relative, poverty). The "dollar a day" poverty line corresponds to a value of 1212 yuan per year in 2002 prices, or the point "12" on the graph. As can be seen, less than 1 percent of people received incomes below this point, so that poverty line is of little use in evaluating changes in the standard of living of the less advantaged Chinese residents. Instead, we focus on two rather arbitrary poverty lines – a broad poverty line of $3 per day and a narrower one of $2. With the $3 poverty line, the proportion of urban people who are poor falls from 36.4 percent in 1988 to 8.5 percent in 2002 (Table 2.1 refers). With the $2 poverty line, the prevalence of poverty falls from 7.3 percent to 2.1 percent. However, Figure 2.1 shows that the conclusion that absolute poverty has fallen is robust to the choice of poverty line because the poverty incidence curve for 1988 dominates that for 2002. Whatever constant poverty line we take, the incidence of poverty is higher in 1988 than in 2002. Furthermore, one corollary of the "first order" dominance revealed in Figure 2.1 is that poverty must also be lower in 2002 than in 1988 on a variety of poverty indicators beyond the simple headcount of the poor. Thus we can say that both the poverty gap (P1) and the squared poverty gap (P2) are lower in 2002 than 1988, irrespective of the poverty line chosen (see Foster, Greer, and Thorbecke, 1984, for definitions of these indicators).

Poverty also falls between each of the surveys for most poverty lines. The most noticeable case where the poverty incidence curves in Figure 2.1 cross is when comparing poverty in 1988 and 1995. Here, for low poverty lines that identify less than 6 percent of people as poor in 1988, we can see that poverty is higher in 1995 than in 1988. This implies that living standards worsened for the poorest 5 percent of the population between 1988 and 1995. This helps to understand the finding that, using the $2 a day poverty line, both the poverty gap, P1, and the squared poverty gap, P2, are estimated to rise between 1988 and 1995.

Very different results would be obtained if we adopted a relative approach to poverty. Since the incomes of poorer urban residents have tended to grow less than average incomes, relative poverty has risen. For example, if we define "poverty line" as half of median income per capita in the survey of that year, relative poverty rises from 3.8 percent in 1988 to 12.8 percent in 2002. Relative poverty rises in all years although the increase is more pronounced between 1988 and 1995 (when relative poverty reaches 9.3 percent) and more modest between 1999 and 2002 (going from 11.8 percent to 12.8 percent).

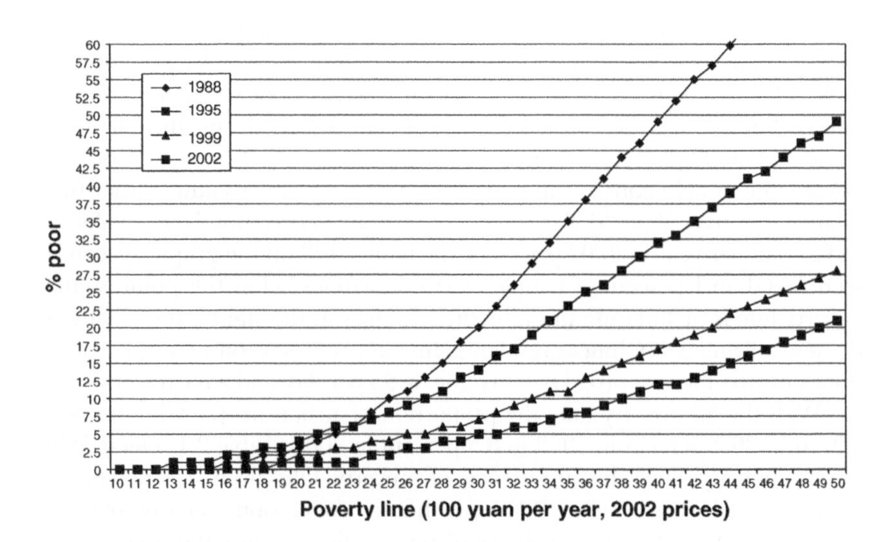

Figure 2.1 Urban poverty incidence curves 1988–2002

Decomposition of poverty changes into growth and distribution components

The problem with focusing on relative poverty is that by construction it does not allow changes in average income to impact on poverty – relative poverty can only change if the distribution of incomes changes. However, it is growth alone rather than redistribution that has raised the living standards of the poor in urban China since 1988. Since inequality has risen, the distributional changes have been unfavorable to the poor. We can decompose the change in poverty into growth and redistribution components following Datt and Ravallion (1992). The strong impact of growth in reducing poverty is evident from the results. For example, using the $3 a day poverty line, we find that if the poor had enjoyed the same rate of income growth between 1988 and 2002 as the means of our samples, then the percentage of urban Chinese who are poor would have fallen by 34.9 percentage points. Since only 36.4 percent lived under $3 a day in 1988, such growth would have implied the virtual elimination of poverty as so defined. In reality, poverty fell by 27.8 percentage points – still very impressive, but short of what would have happened had there been no change in the distribution of income.

Distributional changes in income in the period have generally been unfavorable – as should be expected from the rise in inequality noted previously. More revealingly, the impact on poverty of adverse distributional changes is estimated to have been substantial. For example, had there been no growth in mean income between 1988 and 2002, we calculate that the worsening of the distribution of income would have increased the headcount by 8 percentage points (a rise in poverty of over one-fifth). The only exception to the unfavorable distributional changes is the period 1999 to 2002 when the income distribution improves slightly and so would have implied a reduction in poverty even without growth.

As might be expected, the redistributional component of the decomposition of poverty changes is most marked when considering the interval between the 1988 and 1995 surveys. For the narrower definition of poverty ($2 a day poverty line), the distributional component of the poverty changes is almost fully twice the size of the growth component. This implies that, for the very poor, adverse changes in the distribution of income outweighed the beneficial effects of general economic growth.

The paradox of rising unemployment and falling poverty

Perhaps the most surprising aspect of our results is that poverty in absolute terms has fallen at the same time as mass unemployment has emerged. Some insight into this paradox can be obtained by disaggregating poverty statistics for 1995 and 2002 by the economic activity of the household head. Of particular interest is the comparison between households headed by employed workers and those headed by the unemployed. (Other categories include those headed by the retired and by those who do not participate in the labor market for other reasons, for example because they are attending to domestic duties). Given these mutually exclusive groups, it is possible to decompose the overall change in the proportion living in poverty into the effects of changes in poverty within the groups and changes in the size of each group (Ravallion and Huppi, 1991).

The impact of the program of lay-offs in the state sector in the second half of the 1990s can be seen in the population shares of the various groups. In 1995, only 0.4 percent of individuals in the sample lived in households headed by an unemployed worker. In 2002, this percentage had risen to 6.2 percent. Perhaps even more revealingly, the proportion living in households with employed heads fell from 80 percent to 71 percent. Although one might expect some increase in the proportion living in households with retired heads due to an ageing of the population, our figures are consistent with some of the retrenchment in China having taken the form of early retirement rather than unemployment *per se.*

Other things being equal, the emergence of mass unemployment would be expected to increase poverty. This is born out by the contribution of the population shifts to the sectoral decomposition of poverty. For example, if we consider the $2 a day line, population shifts would increase the poverty headcount by 1.3 points, from its 1995 level of 7 percent. However, this is more than offset by falls in poverty rates within groups. For example, the fall in the headcount among those living in households with employed heads would imply a 4.1 point drop in the poverty headcount. This alone would account for four-fifths of the observed fall in poverty. Moreover, the interaction effects also imply falls in poverty because the groups that have grown in size – those headed by the unemployed and the retired – have also experienced the fastest reduction in poverty. Qualitatively similar results are obtained when using the $3 a day line.

The rise in unemployment is not as disastrous as might be thought because only a minority of households headed by the unemployed are poor. For example, using $2 a day as the poverty line, 7 percent of people in households headed by the unemployed are poor in 2002. This is only a small minority, even though substantially above the 2 percent poverty headcount for the total urban population. If $3 a day was used as the poverty line, 22 percent of those in households headed by the unemployed would be poor. These figures are all the more remarkable because our poverty rates are measured by income, rather than consumption. Clearly, households with unemployed heads are finding sources of income other than their heads' earnings to support themselves. This income is partly earnings from the spouse of the head (or other family members). In this respect, the high rates of female employment in urban China should be acknowledged.

Summary of findings

Using comparable surveys with the best accounting of income available, we have shown that living standards rose across the distribution of income from 1988 to 2002. This truth has been masked by conventional analysis of urban poverty in China which defines only a very small minority of the urban population as poor – for example, the 1 percent obtained using a "dollar a day" poverty line. We find evidence that the withdrawal of subsidies between 1988 and 1995 lowered the real income of the poorest in urban areas. However, this was subsequently outweighed by growth in other sources of income. Perhaps most surprisingly, we find that – despite high unemployment after 1995 – absolute poverty continued to fall, irrespective of where the poverty line was set. This implies that the concern that absolute poverty has risen during urban reform is misplaced.

Discussion and Conclusions

To conclude this chapter, we return to the four theoretical perspectives outlined in the introduction to this book: specifically, modernization theory, dependency theory, the developmental state theory, and market transition theory. We consider which of these four approaches fits best with the trends in poverty we document for urban China from 1988. It should be noted that not all the four theoretical perspectives are mutually exclusive. We will argue that modernization theory perhaps provides the most useful vantage point. However, transitional theory can also contribute some insights. By contrast, the developmental state theory seems to contribute little to understanding our results, while the predictions of dependency theory seem flatly refuted.

In our context, we interpret modernization theory as referring to neoclassical theories of economic growth. From this perspective, urban China's rapid economic growth is to be understood as primarily "catch-up" growth. High rates of saving and investment give rise to a growth "miracle" until capital hits diminishing returns. This process may be compounded by improvements in total factor productivity, as China switches from traditional technology to the best practice technology available. Such theorizing provides a useful way of accounting for rapid economic growth, although it falls short of explaining some of the more fundamental processes at work (Why is China's investment rate high? What factors facilitate the transfer of technology?). More narrowly, in terms of urban poverty, simple neoclassical growth theory predicts that growth will be accompanied by falls in absolute poverty, to the extent that a rising capital stock and improvements in technology raise the return to labor. This prediction does not necessarily follow if we extend the analysis to allow for a rural sector with surplus labor. In such a context, as in the Lewis model, growth will lead to increased rural-urban migration, keeping down the wages of urban residents. In China, we show that – although migration has increased – the living standards (and real wages) of urban residents have nonetheless risen. This may be partly due to residual administrative restrictions on migration and/or labor market segmentation that at least partly insulate urban residents from competition with migrants.

Dependency theory stands in opposition to modernization theory and seems to fit the Chinese experience less well. China's strong growth during a process of opening up to the West is in contradiction to the basic predictions of the dependency theorists. Likewise, far from growth being "immiserating," as some dependency theorists postulated, it has led to reduced absolute poverty among urban residents. Dependency theorists may point to the widening inequalities in urban China, but we question whether this widening is an ongoing process. Instead, the rise in income inequality in urban China may have been largely a one-off response to the removal of some subsidies in the transition from socialism to a market-based inequality.

The theory of the developmental state could potentially be reconciled with modernization theory, to the extent that the state simply substitutes for the market in providing capital accumulation and technological improvements. However, in practice, there tends to be a tension between the two approaches with advocates of a developmental state stressing the importance of government interventions in driving growth. By contrast, modernization theorists tend to assume growth could happen without or even despite such interventions. This debate has tended to be fought most fiercely over East Asian NICs such as Korea. In the case of China, it seems more natural to side with the modernizers, as growth takes off as the state gradually reduces its control over the economy and allows the market to flourish. Indeed, one of the key current challenges for the Chinese government is how to deal with a large but loss-making state sector. The relevance of this debate for analysis of trends in urban poverty is unclear. It is true that China's growth has been less pro-poor and more unequal than that of other East Asian NICs. Nonetheless, this contrast may reflect China's much greater scale and hence internal differentiation rather than anything to do with the role of the developmental state *per se*.

The fourth and final theoretical perspective, transitional theory, is also not mutually exclusive with modernization theory. Indeed, the transition from an inefficient command-economy to a leaner market economy is one reason why we might expect improvements in total factor productivity ("technology") to play a large role in China's economic growth after 1978 – larger than in other East Asian NICs characterized as following the developmental state model. In terms of urban poverty, a transition from socialism is likely to lead to a worsening of inequality and indeed absolute poverty if growth is inadequate to offset this. In China, we do find some initial support for this hypothesis when comparing the data for 1988 and 1994. During this period, the removal of state subsidies and rationing appear to have led to a worsening of the income distribution and a fall in the living standards of the poorest. However, this adjustment was a once-and-for-all one, so it does not necessarily imply a continuing worsening of inequality. Indeed, we find income inequality subsequently to be largely unchanged and hence absolute poverty falls due to strong economic growth.

In sum, we find something useful in three of the four theoretical perspectives discussed in the introduction to this volume (dependency theory being useful only as a straw man). Some readers may find this a bit perplexing, citing the foundational differences among the perspectives. However, to us, the three perspectives provide alternative ways of looking at China's urban development somewhat akin to looking at a mountain from three different angles. While they have differences in emphasis, they do provide insights that are not always mutually exclusive. Certainly it is hard to argue that urban China is not modernizing, does not have a development state and is not in transition.

To conclude, we reiterate that our finding of a fall in poverty among urban residents does not imply that there have been no losers from economic reform. Unemployment obviously adversely affects those subject to it. However, most households whose heads have suffered unemployment have not been driven into absolute poverty, even when the poverty line is defined well in excess of the low "dollar a day" thresholds commonly applied to urban China. Here, a limitation of our money metric approach to poverty and well-being should be noted. We may be understating the costs of unemployment to the extent that it causes suffering not captured purely by the loss of income – for example, social isolation and a loss of self-worth. Nor do our findings deny that relative poverty has risen. Although lower income urban households have experienced improvements in living standards, the rates of growth have been below those enjoyed by more affluent households. Since relative poverty and unemployment are both of concern to people – and hence may be a source for potential social unrest – our results on absolute poverty should not be taken as an argument for complacency when reviewing the distributional aspects of China's economic progress. Nor have we touched upon the rural-urban migrants, whose living standards are much lower than those of urban residents. Nonetheless, it is important to challenge the idea that absolute poverty rose during China's reforms before it becomes a new urban legend.

ACKNOWLEDGMENTS

The authors are grateful to Qingjie Xia for invaluable research assistance. Valuable comments were made by the editors and by an anonymous referee.

REFERENCES

Appleton, S., J. Knight, L. Song, and Q. Xia. (2002) "Labour Retrenchment in China: Determinants and Consequences." *China Economic Review* 13, 2–3: 252–75.

Benjamin, D., L. Brandt, and J. Giles. (2004) "The Dynamics of Inequality and Growth in Rural China: Does higher Inequality Impede Growth?" Paper presented at a conference on "Poverty, Inequality, Labor Market and Welfare Reform in China," August 25–27, Canberra: Australian National University.

Chen, S., and Y. Wang. (2001) "China's Growth and Poverty Reduction: Recent Trends between 1990 and 1999." World Bank Policy Research Working Paper 2651.

Chenery, H., M.S. Ahluwalia, C.L.G. Bell, J.L. Duloy, and R. Jolly. (1974) *Redistribution with Growth.* Oxford: Oxford University Press.

Datt, G., and M. Ravallion. (1992) "Growth and Redistribution Components of Changes in Poverty Measures: A Decomposition with Application to Brazil and India in the 1980s." *Journal of Development Economics* 38: 275–95.

Dollar, D., and A. Kraay. (2001) "Growth *is* Good for the Poor." *Journal of Economic Growth* 7, 3: 195–225.

The Economist. (2001) "To Each According to His Abilities." *The Economist* May 31: 39.

Fang, C., X. Zhang, and S. Fan. (2002) "Emergence of Urban Poverty and Inequality in China: Evidence from Household Survey." *China Economic Review* 12: 430–43.

Fields, G. (2001) *Distribution and Development.* Cambridge, MA: Russell Sage.

Fishlow, A. (1972) "Brazilian Size Distribution of Income." *American Economic Review Papers and Proceedings* 391–402.

Foster, J., J. Greer, and E. Thorbecke. (1984) "A Class of Decomposable Poverty Indices." *Econometrica* 52: 761–5.

Khan, A. (1996) *The Impact of Recent Macroeconomic and Sectoral Changes on the Poor and Women in China.* New Delhi: International Labor Organization.

Khan, A. (1998) "Poverty in China in the Era of Globalization." Issues in Development Discussion Paper 22, Geneva: International Labor Organization.

Khan, A., and C. Riskin. (2001) *Inequality and Poverty in China in the Age of Globalization.* New York: Oxford University Press.

Khan, A.R., K. Griffin, C. Riskin, and Z. Renwei. (1993) "Household Income and its Definition in China". In K. Griffin and Z. Renwei (eds.), *The Distribution of Income in China.* London: Macmillan.

Knight, J., and L. Song. (1999) *The Rural-Urban Divide: Economic Disparities and Interactions in China.* Oxford: Oxford University Press.

Kuznets, S. (1955) "Economic Growth and Income Inequality." *American Economic Review* 45, 1: 1–28.

Meng, X., R. Gregory, and Y. Wang. (2005) "Poverty, Inequality, and Growth in Urban China, 1986–2000." IZA Discussion Paper No. 1452, online at: http://ssrn.com/abstract=644327

Park, A., Y. Du, and S. Wang. (2004) "Is Migration Helping China's Poor?" Paper presented at a conference on "Poverty, Inequality, Labor Market and Welfare Reform in China," August 25–27, Canberra: Australian National University.

Ravallion, M., and S. Chen. (2004) "China's (Uneven) Progress against Poverty." Paper presented at a conference on "Poverty, Inequality, Labor Market and Welfare Reform in China," August 25–27, Canberra: Australian National University.

Ravallion, M., and M. Huppi. (1991) "Measuring Changes in Poverty: A Methodological Case Study of Indonesia during an Adjustment Period." *World Bank Economic Review* 5, 1: 57–82.

Wang, Q.B., G.M. Shi, and Y. Zheng. (2002) "Changes in income Inequality and Welfare under Economic Transition: Evidence from Urban China." *Applied Economics Letters* 9, 15: 989–91.

Wu. F. (2004) "Urban Poverty and Marginalization under Market Transition: The Case of Chinese Cities." *International Journal of Urban and Regional Research* 28, 2: 401–23.

Wong, L. (1998) *Marginalization and Social Welfare in China*. London: Routledge.

3

Class Structure and Class Inequality in Urban China and Russia: Effects of Institutional Change or Economic Performance?

Yanjie Bian and Theodore P. Gerber

Introduction

This chapter examines whether and how class structure and class inequality have changed in urban China and Russia during their transitions in the direction of market-based economies. Our main objective is to derive a set of expectations regarding particular changes in class structure and inequality from broader theories of how marketization and economic growth affect inequality. We also provide some initial empirical tests of our expectations using survey data spanning 1983 to 2003 in China and 1985 to 2001 in Russia. The data permit us to measure patterns of stability and change within each country as market reforms have progressed and to make precise comparisons across national settings at particular stages of the transition process.

Our comparative design is well suited to analyze how stratification processes are and are not affected by market transition. A China-Russia comparison can provide insights that inform a key, if at times implicit, issue in the literature: does market transition affect stratification mainly through *institutional changes* (the introduction of markets, the withdrawal of state planning and redistribution) or mainly through the *changes in economic performance* (rapid economic growth in China, rapid contraction in Russia) that accompany market reforms? Because China experienced institutional change in the direction of markets and rapid economic growth simultaneously, it is hard to disentangle the effects of these two processes on class inequality. In contrast,

Russia's market reforms led to a sustained economic contraction, followed at the end of the 1990s by recovery. Thus, Russia offers a useful comparison case for assessing which factor has been paramount in urban China.

Our empirical analyses address three questions: First, how has the occupational class structure changed in urban China and Russia in the course of their transitions to the market? Second, how has the *level* of class inequality in earnings changed? And third, how has the *structure* of class inequality in earnings changed? That is, which classes benefit most and which suffer most in the course of economic reforms? The empirical answers to these questions based on our survey data from the two countries are relevant for a set of larger theoretical issues involving the effects of market transition on stratification processes, one of the main themes of this volume: Does the magnitude of inequality tend to increase, decrease, or remain the same as a result of market transition? Which aspect of market transition appears more decisive in shaping the emerging levels and patterns of inequality – institutional changes or changes in economic performance (growth vs. stagnation)? Does market transition have any generic effects on stratification in the different countries that undergo it, or are the trajectories of stratification processes in different countries so divergent that no common patterns can be identified?

These questions are important for understanding the changing pattern of class-income inequality, a key form of urban transformations in China. In a simplistic yet useful view, the socialist Chinese city had a dual-class structure consisting of the "socialist elite," given political power, material privileges, or social prestige by the redistributive regime, and the "working-class masses" which included all non-elites. Although they were institutionally secured for lifelong employment in the state sector, the working class nonetheless lived with low incomes in crowded housing space due to policies of economic egalitarianism. For both classes, the redistributive benefits mostly took the form of income-in-kind (allocated housing, free compulsory education, socialized medical insurance, government food coupons, work-unit fringe benefits, etc.), making wage incomes a less important source of class distinction. The post-1980 reforms maintained some of these incomes-in-kind for the very top political and military elite. But, more importantly, they also marketized and commodified consumption, making wage and other cash incomes the more fundamental source of class and lifestyle distinctions. Meanwhile, the transition toward a market system has been reconfiguring the class structure. The socialist elite might have maintained its privileges because most of them could capitalize on their positional power and political networks in the growing market economy. But the working-class masses may well have become more differentiated, because professionals, non-manual, and manual workers held different market positions to compete for jobs with varying returns for their skills. Thus, changing class-based income inequality is an integral part of urban transformations in China.

In the rest of this chapter, we first provide some historical background on China and Russia, comparing similarities and differences of recent economic transformations in the two countries. Next, we derive expectations regarding the emerging patterns of class and income inequalities from two competing perspectives on post-Socialist stratification, emphasizing, respectively, institutional change and change in economic performance. We outline our analytic logic for empirically addressing these explanations with a China-Russia comparison. After describing our data and methods, we present preliminary results obtained to test the expectations derived from the competing explanations. In the conclusion, we use the China-Russia comparison to engage larger theoretical issues about impacts of market reforms on stratification processes in transition economies.

Class and Reforms in China and Russia: Brief Historical Background

Communist regimes sought to eliminate class inequality – at least in principle. Even though class inequalities of some form persisted at the height of Communism in both China and Russia, these inequalities were lower in magnitude than in most market-based societies (Whyte 1984). The growth strategy prioritizing industrial development led to the underdevelopment and exploitation of agriculture in the form of collective Communes, making the gap between rural and urban populations a fundamental form of inequality. But within the urban sector, class differences in earnings and, more broadly, in material standing were kept at lower levels via administrative means. A variety of institutions kept class-based inequalities within limits (Connor 1979; Whyte and Parish 1984): prohibition of private ownership of the means of production, state planning, centralized wage and price setting, non-market distribution of consumer goods, and administrative control over nearly all aspects of economic life (as opposed to reliance on the market for the allocation of labor, production factors, goods, and services).

Communist systems produced their own privileged groups (Djilas 1957). Members of the political and managerial elite constituted one such group (Szelényi 1978; Walder 1985). Professionals, to a lesser extent, formed another (Konrad and Szelényi 1982). At the other end of the scale, agricultural laborers and urban workers in the "collective sector" persistently lagged behind state industrial workers in their standards of living (Bian 1994; Parish and Whyte 1978; Walder 1986). Nonetheless, the absence of "capitalists" in the class structure and the relatively narrow gaps between the classes under Communism meant that class inequality was substantially more limited than in countries with full-blown market economies.

The introduction of market reforms by the Chinese leadership under Deng Xiaoping in the early 1980s and by the newly independent Russian government under Boris Yeltsin in 1992 eliminated the institutional brakes on class-based inequality. At the same time, the reforms created the opportunities for a new class of proprietors to form. By virtually all accounts, these reforms have been accompanied by sharp increases in the levels of economic inequality in both countries (see reviews by Bian (2002) on China and Gerber and Hout (1998) on Russia). But what has been the impact on *class-based* inequality? This question has escaped systematic analysis, despite the considerable scholarly attention devoted to overall levels of inequality and correlates of socioeconomic status such as education, Communist party membership, social networks, and gender. An exception is the study by Nee and Cao (2002) of occupational attainment: they report an increase in occupation-based earnings inequality between 1990 and 1995. However, their analysis does not employ standard class categories used in other national contexts and it is limited to this brief five-year period.

Class inequalities and class conflicts seem to have become a mainstay of Chinese politics and culture. Private owners and wage laborers have become increasingly divided as the Communist factory regime (Walder 1986) has given way to a capitalist pattern of "disorganized despotism" (Lee 1998). New forms of class exploitation, such as the severe oppression of "peasant migrant labor" by management and capitalists, have grown widespread (Pun 1999). Labor protests and official corruption are an everyday phenomenon and are the main theme of political movies and TV plays (Chan 1995). Class boundaries are reflected in housing, car ownership, and whether children are sent to overseas boarding schools (Davis 2000). And class closure is behind class-bound "New Year greeting networks" (Bian et al. 2005). In fact, class conflicts have been substantially more salient in Chinese culture and politics than in Russia, where the working class has done little to challenge the new class of wealthy owners and managers. But this may reflect a deeply ingrained culture of acquiescence and apathy on the part of Russian workers. Class inequalities may have nonetheless increased in Russia.

China and Russia are similar in some respects: both are large, militarily powerful, former state socialist countries characterized (during their Communist eras) by a planned, state-administered economy, Communist Party rule, and relative isolation from the global capitalist economy. Both introduced major reforms beginning in the 1980s with the intention of introducing market mechanisms into their respective economies. However, there are also a number of significant differences between the two countries. In Table 3.1 we present official data from China and Russia that reflect their divergent patterns of institutional change and changes in economic performance during the course of their respective transitions. Pre-1992 data are scarce for Russia, since Russia was part of the Soviet Union until that time. Also, official Russian sources do not provide separate data for urban and rural Russia. Thus, we can only provide national data for 1991 to 2002. Note that

Table 3.1 Indicators of economic development and labor allocation, 1978–2002

Year	*GDP annual growth in %*			*Urban private sector employment in %*		
	Tianjin	*China*	*Russia*	*Tianjin*	*Urban China*	*Russia*
1978	20.9	11.7		0.1		
1979	10.0	7.6		0.0		
1980	10.0	7.8		0.4		
1981	4.8	5.2		0.4		
1982	4.3	9.1		0.6		
1983	8.3	10.9		0.8		
1984	19.3	15.2		1.3		
1985	10.6	13.5		2.2		
1986	5.8	8.8		2.1		
1987	7.6	11.6		2.7		
1988	5.8	11.3		2.9		
1989	1.6	4.1		3.2		
1990	5.4	3.8		3.5		
1991	6.0	9.2	–5.0	5.0		12.6
1992	11.7	14.2	–14.5	6.1	15.8	13.5
1993	12.1	13.5	–8.7	8.4	18.6	18.6
1994	14.3	12.6	–12.7	12.8	21.3	28.5
1995	14.9	10.5	–4.1	15.4	24.5	33.4
1996	14.3	9.6	–3.5	17.4	28.0	35.0
1997	12.1	8.8	0.8	20.2	31.1	36.4
1998	9.3	7.8	–4.9	24.4	46.7	41.1
1999	10.0	7.1	5.4	28.6	51.1	44.8
2000	10.8	8.0	9.0	31.1	58.5	46.1
2001	12.0	7.5	5.0	37.2	62.7	48.8
2002	12.5	8.3	4.3	44.4	66.6	50.2
2003	14.8	9.3	7.3	49.6	69.3	53.7

Sources: Statistical Bureau of Tianjin, *Tianjin Statistical Yearbooks* (1985–2004). Beijing: Chinese Statistical Press. State Statistical Bureau, *Chinese Statistical Yearbooks* (1997, 2001, 2004). Beijing: Chinese Statistical Press. Goskomstat [State Statistical Committee of the Russian Federation], *Rossiiskii Statisticheskii Ezhegodnik* [Russian Statistical Yearbook]. (1996, 2004).

aside from a small number of cooperative owners, most of those formally employed in the "private sector" in Russia in the pre-transition years (prior to 1992) were collective farm workers in the countryside, as Russian official statistics classify collective farmers as such, even though collective farms were legally treated as communal (rather than state) property.

In China, market-oriented reforms have been introduced in a gradual and incremental fashion, with the aim of achieving rapid economic growth without damaging the social-political order of the Communist party-state. Rural reforms started in the early spring of 1979, immediately yielding a dramatic improvement in the harvest that fall. But reforms in the cities, the focus of this book, did not begin until 1983. In a departure from the system put in place under Mao, state factory profits became subjected to tax and factories could retain part of their after-tax earnings (Naughton 1995). The government still determined workers' salaries and raises. Bonuses resulting from retained earnings by the factories were distributed rather equally among workers within a factory (Walder 1987); however, they grew increasingly unequal between factories (Bian 1994, ch. 7). The spread of small commodity markets, the introduction of a managerial responsibility system, and the emergence of labor contracts and "talent markets" in the late 1980s increased income for household business owners and the higher classes with managerial authority and professional skill.

After Deng Xiaoping's 1992 South China Tour spurred a new wave of capitalist-oriented market reforms, private ownership became legitimate, labor markets emerged, inter-firm mobility began rising, "peasant migrant workers" flooded cities and towns, financial markets diversified and grew locally, and foreign direct investments increased rapidly year by year. These trends continued through the first years of the new century. The late 1990s saw significant property rights reforms in the state sector. Subsequently, rising layoffs in state-owned firms were accompanied by re-employments in non-state sectors.

Meanwhile, China has enjoyed a great deal of regime stability. The state has remained strong despite the institutional changes associated with market reforms (Bian and Zhang 2002). Perhaps this very political stability and state strength helped the Chinese economic reforms to stimulate rapid economic growth (Burawoy 1997; Walder 2003). It is also noteworthy that China began its reforms at an earlier stage of development than the centrally planned economies of Eastern Europe and the former Soviet Union: 80 percent of China's population worked in the agricultural sector and the per capita GDP placed China in the ranks of developing countries rather than advanced industrial societies. China's earlier starting point may also help explain its rapid growth rates, particularly as market reforms were initially introduced in the agrarian sector. In any event, market reforms and strong, persistent economic growth have gone hand in hand in China. This is the main theme of a comprehensive review by Wu Jinglian (2003), one of the few respected Chinese economists who contributed to reform policies.

Soviet-era Russia was already a "developed" society, with 74 percent of the population urbanized in 1989 and only 13 percent of the employed population working in agriculture (Goskomstat Rossii 1996). Russia's reforms were introduced in the context of great political instability; the collapse of the Soviet Union meant that the Russian government had to forge new political institutions and economic policies simultaneously, while the entire process was contested. The lack of cohesion within the ruling elite and the nascent character of state institutions have made for a very weak Russian state and rampant corruption.

In contrast to China's gradual market reforms, Russia's institutional changes have been more radical, following the "shock therapy" model first applied to the Polish reforms of the late 1980s. Recommended by foreign advisers, this model was designed to privatize the economy within a short period of time. In early 1992, the Russian government abolished price controls, freed the exchange rate, eliminated planning, opened the economy up to imports, and removed restrictions on private ownership of the means of production. The government quickly privatized small enterprises, and by 1993 began the rapid privatization of medium and large enterprises via a complex voucher scheme. The "shock" of these reforms proved more devastating and lasting than the therapy (Gerber and Hout 1998). Prices skyrocketed as the government printed money to finance its growing deficits. Supply chains and distribution networks were disrupted by the chaos that ensued after the collapse of the Soviet Union. Heavily subsidized industries like defense, mining, and steel production could not stand on their own feet. In other sectors, domestic producers lost out in the competition with cheap foreign imports (including many from China). Firms could not obtain capital needed for restructuring, and the state could not collect its taxes. In the scramble to survive and profit in the new environment, managers engaged in asset stripping for their personal benefit, and banks took part in speculative schemes rather than provide capital for investment. Wage arrears and barter reached historically unprecedented heights. While most of the population saw its wages eroded by steep inflation and arrears, a small minority reaped vast profits from insider deals for state property.

The economy quickly plunged into a deep crisis, which lasted until 1997, the first year the Russian economy did not contract. August 1998 dealt another blow when the government defaulted on high interest bonds. But beginning in 1999, a combination of relative stability, some successful restructuring, and high oil prices on the international market produced several years of substantial economic growth. Thus, in contrast to China, Russia's more rapid market reforms were initially accompanied by a steep decline in economic performance. At the end of the 1990s, however, economic growth resumed in Russia, though by then the process of institutional change had essentially been completed.

In sum, China and Russia have diverged in terms of their initial "starting point" at the time reforms were introduced, the rapidity and scope of the institutional changes, and the growth trajectories that have accompanied the transition process. These contrasts mean that a China-Russia comparison has potential to yield insights into the respective roles played by institutional reforms and changes in economic performance in shaping class inequality in transition societies.

Competing Explanations: Institutional Change or Economic Growth?

Institutional change: The impact of market reforms

A lively and sizable literature has emerged examining the impact of market transition on stratification processes in China, Russia, and Eastern Europe. Victor Nee's "market transition theory" (Nee 1989, 1991, 1996; Cao and Nee 2000; Nee and Cao 1999, 2002) has had a formative influence on this literature. Nee argues that as markets come to play a greater role in allocating economic resources and, correspondingly, state redistribution recedes in importance, the economic returns to human capital and entrepreneurship will increase. Meanwhile, the returns to political capital (operationalized as Communist Party membership and cadre status) will decrease. In his initial formulations, Nee believed that these two trends would offset each other, and thus overall inequality would likely remain stable, if not to decline.

Nee's theory has generated controversy, as other scholars have challenged or refined various aspects of his claims, both theoretically and empirically (Bian and Logan 1996; Bian and Zhang 2002; Parish and Michelson 1996; Walder 1996; Xie and Hannum 1996; Zhou 2000). This literature has focused on issues first put forward by Nee: whether the growth of market economies increases the returns to education (human capital) and decreases the returns to cadre status (political power). Some studies have examined changes in job mobility (Zhou, Tuma, and Moen 1996, 1997), elite formation (Walder 1995; Walder, Li, and Treiman 2000), and gender inequality (Shu and Bian 2002, 2003), but these offer fewer broad theoretical guidelines for the topic of class inequality. We have seen no sustained analysis of patterns and trends in class structure and class inequality in the course of China's transition. The literature on post-Socialist stratification in Eastern Europe and the former Soviet Union has also grown (Gerber 2000, 2001a, 2001b, 2002, 2003; Gerber and Hout 1998, 2004; Gerber and Schaefer 2004; Rona-Tas 1994). These studies have pointed away from the marketization effects predicted by Nee. But with few exceptions (Gerber and Hout 1998, 2004), these studies also do not focus on class inequality.

Nee and Cao (2002) note that their descriptive data shows increasing earnings inequality across seven occupational groups between 1990 and 1995 in their study of occupational attainment. They comment (2002, p. 23): "As the Communist party's promise of an egalitarian utopia continues to fade, landing on the right career track becomes increasingly crucial for the Chinese urbanites." This statement clearly suggests that the retreat from socialist redistribution is the driving force behind increased inequalities based on occupational class. In other words, consistent with Nee's arguments regarding returns to education and political capital, *institutional changes* (the retreat from state-based economic institutions and the rise of market institutions) are the driving force shaping patterns of class inequality in transition societies. In fact, Nee's argument regarding the enhanced returns to productivity due to marketization implies predictions about trends in specific class-based inequalities (see Gerber and Hout 1998). *Professionals* and (relative to unskilled manual workers) *skilled workers* typically have high human capital. According to Nee's logic, their earnings advantages over unskilled manual and non-manual workers should increase. The new class of *entrepreneurs* should have large and increasing advantages stemming from the rewards to innovation, creativity, and risk taking.

Researchers have not proposed explicit theories regarding the impact of marketization on class structure (i.e., on the distribution of the workforce across occupational class categories). The most straightforward prediction would clearly be an increase in the class of proprietors, which virtually does not exist under state socialism. Also, the shrinking of the state sector may have driven recently laid off workers into small-scale business ventures (Wu and Xie 2003). Another possible effect of institutional change is more subtle, and depends not only on the introduction of market reforms but also on the investment policies of socialist-era planners. For both political and economic reasons, the leaders of state socialist economies tended to emphasize investment in heavy industry and defense production at the expense of investment in consumer goods and in consumer services. The introduction of markets implies that demand, rather than political decisions, should determine the flow of capital and resources to various sectors of the economy. Thus, when market reforms are introduced in a context of both pent-up demand for consumer goods and services and oversupply of heavy industrial and defense production, the expected result of this institutional change would be a structural shift of the workforce out of heavy industry and into consumer goods and services (Gerber 2002). Although this shift is largely sectoral in nature, it implies a change in the class structure as well: a *decline in the share of skilled and unskilled manual* occupations and a corresponding *increase* in the share of *routine non-manual* occupations.

Growth and class inequality

An alternative perspective for anticipating developments in class inequality in contemporary urban China focuses on economic growth rather than institutional change as the chief factor motivating changes in stratification. Xie and Hannum (1996) found higher returns to education in regions marked by more rapid economic growth. Incorporating temporal and regional variation, Hauser and Xie (2005) repeated this analysis and obtained about the same findings. Walder (2002) sought to disentangle the effects of growth from the effects of market reforms *per se*, and he found that returns to education and to position of political power both increased in villages of higher growth but not necessarily greater privatization. Zhou et al. (1996, 1997), Walder et al. (2000), and Zhao and Zhou (2002) point to long-term changes in the returns to education and elite formation processes that they attribute not to institutional change but to China's economic development and associated policies.

How is growth related to class structure and class differences in earnings? The growth process itself often involves a dramatic shift of the class structure. According to the influential model of Kuznets, countries experience a shift from agricultural to industrial classes when they undergo modernizing growth (Kuznets 1955; Nielsen 1994). This would obviously be relevant for all of China. But our focus is on urban China, where the agricultural classes were negligible to begin with. Nonetheless, the developmental shift accompanying industrialization is relevant, as it pushes migrants from the countryside into urban areas. Given their low education and dubious legal status, migrants would probably swell the ranks of the unskilled manual and non-manual classes. Thus, China's rapid economic growth seems likely to produce an increase in the share of these classes in urban areas. Moreover, in developed countries, recent periods of growth driven by technological changes and globalization have seen a decline in the manual workforce and expansion of both lower-level service occupations and professionals (Alderson 1999; Nielsen and Alderson 1997). These developments have led some scholars to reject the Kuznets model altogether, and others to recast Kuznets' basic insight that growth entails population redistributions across sectors and categories that, in turn, have implications for the magnitude and pattern of inequality (see Korzeniewicz and Moran 2005). Although China should not be considered a developed country and tends to export goods rather than capital, technological advances associated with its recent growth may nonetheless also drive an increase in the size of the professional classes.

As for the impact of growth on class-based earnings inequalities, Chiswick (1971) suggested that economic growth increases the returns to human capital because individuals who allocate resources efficiently are better able to profit from new opportunities. This implies that growth should increase the earnings advantages of *professionals* and *managers*. Growth, in other words, puts a greater premium on technical skills and training, which many labor economists see as the source of increased returns to college education in the US during the economic boom of the 1980s and 1990s (see Morris and Western 1999).

Another approach can be found in the original argument of Kuznets (1955): at early stages of development economic growth is accompanied by increasing inequality, while at more advanced stages of development, growth is accompanied by decreasing inequality (see the contribution of Appleton and Song to this volume). Although Kuznets focuses mainly on the shift from the traditional to the modern sectors, his discussion also notes a tendency for the middle classes to gain income share in the urban areas following this initial transformation. This claim can be interpreted to imply a growth in the advantages (relative to those at the bottom) of those classes in the middle of China's earnings distribution. Of course, the impact of growth on class inequality may defy generalization; it may depend on whether and what kinds of changes in economic and political institutions accompany growth.

Logic of Comparison

The effects of growth and institutional change are difficult to disentangle in China, because the two factors co-vary regionally and temporally (Hauser and Xie 2005). This makes the Russian case useful for elucidating which factor is more decisive in shaping class structure and class inequality: Russia has experienced a period of slow reform and slow contraction (the late Soviet era: 1985–91), a period of rapid institutional change and rapid contraction (1992–6), and a period of very little new institutional change and moderate growth (1997–2002, with the exception of 1998). Thus, economic reforms and economic performance have not co-varied in Russia the way they have in China. By comparing the patterns and trends in class structure and inequality in Russia over these three periods with the patterns observed in urban China throughout the transition era, we can gain insight into the relative influence of institutional change and economic performance in shaping patterns of inequality.

We schematically illustrate our initial logic of comparison in Table 3.2. We proceed by obtaining empirical estimates for changes in the distribution of the work force across class categories (class structure) and both the magnitude (overall class-based inequality) and form (advantages of particular classes) of class inequality in earnings in urban China and Russia for the periods in question. Unfortunately, we do not have the data we would need to model class differences in earnings in pre-1992 Russia, so the comparison of this period with China will be limited to patterns of class structure. Based on our empirical results, we ask: How do the patterns observed in Russia during each of the three periods compare to the patterns observed in China? The observed patterns of similarity and difference might yield conclusions about the relative impact of institutional changes and changes in economic performance. For example, if (as Nee's theory implies) marketization increases the returns to human capital, we would expect to see an increase in the advantages of professionals in Tianjin and in early reform Russia, but not in late reform Russia (when there was little additional movement in the direction of markets.) If economic growth improves the relative standing of the middle classes, we should see increased middle class advantages in Tianjin and in late reform Russia, but decreased advantages in early reform Russia. Our reasoning is reflected in the entries in the final two columns of Table 3.2.

Table 3.2 Logic of comparison

	Pace of economic reforms	*Change in economic performance*	*Russia/China similarity implies...*	*Russia/China difference implies...*
Tianjin, 1983–2003	Gradual	Rapid growth		
Russia, 1985–91	Gradual	Moderate contraction	Institutional changes decisive	Economic performance decisive
Russia, 1992–6	Rapid	Rapid contraction	Institutional changes decisive	Economic performance decisive
Russia, 1997–2002	None	Moderate growth	Economic growth decisive	Institutional changes decisive

Of course, we may encounter patterns of change that are too complex to analyze with a simple schema like that in Table 3.2. Our initial empirical findings – presented below – already point in that direction. The logic of the Kuznets curve may imply, for example, that growth will have a different effect on inequality in Russia than in China, since Russia began the process at a more advanced level of development. In any case, we now examine patterns in class structure and in the effects of class on earnings empirically. Then we attempt to make theoretical sense of what they reveal about the relative importance of institutional change and growth as factors shaping class inequality in urban China and other post-Socialist contexts.

Data and Measures

Our China data come from a single but very large city, Tianjin. Starting in 1988, the first author conducted his Tianjin household surveys based on the same multi-stage probability sampling procedure every five years, and for this chapter we analyzed the surveys of 1988, 1993, 1999, and 2003. All these surveys collected information necessary for analysis of earnings inequalities at the individual level. Because the year 1983 is important as a starting reference point, the year's retrospective data from the 1988 survey are also analyzed. While Tianjin is certainly not representative of urban China and may be placed in the lower end of the top one-tenth of all Chinese cities in terms of pace of market reforms and economic growth, Tianjin's surveys combined with official statistics are a valuable source of data for scholarly analysis.

The advantage of using the Tianjin data is that they cover a long time span. While no national data of this time span are available from China, this large city's series of sample surveys permits a comparison with Russia.

Our Russian data come from surveys conducted in 1993 (with retrospective data going back to 1988), 1996, 1998 (with retrospective data going back to 1991), 2000, and 2001–2 (with retrospective data going back to December 1984). These surveys were conducted with nationally representative, multi-stage samples, including households in both urban and rural areas. Because all our Chinese data come from an urban area, we include in our analysis only the urban portion of the Russia surveys. Such a data manipulation suits our interest in the changing occupational class structure and earnings inequality over time.

To measure social class in China and Russia, we use a version of the Erikson-Goldthorpe class schema (Erikson and Goldthorpe 1992), which treats ownership, authority, and skill as the bases of class distinction in modern economic systems. This is the most widely used class schema in international research on stratification and mobility; thus, it helps us generate results that can be replicated in other countries without great difficulty. Gerber and Hout (1998; 2004) adapted the schema to Soviet and post-Soviet Russia by aggregating skilled manual workers and supervisors of manual workers (classes V and VI), separating managers from the upper and lower professional classes, and combining all proprietors (with or without employees) into a single category. We implement these adaptations, consistent with Gerber and Hout (1998, 2004) who show that the managerial class is distinct, both in terms of class effects on earnings and in terms of the social distance between classes as measured by intergenerational class mobility. While past Chinese studies have been based on various schema of classifying occupations into some kind of categories, we believe that the modified Erikson-Goldthorpe schema is adequate for China because the country is increasingly capitalistic, globalized, and market-oriented.

Results

Class structure

Table 3.3 shows change and stability in class structure (overall and by gender). For the most part, the class structures of both countries persist across the entire transition period, with some exceptions consistent with the institutional perspective. In China (at least in Tianjin), the proportion of managers decreased from 1993 onward. This probably reflects institutional changes: the dismantling of small and medium-sized state and collective enterprises after 1993 most likely pushed managers of these enterprises into other classes. Second, as the institutional perspective anticipates, the number of proprietors grew starting in 1999. Although 1993 was the year when Tianjin, consistent with a national trend, saw the rise of household businesses and self-employed, it was after 1998 when a growing number of state workers went jobless, some of whom were forced to become self-employed or owners of very small business operations. By 2003, some either went bankrupt or were forced to give up, while others solidified a foothold in the ownership class. Third, where skilled workers experienced some decline as a result of the disorganization of

Table 3.3a Class structure in Tianjin, selected years[a]

	1983 (%)	*1988 (%)*	*1993 (%)*	*1999 (%)*	*2003 (%)*
Men and Women					
I+II(M). Managers	12.7	11.2	5.3	8.5	5.7
I(P). Higher professionals	4.8	5.1	5.6	7.2	5.1
II(P). Lower professionals	15.4	15.7	15.6	13.8	17.8
IIIa. Upper routine non-manual	8.9	8.8	14.5	9.5	12.3
IIIb. Lower routine non-manual	6.7	6.9	4.3	7.5	9.2
IV. Proprietors	0.9	0.9	0.6	19.1	6.1
VI. Skilled manual	33.4	33.1	25.5	15.9	25.8
VII. Semi- and unskilled manual	17.4	18.2	28.6	18.6	18.1
N	812	845	788	923	741
Men					
I+II(M). Managers	18.9	17.4	7.2	10.1	5.8
I(P). Higher professionals	3.7	3.9	4.1	2.9	3.2
II(P). Lower professionals	13.6	14.3	15.9	12.0	19.1
IIIa. Upper routine non-manual	9.4	9.1	14.9	9.7	10.4
IIIb. Lower routine non-manual	5.3	5.2	3.5	6.0	6.0
IV. Proprietors	0.9	1.1	0.7	22.8	6.2
VI. Skilled manual	34.2	33.9	28.6	18.8	31.8
VII. Semi- and unskilled manual	14.0	15.0	25.1	17.8	17.5
N	456	460	542	517	434
Women					
I+II(M). Managers	4.8	3.9	1.2	6.4	5.5
I(P). Higher professionals	6.2	6.5	8.9	12.6	7.8
II(P). Lower professionals	17.7	17.4	15.0	16.0	16.0
IIIa. Upper routine non-manual	8.1	8.3	13.4	9.4	15.0
IIIb. Lower routine non-manual	8.4	8.8	6.1	9.4	13.7
IV. Proprietors	0.8	0.8	0.4	14.3	5.9
VI. Skilled manual	32.3	32.2	18.7	12.3	17.3
VII. Semi- and unskilled manual	21.6	22.1	36.2	19.7	18.9
N	356	385	246	406	307

[a]Currently employed women aged 18–55 and men aged 18–60.

Table 3.3b Class structure in Russia, selected years[a]

	1984 (%)	*1988 (%)*	*1993 (%)*	*1998 (%)*	*2001/2 (%)*
Men and Women					
I/II(M). Managers	2.2	4.5	5.3	5.6	3.0
I(P). Higher professionals	6.2	7.9	8.5	7.7	6.7
II(P). Lower professionals	16.5	16.3	15.2	16.0	15.0
IIIa. Upper routine non-manual	13.4	13.3	12.0	11.9	13.3
IIIb. Lower routine non-manual	4.9	5.4	6.1	10.8	10.3
IV. Proprietors	0.0	2.0	4.5	5.3	5.2
VI. Skilled manual	28.9	27.7	25.6	21.5	22.9
VII. Semi- and unskilled manual	27.9	22.8	22.9	21.4	23.6
N	2,858	2,470	2,430	1,968	2,756
Men					
I/II(M). Managers	3.6	6.7	7.3	7.6	4.4
I(P). Higher professionals	3.2	4.7	4.9	4.6	3.7
II(P). Lower professionals	12.1	11.9	10.3	11.3	10.8
IIIa. Upper routine non-manual	3.3	3.8	3.5	4.2	4.9
IIIb. Lower routine non-manual	2.1	1.9	3.9	6.9	7.5
IV. Proprietors	0.0	2.4	6.2	6.9	7.0
VI. Skilled manual	42.7	40.2	36.2	31.3	32.9
VII. Semi- and unskilled manual	33.0	28.5	27.6	27.2	28.9
N	1,067	1,056	1,115	941	1,225
Women					
I/II(M). Managers	1.0	2.5	3.2	3.2	1.5
I(P). Higher professionals	8.8	11.0	12.4	11.2	10.0
II(P). Lower professionals	20.4	20.4	20.3	21.3	19.7
IIIa. Upper routine non-manual	22.1	22.1	21.0	20.8	22.9
IIIb. Lower routine non-manual	7.3	8.6	8.3	15.2	13.5
IV. Proprietors	0.0	1.6	2.6	3.3	3.1
VI. Skilled manual	17.0	16.3	14.4	10.3	11.7
VII. Semi- and unskilled manual	23.5	17.6	17.8	14.8	17.6
N	1,791	1,414	1,315	1,027	1,531

[a]Currently employed women aged 18–55 and men aged 18–60.

state-owned enterprises in mining and manufacturing industries, upper routine non-manual staffers grew as market reforms widened and deepened. We see no expansion of professionals due to technological advances or growth in the unskilled manual class due to in-migration from rural areas, as predicted by the growth perspective. Overall, institutional change seems to offer a better account of developments in Tianjin's class structure during the transition.

The robustness of the institutional explanation is confirmed by the fact that urban Russia experienced a very similar pattern, even though it experienced contraction rather than growth after market reforms were introduced. There, too, the routine non-manual (especially lower) and proprietor classes have grown in size, reflecting structural change driven by the reforms (growth of trade and services) and new entrepreneurial opportunities associated with market transition. Second, the proportion of manual workers (especially skilled) has declined, as market reforms and global competition spurred deindustrialization. Professional classes essentially maintained their relative sizes throughout the transition.

Although this is not our focus, we note that the gender-specific class distributions reveal interesting China-Russia differences, which point to the different stages of development in the two countries (see also Fan and Regulska's contribution to this volume). In the less developed Chinese context, male managers outnumbered female managers by several times at the beginning of reforms, but this gender gap decreased in the most recent years as the economy has become more developed and the service industries have grown. In the more developed Russian context, the relative proportions of male and female managers have remained stable throughout the transition period. Furthermore, Russia's skilled and unskilled workers have declined proportionally for men and women, a sign that men and women experienced about the same effects of change of the occupational structure. In China, however, men seemed to maintain their skilled jobs while increasingly taking up unskilled jobs, and women, while maintaining unskilled jobs, lost their skilled jobs by half from 1988 to 2003. At the same time, women took up jobs as both upper and lower routine non-manual staffers. These newer jobs might not have given them more occupational prestige, but improved income levels compared to men.

We must interpret our findings regarding class structure with caution, because of the small samples and different sampling frames employed in different surveys. Tianjin's surveys were kept within a sample of 1,000 households in each survey, and Russia's surveys, although 2 or 3 times larger, are not sufficiently large to offer reliable estimates of changes in labor force structure. Data from censuses and labor force surveys should be used to verify our findings. Nonetheless, our survey figures are illustrative and suggestive, particularly since they permit us to apply our adapted version of the Erikson-Goldthorpe class schema. They suggest that market reforms, more than economic growth, have been the key force shaping trends in the class structure of urban China.

Class inequality in earnings

Table 3.4 shows OLS regressions for class inequality in earnings, net of gender and age, for multiple years in both countries. Because education and party membership are highly correlated with the class categories, we purposefully

Table 3.4a OLS regressions of logged personal annual earnings, Tianjin

	1983	*1988*	*1993*	*1999*	*2003*
I+II (M) Manger	.227***	.299***	.464***	.495***	.682***
I. Higher professionals	.133*	.300***	.365***	.535***	.651***
II. Lower professionals	.129**	.276***	.314***	.609***	.537***
IIIa. Upper routine non-manual	.128*	.232***	.174***	.455***	.410***
IIIb. Lower routine non-manual	–.067	.062	.056	.091	–.115
IV. Proprietors	.022	.194	.690***	.710***	.043
VI. Skilled manual	.014	.068*	.096*	.042	.050
SEX	.200***	.218***	.287***	.310***	.227***
AGE	.041***	.052***	.031***	–.010	–.024*
AGE2	.000***	–.001***	.000***	.000	.000
Constant	5.748***	5.713***	7.175***	8.682***	9.505***
R Square	.243	.270	.190	.194	.189
Kappa	.099	.103	.224	.255	.325
N	868	963	1024	947	921

Two-tailed significance * <.05, ** <.01, *** <.001.

Reference group: Semi- and unskilled manual.

Table 3.4b OLS models for logged earnings in urban Russia, selected years

	March 1993	*January 1996*	*September-November 1998*	*September 2001–January 2002*
I/II(M). Managers	.479**	.616**	.633**	.583**
I(P). Higher professionals	.308**	.512**	.509**	.482**
II(P). Lower professionals	.285**	.398**	.308**	.274**
IIIa. Upper routine non-manual	.201**	.277**	.136*	.216**
IIIb. Lower routine non-manual	.141*	.306**	.253**	.179**
IV. Proprietors	.795**	1.016**	.720**	.778**
VI. Skilled manual	.173**	.265**	.255**	.288**
Age	.047**	.037**	.061**	.024**
Age-square/100	–.061**	–.047**	–.082**	–.032**
Woman	–.383**	–.516**	–.514**	–.491**
Constant	9.181**	12.377**	5.526**	7.371**
R-square	.130	.212	.160	.164
N	2,250	903	1,499	2,368
Kappa	.230	.268	.220	.222

Two-tailed significance * <.05, ** <.01, *** <.001.

All models include dummy variables for months to capture monthly inflation effects, but these are not shown.

exclude these common stratification variables from our models designed to estimate the gross effects of class categories on income inequality. In preliminary analyses we included a control for urban *hukou* in our models for China. However, we found it had no impact on the findings, so we decided to omit it for the sake of clear comparability across countries.

We use *kappa*, the standard deviation of the coefficients on the class dummy variables, as a global measure of class inequality in a given year (see Hout, Brooks, and Manza 1995). In China, class inequality increased steadily throughout the entire period. Compared to the reference group of unskilled workers, managers increased their income advantages from 25 percent (anti log of .227) in 1983 to 98 percent (anti log of .682) in 2003; higher professionals from 14 percent (anti log of .133) to 92 percent (anti log of .651); lower professionals from 14 percent (anti log of .129) to 71 percent (anti log of .537); and upper routine non-manual staff from 13 percent (anti log of .128) to 51 percent (anti log of .410). In the same time span, lower routine non-manual staff and skilled manual workers had comparable income levels as compared to unskilled manual workers. Proprietors experienced a drastic increase in relative income, going from no clear income advantage over unskilled workers in the early stage of market reforms, when they constituted less than 1 percent of the labor force, to a very high advantage of twice the income of unskilled workers. Overall, the pattern and magnitude of class inequalities in earnings in urban China grew steadily more similar to the patterns and magnitudes observed in advanced market societies.

The pattern of class-income inequality in Russia is more complicated. In the initial phase of the post-Soviet era (when economic decline was steepest) class differences in income grew dramatically (from 1993 to 1996). Then, as the economy stabilized (1996 to 1998, since our 1998 survey is pre-August 1998 crisis) class inequality in earnings actually began to subside, a pattern that continued as growth resumed at the end of the decade. This pattern holds both for the overall change in degree of class-income inequality across these survey years and for the changes in specific between-class inequalities. For example, the relative income advantages for managers, higher professionals, lower professionals, and proprietors all increased from 1993 to 1996 and gradually decreased after 1998. Thus, it would seem that in broad strokes economic growth and class inequality in earnings are *inversely* related in Russia, while the impact of market reforms is hard to discern.

Two differences in trends of class-based income inequality should be noted. First, proprietors in Russia enjoyed a consistently strong income advantage throughout the reform era, which was marked by the immediate recognition of private ownership. In contrast, Chinese proprietors did not have an income advantage until 1993, the year when private ownership began to be legitimized in Tianjin and on a national level. Second, in Russia lower routine non-manual staff and skilled manual workers have maintained earnings advantages over unskilled workers. This was not the case in China, where the three classes had about the same income throughout the reform era. In China, income inequality is basically between the minority of "high classes" and the majority of the "working masses," while the pattern of class income inequality in Russia is more variegated.

China and Russia also differ in sex and age inequalities in income. Both countries have gender gaps throughout the reform era, but they are smaller in China than in Russia. While Chinese women earned 20–27 percent less than their male counterparts, Russian women earned 32–40 percent less than Russian men. In China, market reforms seem to have removed the income advantages of older persons, perhaps by eliminating the seniority rule.

Conclusion

Our China-Russia comparison points to similar patterns of change in the class structures of the two countries during their periods of market reforms. This is particularly noteworthy, given the different (modernized versus agricultural) starting points in the two countries (we might have expected a more dramatic change in China than in Russia), and their sharp divergence in terms of growth rates during this period. Perhaps because the China sample is strictly urban (and thus more "modern" in the old "modernization" sense) it is not so surprising. The urban areas in both countries increasingly exhibit class structures typical of modern industrialized economies, regardless of their divergent growth trends. The similarities of the changes that have occurred suggest that institutional reforms, not growth patterns, play the most important role in shaping the development of urban class structures during market transition.

Second, our preliminary results regarding the magnitude and patterns of class-based earnings inequalities point to a very interesting contrast that raises more questions than answers. In Russia, economic growth appears to have been inversely related to the level of class inequality and market reforms would seem not to have much of a relationship, though one might make some kind of argument about lagged institutional effects. The apparent lack of relationship between the pace of reforms and the level of class inequality in Russia suggests that institutional factors may play less of a role than growth in China. Yet it is puzzling that growth appears to have a positive relationship with class inequality in China, while it has negative relationship in Russia. Perhaps the different starting points come into play here – China may have begun at an early point on the Kuznets curve, so to speak. The sharp contraction of earnings in Russia may have affected the middle classes disproportionately, so that they recovered disproportionately when growth resumed. Another possible explanation is that the effects of institutional change and growth interact: the Chinese pattern reflects the combination of marketization and growth, while the Russian pattern shows the effects of marketization accompanied by contraction.

In any case, our results suggest that the relationship between institutional change, growth, and class inequality is complex. Based on our comparison between China and Russia, we conclude that it would be too simplistic to attribute patterns of class inequality in contemporary urban China to either market reforms alone or to economic growth alone. Of course, our empirical findings are very provisional, as they are based on relatively small samples and data that are not, strictly speaking comparable. We need to refine and confirm these findings by analyzing multiple and larger Chinese surveys (rather than focusing on a single city), and perhaps also expanding the samples from Russia. Clearly, much work remains to be done to elucidate the relative roles played by institutional changes and economic growth in shaping class structure and class inequality in urban China, as well as in other former state socialist societies. We believe that fruitful insights can be gained from more detailed and robust comparisons of patterns in China, Russia, and potentially other countries as well. We hope that some of the theoretical ideas we propose may be of use in future endeavors to identify the particular influences of institutions and economic performance on class inequality in China and elsewhere.

ACKNOWLEDGMENTS

We acknowledge grants to Bian from Hong Kong's Universities Grants Committee (HKUST6052/98H, HKUST6007/00H, HKUST6007/01H) and to Gerber from the National Science Foundation (SBR9729225 and SBR0096607) and the Spencer Foundation. We also thank Li Yu, Zhang Lijuan, and Zhang Wenhong for research assistance and John Logan, Fulong Wu, Neil Brenner, and an anonymous reviewer for helpful comments.

REFERENCES

Alderson, A.S. (1999) "Explaining Deindustrialization: Globalization, Failure, or Success?" *American Sociological Review* 64: 701–21.

Bian, Y. (1994) *Work and Inequality in Urban China.* Albany: State University of New York Press.

Bian, Y. (2002) "Chinese Social Stratification and Social Mobility." *Annual Review of Sociology* 28: 91–116.

Bian, Y., and J.R. Logan. (1996) "Market Transition and the Persistence of Power: The Changing Stratification System in Urban China." *American Sociological Review* 61: 739–58.

Bian, Y., and Z. Zhang. (2002) "Marketization and Income Distribution in Urban China: 1988 and 1995." *Research on Social Stratification and Mobility* 19: 377–415.

Bian, Y., R. Breiger, D. Davis, and J. Galaskiewicz. (2005) "Occupation, Class, and Networks in Urban China." *Social Forces* 83, 4: 1443–68.

Burawoy, M. (1997) "Review Essay: The Soviet Descent into Capitalism." *American Journal of Sociology* 102: 1430–44.

Cao, Y., and V. Nee. (2000) "Comment: Controversies and Evidence in the Market Transition Debate." *American Journal of Sociology* 105: 1175–88.

Chan, A. (1995) "The Emerging Patterns of Industrial Relations in China and the Rise of Two New Labour Movements." *China Information* 9: 36–59.

Chiswisk, B.R. (1971) "Earnings Inequality and Economic Development." *Quarterly Journal of Economics*. 85: 21–39.

Connor, W.D. (1979) *Socialism, Politics and Equality: Hierarchy and Change in Eastern Europe and the USSR*. New York: Columbia University Press.

Davis, D.S. (ed.). (2000) *The Consumer Revolution in Urban China*. Berkeley: University of California Press.

Djilas, M. (1957) *The New Class: An Analysis of the Communist System of Power*. New York: Praeger.

Erikson, R., and J.H. Goldthorpe. (1992) *The Constant Flux: A Study of Class Mobility in Industrial Societies*. Oxford: Clarendon Press.

Gerber, T.P. (2000) "Membership Benefits or Selection Effects? Why Former Communist Party Members Do Better in Post-Soviet Russia." *Social Science Research* 29: 25–50.

Gerber, T.P. (2001a) "The Selection Theory of Persisting Party Advantages in Russia: More Evidence and Implications." *Social Science Research* 30: 653–71.

Gerber, T.P. (2001b) "Paths to Success: Individual and Regional Determinants of Entry to Self-Employment in Post-Communist Russia." *International Journal of Sociology* 31: 3–37.

Gerber, T.P. (2002) "Structural Change and Post-Socialist Stratification: Labor Market Transitions in Contemporary Russia." *American Sociological Review* 67: 629–59.

Gerber, T.P. (2003) "Loosening Links? School-to-Work Transitions and Institutional Change in Russia Since 1970." *Social Forces* 82: 241–76.

Gerber, T.P., and M. Hout. (1998) "More Shock than Therapy: Market Transition, Employment, and Income in Russia, 1991–1995." *American Journal of Sociology* 104: 1–50.

Gerber, T.P., and M. Hout. (2004) "Tightening Up: Declining Class Mobility during Russia's Market Transition." *American Sociological Review* 65: 677–703.

Gerber, T.P., and D.R. Schaefer. (2004) "Horizontal Stratification of College Education in Russia: Temporal Change, Gender Differences, and Labor Market Outcomes." *Sociology of Education* 77: 32–59.

Goskomstat Rossii [State Committee on Statistics of the Russian Federation]. (1996) *Rossiiskii Statisticheskii Ezhegodnik 1996* [Russian Statistical Annual, 1996]. Moskva: Goskomstat.

Goskomstat Rossii [State Committee on Statistics of the Russian Federation]. (2004) *Rossiiskii Statisticheskii Ezhegodnik 2004* [Russian Statistical Annual, 2004]. Moskva: Goskomstat.

Griffin, K., and R. Zhao. (1992) *Chinese Household Income Project, 1988 (MRDF)*. New York: Hunter College Academic Computing Services (Producer). Distributed by Inter-University Consortium for Political and Social Research, Ann Arbor, Mich.

Hauser S.M. and Y. Xie. (2005) "Temporal and Regional Variation in Earnings Inequality: Urban China in Transition between 1988 and 1995." *Social Science Research* 34: 44–79.

Hout, M., C. Brooks, and J. Manza. (1995) "The Democratic Class Struggle in the United States, 1948–1992." *American Sociological Review* 60: 805–28.

Khan, A.R., and C. Riskin. (1998) "Income Inequality an Its Distribution and Growth of Household Income, 1988 to 1995." *The China Quarterly* 154: 221–53.

Konrad G., and I. Szelényi. (1982) *The Intellectuals on the Road to Class Power: A Sociological Study of the Role of the Intelligentsia in Socialism.* New York: Harcourt Brace Jovanovich.

Korzeniewicz, R.P., and T.P. Moran. (2005) "Theorizing the Relationship between Inequality and Economic Growth." *Theory and Society* 34: 277–316.

Kuznets, S. (1955) "Economic Growth and Income Inequality." *American Economic Review* 65: 1–28.

Lee, C.K. (1998) *Gender and The South China Miracle.* Berkeley: University of California Press.

Morris, M., and B. Western. (1999) "U.S. Earnings Inequality at the Close of the 20th Century." *American Sociological Review* 25: 623–57.

Naughton, B. (1995) *Growing Out of the Plan: Chinese Economic Reform, 1978–1993.* New York: Oxford University Press.

Nee, V. (1989) "A Theory of Market Transition: From Redistribution to Market in State Socialism." *American Sociology Review* 54: 663–81.

Nee, V. (1991) "Social Inequalities in Reforming State Socialism: Between Redistribution and Markets in State Socialism." *American Sociological Review* 56: 267–82.

Nee, V. (1996) "The Emergence of a Market Society: Changing Mechanisms of Stratification in China." *American Journal of Sociology* 101: 908–49.

Nee, V., and Y. Cao. (1999) "Path Dependent Societal Transformation: Stratification in Hybrid Mixed Economy." *Theory and Society* 28: 799–834.

Nee, V., and Y. Cao. (2002) "Postsocialist Inequality: The Causes of Continuity and Discontinuity." *Research in Social Stratification and Mobility* 19: 3–39.

Nielsen, F. (1994) "Income Inequality and Industrial Development: Dualism Revisited." *American Sociological Review* 59: 654–77.

Nielsen, F., and A.S. Alderson. (1997) "The Kuznets Curve and the Great U-Turn: Inequality in US Counties, 1970 to 1990." *American Sociological Review* 62: 12–33.

Parish, W.L., and E. Michelson. (1996) "Politics and Markets: Dual Transformations." *American Journal of Sociology* 101: 1042–59.

Parish, W.L., and M.K. Whyte. (1978) *Village and Family in Contemporary China.* Chicago, IL: University of Chicago Press.

Pun, N. (1999) "Becoming Dagongmei: The Politics of Identity and Difference in Reform China." *China Journal* 42: 1–19.

Rona-Tas, A. (1994) "The First Shall be Last? Entrepreneurship and Communist Cadres in the Transition from Socialism." *American Journal of Sociology* 100: 40–69.

Shu, X., and Y. Bian. (2002) "Intercity Variation in Gender Inequalities in China: Analysis of a 1995 National Survey." *Research on Social Stratification and Mobility* 19: 269–309.

Shu, X., and Y. Bian. (2003) "Market Transition and Gender Gap in Earnings in Urban China." *Social Forces* 81, 4: 1107–45.

Szelényi, Ivan. (1978) "Social Inequalities in State Socialist Redistributive Economies." *International Journal of Comparative Sociology* 19: 63–87.

Szelényi, I. (1983) *Urban Inequality Under State Socialism*. Oxford: Oxford University Press.

Walder, A.G. (1985) "The Political Dimension of Social Mobility in Communist States: China and Soviet Union." *Research in Political Sociology* 1: 101–17.

Walder, A.G. (1986) *Communist Neo-Traditionalism: Work and Authority in Chinese Industry*. Berkeley, CA: University of California Press.

Walder, Andrew G. (1987) "Wage Reform and the Web of Factory Interests." *The China Quarterly* 109: 22–41.

Walder, A.G. (1995) "Career Mobility and the Communist Political Order." *American Sociological Review* 60: 309–28.

Walder, A.G. (1996) "Markets and Inequality in Transitional Economies: Toward Testable Theories." *American Journal of Sociology* 101: 1060–73.

Walder, A.G. (2002) "Markets and Income Inequality in Rural China: Political Advantage in an Expanding Economy." *American Sociological Review* 67: 231–53.

Walder, A.G. (2003) "Elite Opportunity in Transitional Economies." *American Sociological Review* 68: 899–916.

Walder, A.G., B. Li, and D.J. Treiman. (2000) "Politics and Life Chances in a State Socialist Regime: Dual Career Paths into the Urban Chinese Elite, 1949–1996." *American Sociological Review* 65: 191–209.

Wu, J. (2003) *China's Economic Reform* (Dangdai Zhongguo Jingji Gaige). Shanghai: Shanghai Fareast Press (Shanghai Yuandong Chubanshe).

Wu, X., and Y. Xie. (2003) "Does the Market Pay Off? Earnings Inequality and Returns to Education in Urban China." *American Sociological Review* 68: 415–22.

Whyte, M.K. (1984) "Sexual Inequality under Socialism: The Chinese Case in Perspective." In J. Watson (ed.), *Class and Social Stratification in Post-Revolution China*." New York: Cambridge University Press: pp. 198–238.

Whyte, M.K. and W.L. Parish. (1984) *Urban Life in Contemporary China*. Chicago, IL: University of Chicago Press.

Xie, Y., and E. Hannum. (1996) "Regional Variation in Earnings Inequality in Reform-Era Urban China." *American Journal of Sociology* 101: 950–92.

Zhao, W. and X. Zhou. (2002) "Institutional Changes and Returns to Education in Urban China: An Empirical Assessment." *Research in Social Stratification and Mobility* 19: 337–73.

Zhou, X. (2000) "Economic Transformation and Income Inequality in Urban China: Evidence from a Panel Data." *American Journal of Sociology* 105: 1135–74.

Zhou, X., N.B. Tuma, and P. Moen. (1996) "Stratification Dynamics under State Socialism". *Social Forces* 28: 440–68.

Zhou, X., N.B. Tuma, and P. Moen. (1997) "Institutional Change and Job-Shift Patterns in Urban China." *American Sociological Review* 62: 339–65.

4

Gender and the Labor Market in China and Poland

C. Cindy Fan and Joanna Regulska

Introduction

Gender is an important dimension for understanding labor-market experiences throughout the world. In this chapter, we examine China and Poland. Specifically, we ask which theoretical perspectives can best explain the persistence of gender differentials in the labor market and their changes in these two countries since socialism.

Both China and Poland experienced state socialism, imported from or imposed by the former Soviet Union, and have witnessed profound structural changes in recent decades. China, while sustaining a certain degree of central control, opened and restructured its market in unprecedented fashion and continues to aspire to join ranks of capitalist industrialized economies (for example via the World Trade Organization (WTO)). Poland not only broke away (as the rest of the Central and East Europe (CEE) countries did) from the Soviet regime, but a decade later realigned itself politically, economically and socially by joining the European Union (EU) and North Atlantic Treaty Organization. In both countries, state socialism and recent economic liberalization have had direct impacts on women's experiences in the labor market, and gender ideologies continue to undervalue women and their work.

The rationale for studying China and Poland side by side is to examine the effects of state socialism and the subsequent economic restructuring on gender inequality. State socialism, despite its variations across countries, subscribed to a strikingly similar philosophy toward women's labor. Also, recent market reforms, despite their varied levels of success, have had some similar impacts on women's labor-market experiences. This is not to say, however, that all post-socialist economies are the same (see also Chapter 3 in this volume). Far from it, by examining China along with another transitional economy with different historical and economic trajectories, we wish also to highlight contextual differences that contribute to the persistence of gender inequality. We do not claim that post-socialist or transitional economies constitute the best comparisons for China. Indeed, as Read and Chen (Chapter 14, this volume) convincingly argue, societies in East Asia that never experienced socialism may very well be good comparative cases for certain phenomena in China; and other chapters in this volume illustrate meaningful comparisons between China and Western societies. Yet because issues of inequality, including gender inequality, were fundamental to state socialism and because socialist policies and market reforms had pronounced effects on women's labor-market experiences, a study of two post-socialist economies would shed light on the effects of macro restructuring as well as other structural and contextual factors on gender inequality.

China and Poland differ in many respects. China is much bigger in size and its economy is less urbanized than Poland's. In terms of political history, the Chinese did not resist socialism as much as the Polish, who consistently displayed a civil disobedience spirit that culminated in the emergence of the Solidarity movement. Economic liberalization in China occurred gradually and incrementally, which proved to be an overall successful formula for sustaining high rates of economic growth. Poland, on the contrary and like a number of other CEE countries, introduced economic reforms as a "shock therapy" (see also Chapter 3, this volume). Despite these differences, they share important similarities in terms of women's experiences in the labor market and above all theoretical explanations for these experiences.

In this chapter, we wish to explain not only the persistence of gender differentials, but also new opportunities and the agency of women in the labor market. The realities of gender differences are complex and thus approaches that are concerned solely with whether women's situations have improved or worsened are often less than satisfactory. Rather, we wish to critique and analyze several theoretical perspectives that have been proposed as explanations of gender differentials. Here again, singular explanations are insufficient for grasping the complexity of gender differences, especially since shifts at the global scale have produced new conditions for a variety of actors that implicate the gendering of labor markets. In the analysis, we do not distinguish specifically between rural and urban parts of the countries for two reasons. First, theoretical explanations for gender differences primarily deal with structural issues that affect both urban and rural populations. Second, the data and information we rely on do not permit a systematic disaggregation along rural-urban lines. However, the indicators that we use, including gender differentials in employment and unemployment, occupation and wages, and new women's organizations, reflect more directly urban restructuring than rural changes in both China and Poland. In addition, in our review of existing findings, the emphasis is on the urban economy more so than on the rural one.

We use the four perspectives outlined by Logan and Fainstein as points of reference rather than rigid categories of explanation. This is because, first, as several other contributors to the volume (e.g., Huang and Low; White et al.) also suggest, none of the four perspectives can fully explain social and economic inequalities. And, second, these perspectives are not exhaustive in explaining all aspects of society and economy. To focus on gender differentials, we review three theoretical approaches which have some parallels with Logan and Fainstein's four perspectives. We begin by commenting on the neoclassical perspective, which subscribes to the same reasoning as modernization theory and highlights human capital as the basis of gender differentials. Second, we focus on state socialism, which sought to correct dependency via a strong central state. We show how the effects of state socialism are mixed, in that it advocated a degree of equalization and mobilized women to enter the labor force in mass numbers, but it did not challenge the deep-rooted and gendered social order. Third, we examine the recent repositioning of the state, which is marked by the promotion of both a developmentalist agenda and market transition. The developmentalist state repositioned itself *vis-à-vis* different social groups and withdrew support for marginalized groups. New market conditions led to layoffs and unemployment, which legitimized gender discrimination in hiring, compensation, and layoffs and deepened inequality. At the same time, market transition opened new opportunities for individuals and groups.

We argue that the above theoretical perspectives can only partially explain the persistence of gender differentials and their changes. Similar to much of the mainstream discourse, these perspectives are masculinist as they do not emphasize how social categories are constructed and they tend to reduce women to passive victims (Nagar et al. 2002). We show that feminist theory is a useful alternative to and complements these perspectives because the former emphasizes how gender is socially and culturally constructed, how work is gendered, and the process and outcome of women's agency. We focus specifically on how patriarchal ideology and societal and cultural norms have shaped and are shaping the social and gendered practices at home, work, and in the public sphere; and how women's agency has gained strength and has created new spaces of engagement such as non-governmental organizations (NGOs).

Neoclassical Perspective

The neoclassical economic literature subscribes to modernization theory, whose premise is that diffusion of capital, technology, and education will narrow inequality. In this view, labor-market outcomes are functions of individuals' human capital and productivity differentials. Thus, neoclassical theorists point to improving women's education and human capital as the main instrument to close the gender gap in the labor market (Stiglitz 1998). In societies where women have lower level of education than men, improving the competitiveness of women is indeed a means to advance their labor-market positions. Yet, there is plenty of evidence that show that women's lower wages and under-representation in powerful positions cannot be fully explained by gender differentials in education and experience (e.g., Bauer et al. 1992).

In China, the gender gap in educational attainment remains large but has gradually narrowed over time. In 2004, 63.6 percent of men in China, compared with 53.2 percent of women, had received education at or beyond the junior secondary level, compared with respectively 34.5 percent and 21.5 percent in 1982 (NBS 2005, pp. 105–6; SSB 1985). In Poland, women are better educated than men in terms of years of schooling and women outnumber men at the post-secondary level of education (Bialecki and Heynes 1993). Yet, greater access to education and higher level of educational attainment did little to eradicate gender differences in the economic sphere. The notion that women are not as productive and efficient as men and are only suitable for certain jobs is deeply ingrained in both societies, with the result that women are consistently tracked into low-skilled and low-paid jobs (Hershatter 2004; Ingham, Ingham and Dománski 2001). Similar to observations of labor-market segmentation elsewhere (Harris and Todaro 1970), gender segregation in the Chinese and Polish labor markets is explained not only by human capital differences but by women's lower social status and barriers that block them from certain jobs (Huang 1999).

Explanations for women's higher rates of unemployment often include human capital considerations. While differentials in educational attainment do in part explain gender disparities in the Chinese labor market (Gustafsoon and Li 2000), women in Poland with higher education continue to be worse off than men. In 2003, over 50 percent of unemployed women in Poland had at least a secondary level of education, compared with 32 percent for unemployed men (GUS 2004). In addition, unemployment of women is highest where the proportion of women with college and higher education degrees is also highest (Knothe and Lisowska 1999). These findings show that education does not protect women from unemployment.

The Socialist State

The advent of socialism in China and Poland not only fostered a new political ideology but also legitimized a strong central state, one that dependency theorists (see the Introduction to this volume) advocated. A premise of state socialism is that the state should and could equalize. In an attempt to create a classless society, the Chinese as well as Polish state eliminated landlords, centralized labor allocation, kept wages even, and officially maintained near full employment. The Marxist perspective contends that capitalism's inevitable outcome is a dualistic and polarized society, where women's entering the labor market and taking up unskilled, low-paid jobs would further aid capital accumulation (Beechey 1986). Instead, state socialism places women at work side by side with men and promises them equal opportunities and rewards.

The socialist attempt to promote gender equality was nowhere more profound than the state's massive mobilization of women to enter the labor force. Drawing from the Hungarian experience, Szelényi (1992, p. 576) went as far as concluding that "the only effect socialism had on the gender division of labor was with respect to labor-force participation." In the 1980s and beyond, more than 80 percent of Chinese women between the ages of 15 and 54 were in the labor force (ACWF 1991, pp. 234–7; SSB 1985, pp. 440–3). By 1989, women's labor force participation in Poland was almost 70 percent (Regulska 1992). Ideologically, the state adopted the notion that emancipation of women depends on increasing their economic contribution to the non-domestic sphere. Mao's famous statement that "women carry half of the heavens on their shoulders" summarizes a Marxist version of gender ideology, one that emphasizes economic production. The construction of the "iron girls" or "women on the tractor" images that were heavily promoted in China and Poland aimed at advancing the idea that women were capable of performing jobs commonly done by men (Haney and Dragomir 2002; Loscocco and Wang 1992). This interpretation of equality, however, emphasized physical strength and the sameness between men and women rather than women's distinct and differing social expectations, circumstances, and contribution; it prescribed that women should improve their status by aspiring to become men (Honig 2000). Thus, in both China and Poland, "emancipation of women" did not necessarily enrich women's identity with the desire for professional success and economic independence. Rather, a reductionist theory of women's liberation, focusing almost exclusively on bringing women into economic production, was promoted (Hershatter 2004). In addition, mass mobilization was imposed from above and "was done without women's will and participation" (Titkow 1992, p. 6). And, feminism – a label connoting Western bourgeois influence – and along with it feminist perspectives that address gender differences and identities, were shunned.

Women were mobilized to work not only on ideological grounds but also for pragmatic reasons. In the aftermath of the civil war in China and the massive destruction during World War II in Poland, both economies badly needed to be rebuilt. Women's entering the labor force was expected to boost economic development, especially during periods of labor shortage, and to meet financial needs of families (Lake and Regulska 1990). However, during periods of slow growth or economic crises, women were under pressure to return to the home. In China, economic disasters in the late 1950s and early 1960s coincided with the official promotion of "good socialist housewives" (Loscocco and Wang 1992). Likewise, during slower growth periods in the late 1950s and 1960s in Poland, the state's attitude toward women's labor-force participation shifted and women's "homecoming" was officially celebrated.

The focus on labor-force participation and the sameness between men and women did little to address gender as a focal point of social process or to challenge patriarchal ideology in the home (Titkow 1998). On the contrary, the "ideal" woman had to combine her traditional duties in the private sphere and her production roles outside the home (Bauer et al. 1992; Środa 1992, p. 15). In both China and Poland, women suffered from a double burden as they worked outside the home but continued to shoulder the bulk of housework (Bauer et al. 1992; Nowakowska 2000). The state provision of services and benefits attempted often to achieve both: to retain women at work, but at the same time to encourage them to continue to bear and rear children and perform other home-related duties. So for example, in China social services such as childcare were organized at the place of work ("work units" or *danwei*) and by local-level government units (e.g., residents' committees in cities) (Stockman 1994). The Polish state simultaneously and repeatedly reaffirmed women's role as mothers, producers, and reproducers through enactment of legislation (maternal leaves) and by offering a variety of social entitlements including housing, childcare facilities, subsidized transportation, vacation, and a variety of other services. Yet, this process sustained and enforced the traditional pattern, according to which "the burdens of family life should be borne by women and their professional careers must be subordinated to family needs," and it created public patriarchy where women relied predominantly on the state (Ingham et al. 2001).

Evidence on the socialist state's commitment to and effectiveness in achieving gender equality is mixed. Despite socialist centralization of labor allocation, women continued to be concentrated in less prestigious sectors (Ingham et al. 2001; Stockman 1994). Based on China's 1982 Census, women concentrated in the following occupations: childcare, nursery school teachers, nurses, embroiderers, knitters, housekeepers, sewers, weavers, washers and darners, electronic assembly, heads of neighborhood committees, kindergarten teachers, babysitters, housemaids, and hotel attendants (Loscocco and Wang 1992). Men constituted 56 percent of the labor force, but they accounted for over 75 percent of the jobs in government, construction, transportation, and geological survey and exploration, all sectors that had traditionally enjoyed high prestige and/or good earnings (Bauer et al. 1992; Loscocco and Wang 1992). Compared to Chinese women, those in Poland appeared to have made more inroads into occupations previously taken on only by men. In the late 1970s, labor-force participation in state-controlled enterprises in Poland was similar for women and men (46 percent) and the average occupational prestige of working men and women was "the same" (Titkow 1984, p. 561). Still, Polish women concentrated in services, clerical work and sales (60 percent in 1979), and 70 percent of teachers were women. Women were also dominant in health and social work (Titkow 1984). These observations from China and Poland echo Szelényi's (1992) finding in Hungary that state socialism did not change the occupational segregation between men and women. This pattern of occupational segregation continues to persist during the period of transformation (Ingham and Ingham 2001) (see also Chapter 3 in this volume, which suggests that gender gaps in occupation differed between China and Russia because of their different stages of development).

In both China and Poland, the state did not eliminate the gender gap in work compensation, despite their respective constitutions that provided equal rights to women and guaranteed them "equal pay for equal work." Plenty of evidence documents that Chinese work units and communes discriminated against women in rewards (Meng and Miller 1995). Though little data on wages and earnings during the Polish socialist period is available, existing evidence reinforces the notion that women received lower pay than men in all industries and occupations (Inghan and Inghan 2001, p. 56).

Marxist theory is fundamentally a class theory that emphasizes capital-labor relations rather than gender ideology, and the exploitation of proletariats more than the historical and social oppression of women. Thus, the state was inconsistent in its pursuit of gender inequality, which was given a lower priority than class conflict or production goals (Honig 2000). Changes introduced under socialism did not eliminate gender discrimination as reflected in wages, occupational attainment, and promotion, and at the same time they reinforced women's double burden in and outside the home. Yet, women's massive mobilization into the labor force did have long-lasting implications for their opportunities in the labor market. A recent International Labor Organization (ILO) (2004) report, for example, argued that one of the reasons why women's overall share of professional jobs in 2000–2002 was highest in East Europe (60.9 percent in Poland), was precisely because of long-standing policies supporting working mothers.

Market Transition and the Developmentalist State

Since the late 1970s in China and the 1990s in Poland, both states have pursued economic liberalization and repositioned themselves as developmentalist states (see also the Introduction in this volume). Similar to other transitional economies, China and Poland introduced market mechanisms as a means toward economic growth. In China, such transition is legitimized by the invention of a new model of development known as "socialist market economy." Mass layoffs, privatization of state enterprises, state withdrawal of support, and rapid growth of the private sector are among the new market conditions. At the same time, both states retreated from the previous gender-equality agenda, despite legislative changes and new labor requirements due to China's admission to the WTO in 2001 and EU pressures on Poland during the accession negotiations in the 1990s (Fan 2004; Ingham et al. 2001; Titkow 1998). There are many indicators that as a group, women in China since the 1980s and women in Poland since the 1990s have lost ground in the labor market relative to men (Hare 1999; Maurer-Fazio, Rawski, and Zhang 1999; Millimet and Wang 2006).

Market reforms give employers greater autonomy to hire and fire. In China, the lack of proper legislation and the inconsistent enforcement of existing legislation have opened unlimited possibilities for discrimination in the labor market (Bauer et al. 1992; Huang 1999). Overt discrimination by gender in hiring is common, as seen in advertisements that exclude women, especially mothers. A survey of Chinese newspapers in 2000 finds that 85 percent of advertisements for managers and senior clerical staff specified that only male applicants needed to apply and 85 percent of advertisement for restaurant and service work, which typically is of lower pay and status, targeted only female workers (Li 2002, p. 55). In Poland, the existence of legal protection of women's work (especially now after the entrance to EU as the labor code was strengthened) has potential to place them in a better position (Fuszara and Zielińska 2006), but women's knowledge and awareness of their constitutional and labor rights are often limited. While not necessarily coded in advertisements, in Poland women are expected to be better educated than men when applying for the same jobs.

The emergence of unemployment in China and Poland reflects new market practices that endorse gender discrimination. In 2000, the unemployment rate (15 years and older) in Chinese cities was 10.4 percent for women and 8.7 percent for men (NBS 2002, pp. 1241–4). Between 1990 and 2000, the gender gap in employment (ages 15–54) in China had widened from 6.5 percent to 9.5 percent (ACWF 1998, pp. 22, 324; NBS 2002, pp. 1237–42), hinting at a setback to the gains in women's labor-force participation made earlier. In Poland, the proportion of women among the unemployed increased from 50.9 percent in 1990 to 60.4 percent in 1997 (Kotowska 2001), and the majority of the urban employed were women (Pater 1998). In 2003, the unemployment rate was 18.4 percent for men and 20.3 percent for women (GUS 2004). Women's unemployment grew despite high economic growth in the second half of the 1990s, indicating that the former was related to not only economic but also structural and institutional causes. Moreover, Polish women remain unemployed for longer periods and have greater difficulties reentering the labor market than men (World Bank 2004).

Gender segregation in occupations is believed to have accelerated since market reforms. In China, agriculture is increasingly feminized (Jacka 1997), while women's relative lack of access to leading administrative positions has continued (Loscocco and Wang 1992). Data from 2000 show that urban women concentrated more so than men in commerce, health and education – all sectors of rapid growth but relatively low earnings – but men accounted for 83.2 percent of the "heads of governments, parties, social organizations, enterprises and institutions" category (NBS 2002, pp. 935–1096; NBS 2004, p. 44). Reflecting also the urban-rural divide, rural women migrants are squeezed into positions, mostly in manufacturing and low-end services, that confer less dignity and bring less reward (Huang 1999; Knight and Song 1995). In Poland, likewise, men move out from agriculture at a faster rate then women, and women are predominantly in health and education. Even though women's employment in government administration increased from 2.2 percent in 1982 to 7.8 percent in 2000, the role of the public sector in the economy has been diminishing. Women's employment in finance and insurance also increased, from 1.9 percent in 1982 to 4.8 percent in 2000, but Titkow (1998, p. 25) pointed out that "even in the better-paid sectors women are primarily lower-status administrative employees or semi-skilled workers."

Market reforms opened the possibilities for self-employment and entrepreneurship, which are seen by many, especially those in urban areas, as a means to escape unemployment and respond to the threat of losing jobs in the public sector. In China, privatization has increased rapidly but women remain underrepresented in self-employment and among entrepreneurs. The 2002 urban employment figures indicated that 6.8 percent of men and 5.5 percent of women were employers in the private sector and 13.4 percent of men and 11.9 percent of women were self-employed (NBS 2003, p. 68). In Poland, between 1989 and 1998, the rate of women ownership of small and medium size enterprises increased from 27 percent to 40 percent, and the number of women entrepreneurs grew from 3.7 percent to 22 percent (Lisowska 2005). These reflect social changes including a growing desire by women for greater independence and self-sufficiency (Nowakowska 2000). Nevertheless, men began to own their first businesses earlier and women are still less likely to be self-employed then men (Ingham and Ingham 2001). The biggest challenges for women in Poland to become entrepreneurs appear to be the persistent stereotypes about their family role and the numerous structural barriers such as instability of tax regulations and lack of operating capital (Lisowska 2005; Reszke 2001). Yet, the respondents in Reszke's (2001) survey saw business ownership as equally suited for their sons and daughters, hinting further attitudinal changes to be expected.

The findings about gender wage gap since market reforms are inconclusive. In China, female wages are placed at anywhere between 56 percent to 93 percent of male earnings (Meng and Miller 1995; NBS 2004: 52; Shu and Bian 2003), and Gustafsson and Li (2000) show that the gender wage differential grew from 15.6 percent in 1988 to 17.5 percent in 1995 (see also Maurer-Fazio et al. 1999). Yet, Whyte (2000) argued that most studies on China lacked comparable data and could not validate the claim that women had lost ground. What is clear, however, is a persistent gender gap in earnings (Shu and Bian 2003). In Poland, while Ingham et al. (2001, p. 11) argued that the gender gap in earnings narrowed over the period 1992 to 1995, in 1998 women's salaries constituted only 69.5 percent of an average man's salary (Nowakowska 2000, p. 63) and in 2002 women's average wages and salaries were lower by 16.9 percent than those of men (GUS 2004). Kotowska (2001, p. 70) questioned if gender earnings differentials would hold when controlled by hours of work, as women work fewer hours. High concentration of women in sectors with lower wages has clearly contributed to the persistent gender wage gap in both countries (Loscocco and Wang 1992; Maurer-Fazio et al. 1999; UN 2000). Yet, Hare's (1999) survey in Guangdong found that in 1989 women earned 29 percent less than men even after controlling for differences in economic sector, indicating still other factors at work, including discrimination, that depress women's wages. The persistence of a gender wage gap is also found in Russia, another post-socialist economy (Gerber and Hout 1998).

As a whole, therefore, the actual impacts of market reforms on women's position in the labor market are rather mixed. Market reforms have indeed introduced new economic opportunities for women, as seen in their moving into growing urban sectors and entrepreneurship. Shu and Bian (2003) found that although a gender earnings gap persisted, market transition had accelerated the returns to human capital in Chinese cities with the highest degrees of marketization. At the same time, the retreat of the state from a gender-equality agenda has legitimized discriminatory practices in the labor market, as seen in sexist hiring policies and higher unemployment among women. Gender segregation in the labor market persists and the gender wage gap remains large. The state's prioritizing productivist goals over equality has certainly endorsed, if not exacerbated, persistent patriarchal attitudes towards gender division of labor and stereotypes of women as less valued workers. The state's pursuit of developmentalist goals is, therefore, gendered, as seen also in China's one child policy, which legitimizes public surveillance of women's bodies, invades their privacy, and penalizes fertility, all in the name of economic development. Rather than improving the status of women, the severe sex imbalances at birth (due to sex-selective abortion, hiding and abandonment of girls, and girl infanticide) and shortage of marriage-age women in China have in fact further aggravated women's hardship, as seen in for example incidences of incest and women trafficking (Riley 2004).

Feminist Framework

The theoretical explanations reviewed above, while partially explaining the persistence of gender differentials and the emergence of new opportunities for women in the labor market, downplay the roles of gender ideology and women's agency. In most mainstream theories for analyzing economic questions, gender is absent or treated as merely one of many categories. We argue that feminist perspectives can shed important light on understanding the situation of women in the labor market. Feminist theories focus on analyses that unfold the construction of gender, class, and race and conceptualize subjects as embodied actors in social relations (Nagar et al. 2002). For example, some jobs are regarded as or have become female jobs not only because of the biological argument but because they are considered of low value, which reflects gender hierarchy rather than the characteristics of the jobs themselves (McDowell 1999, p. 127). In the following, we articulate the contribution of feminist perspectives via two prisms: (1) how women's role and work are gendered by patriarchal ideology; and (2) how women claim their agency through their engagement with NGOs.

Patriarchal ideology

Confucian thought in China and Catholicism in Poland have, over the centuries, molded in fundamental ways societal norms and practices in the two countries. Both of these philosophies prescribe clear gender differences and position women in the society and family as inferior to men. In China, Confucianism designates specific roles to individuals based on their gender, social class, and position relative to others; it erects boundaries and defines power relations between members of society. Thus, women are subordinate to other male members of the family and are expected to be submissive and to sacrifice their interests to those of men (Fan 2003). This patriarchal ideology underlies age-old practices whereby marriage connotes the transfer of a woman's membership and labor to the husband's family (Croll 1984). Women's ultimate membership in the husband's family discourages the natal family from investing in girls' education, even when socialism began investing in education, and perpetuates parents' strong preference for sons. These limitations of access set off a vicious cycle in which women are poorly educated and thus have low returns in the labor market, which further discourages investment in their education.

With limited access to education and constrained by subordinate social position, women in China are told that their place is "inside" the household while their husbands are responsible for the "outside" sphere, including making the earnings to support the family (Jacka 1997). Housework continues to be the primary reason why some women in China are not in the labor force (NBS 2002, pp. 1237–42). Women are coded as less important family members who are responsible for a limited sphere that is informal and domestic; men, on the other hand, access a much broader sphere outside the home which is important (as the family's economic well being hinges on it) and more formal. Feminist scholars have shown that power relations in the home and at the work place are key to explaining gender differentials in the labor market.

Catholicism in Poland can be perceived as not only the religion but also as "the lifestyle, the worldview, the educational model" (Środa 1992, p. 13). In the Catholic tradition, the female identity rests not only on submission but also on obedience, religiousness, and passiveness; from this perspective, women's agency is limited. The Catholic ideal demands from women a family and child-centered identity and compliance as wives and mothers (Środa 1992, p. 15). In addition, the Church and the state repeatedly called upon women to preserve the nation when under threat, to safeguard the family and to use all their capabilities to care, manage, and sacrifice. Thus, women were critical to the survival of the family and the nation, but their role was symbolic and invisible in the public sphere and reflected a *de facto* pseudo-agency. This system of beliefs goes beyond religious stand and as such, it signifies the existence of a far deeper patriarchal order. So while Catholic traditions cultivate "woman" as a national and religious symbol, they in the end constrain, similarly as in Confucianism, women to the private sphere of family duties. While the stereotyping of women to be the nurturing members and primary caregivers in the home is not unique to China and Poland (McDowell 1999, p. 126), the deep-rootedness of patriarchal ideology in both societies explains the persistence of these cultural norms despite state socialism.

These cultural norms have repeatedly constrained women's ability to become active economic subjects, and the underlying patriarchal ideology is a strong, fundamental, and persistent force that interacts with macro structural changes to produce varied labor-market outcomes for women. As described earlier, state socialism did mobilize women massively into the labor force, but it did little to mitigate gender discrimination as reflected in wages and occupational attainment. At the same time, socialist policies reinforced social norms by expecting and exacerbating women's double burden in and outside the home.

The retreat of the developmentalist state from addressing gender inequality, in essence, allowed for more explicit practices of traditional gender ideology in the labor market. Patriarchal ideology has consistently undervalued women's work and constructed women as less productive labor. Women in assembly line-type manufacturing, for example, are stereotyped as docile, passive, obedient, and disciplined workers and their work is perceived as secondary to familial responsibilities (Lee 1995). Thus, work processes are imbued with notions of appropriate femininity, which exemplifies the conventional view that women belong to domestic, informal, and above all, less important spheres (Mohanty 1997). Mandal's (1998, p. 71) claim of "sexism as an ideology that 'explains' and 'justifies' women's inferior situation on the job market" succinctly summarizes the ideological basis of women's disadvantages. She argued that gender not only hinders women's ability to advance professionally, but also represents a significant barrier to women's attempts to leave the ranks of the unemployed. The traditional gender roles in the family as well as negative stereotypes of women being less reliable, less committed and effective workers, are among chief factors that perpetuate sexism in the Polish job market (Titkow, Budrowska, and Duch 2003).

Since market reforms, in both China and Poland management decisions increasingly manifest the sociocultural perceptions of who is a valuable worker and the traditional gender roles (Titkow et al. 2003), as reflected especially in the joint discrimination by gender, age, and marital status. Chinese migrant women in their late teens and early twenties are more likely to get work than older women because the former are mostly single and are perceived to be physically fit for detailed work (Tam 2000). Married women are discriminated against because they are seen as less productive workers as a result of family demands. These are, in fact, not new practices as marriage and pregnancy were indeed causes for firing women workers in pre-socialist Chinese cities (Hershatter 1986). Similarly, in Poland, women aged 18 to 24 show high unemployment as they are expected to marry and have children in the near future. In addition, women aged 30 to 44 are also disadvantaged as they are perceived to have acquired bad work habits – those associated with low efficiency, poor work ethics and frequent sick leaves – under Communism (GUS 2004). When job shortages emerge, in particular, employers generally believe that men rather than women should be given jobs. In China, the layoffs that are due in part to the bankruptcy of state-owned enterprises disproportionately affect women, especially women of peak childbearing ages, and urban women are under greater pressure than their male counterparts to take early retirement (Li 2002, p. 56; Maurer-Fazio et al. 1999). While the five-year difference in mandatory retirement age for state employees (55 for women and 60 for men) has been a socialist practice since the 1950s, it is now increasingly used as a means to force women to leave the labor market, despite their marked improvement in health and life expectancy over the past decades.

Women's agency and NGOs

State socialism in China and Poland, as described earlier, was imposed from above. Though it appeared that women had power to act on their behalf, in reality what their pseudo-agency amounted to was restricted freedom. They had little power to influence the political decision-making process or to control the conditions of their work; and *de facto* their roles as mothers or workers were often privileged over that of independent agents. Economic liberalization, on the other hand, has created new conditions for women to be more assertive in drawing attention to their economic position. First, market reforms and discriminatory labor practices render the marginalized segments of society ever more vulnerable. Second, the state's social functions, such as those previously carried out by *danwei* in China or by state-owned enterprises in Poland, have gradually but substantially reduced. Third, partly because of the above, a collective awareness and heightened consciousness among women about gender inequality and women's interests have appeared. According to a survey on the social status of women conducted in 2000, 74.7 percent of Chinese women agreed that women should constitute at least 30 percent of high-level leaders in the government (Fuszara 2001; NBS 2004, p. 109). Finally, an increasing number of Chinese and Polish women are active in mobilizing themselves and others to engage feminist perspectives – perspectives that are independent of state socialism and that recognize women's marginalization but also their agency to act. In China, for example, women scholars led the feminist discourses and translated the term "gender" – which does not have a direct equivalent in the Chinese language – into "social sex" (*shehui xingbei*), which emphasizes the social contexts in which gender identity is constructed (e.g., Li 2002). Similarly, in Poland, gender was introduced as "cultural identity of the sexes." In both countries, women began to analyze actively the legacies of socialism and neoliberal policies, engage in efforts to increase their representation and participation in the public sphere, and monitor existing legal provisions in order to bring about real impacts on the status and position of women.

Women's collective agency can be best seen in how they organize in order to advance their positions and interests and enhance their skills. Yet, both countries had to struggle with the legacies of mass-membership women's groups that wanted to retain their power. Established in 1949, the All China Women's Federation's (ACWF) mission is "to represent and to protect women's rights and interests and to promote equality between men and women" (ACWF 2004). Like other state-sponsored organizations, ACWF has a vertically integrated structure consisting of local women's federations at every level of government and administrative division. It remains a product of the socialist experiment and thus faces challenges of how to adjust to new social and economic realities and of working with NGOs. In Poland, there were two such organizations – the Circles of Rural Women and the League of Polish Women (LPW). While the former had a long tradition focusing on the peasant family (Malinowska 2001, p. 194), the latter, created in 1945, reflected socialist ideology tenets and was mainly operating in urban areas. With the transition of 1989, both organizations initially lost power, finances, and prestige, although they (especially LPW) did begin to rebuild their networks and influence during the 1990s.

In both countries, the development of numerous and diverse NGOs in social and economic fields, since economic liberalization, is noticeable. This should be seen as a response to new pressures that result from the repositioning of the state *vis-à-vis* women, to changing economic conditions, and to new ambitions expressed by women that reflect their desire to increase their skills, knowledge, and independence and recognition that collaborative efforts could advance their goals. The specific contexts that conditioned the actual development of women's NGOs differed between the two countries. In China, where the reforms were more gradual, this process was instigated during the United Nations Fourth World Conference on Women that took place in 1995, when the notion of NGO began to demand attention from Chinese women and society at large. At the center of debate were two issues. First, the UN main conference was held in Beijing while the NGO Forum was held in Huairou, a small town about an hour from downtown Beijing, which sparked criticisms that the Chinese government wanted to undermine the visibility of NGOs. Second, though ACWF had since its establishment been seen as a part of the government, it demanded to be recognized as an NGO (and the largest women's NGO in China), in part to legitimize its representation of women and in part to emphasize that the government was in partnership with NGOs (Liu 2003).

The NGO debate in China has indeed motivated the founding of new women's groups, especially in urban areas, that are organized strictly outside the government structure. Though most of these NGOs are in their infancy, they have already become a force behind the passing of legislations that protect women's rights, including labor rights. Some of these new NGOs are research oriented and are affiliated to universities and research institutes, and some are grass-root organizations that provide services and networking opportunities. A well-known example of the latter is the Migrant Women's Club, a Beijing-based organization that provides a physical venue and organizes a magazine and other resources for migrant women to network, and for their voices to be heard (Gaetano 2005; Gong 2002; Jacka 2006). Like many women's NGOs in large cities, the Migrant Women's Club focuses in particular on issues of marriage, family, and employment. Though as a whole the impacts of NGOs on women's labor-market experiences are still limited, their increased number and expanded roles suggest that Chinese women are increasingly exerting their agency and advancing their goals via these collective efforts.

In Poland, the drastic political changes of 1989 immediately opened opportunities for NGOs to grow and compelled women to self-mobilize. In 1995, there were 68 women's formal and informal groups that addressed wide-ranging issues such as women's rights, family planning, health, and political participation (Malinowska 2001). By 2004, the number of groups increased to almost 300, of which almost 50, most of which were in urban areas, dealt specifically with economic issues, especially unemployment and economic activization of women (CPK 2005). These groups were created in response to increased unemployment and layoffs, and thus focus on basic skills enhancement and social and legal services and often target specific groups such as women above 40 and single mothers (CPK 2005).

Numerous new support groups and networks have also emerged in Poland for women who have jobs and who desire to improve their skills and qualifications and learn about their economic rights (e.g., the Federation of Business and Professional Women, European Forum of Business Women Owners, Polish Federation of Business and Professional Women Clubs). Rumińska-Zimny (2002, p. 89) observed that women's relative lack of networking is related to their concentration in the service sector (food, health, education, consumer services), which remains "outside the most influential bureaucratic and technocratic networks." The emergence of new business and professional groups helped women to overcome their marginalization, establish personal contacts, and obtain information and much needed advice. In fact, Rumińska-Zimny argued that "women's business associations and NGOs played a key role in the development of women's entrepreneurship in all transitional countries" (2002, p. 94). These arguments are echoed by others, including Mandicova (2002), who highlighted the role of women's groups and professional associations in creating more socially responsible leadership and in lobbying such bodies as UNCTAD, WTO or Human Rights Commission on wide ranging issues affecting women. Indeed, connecting the specific Polish context of women's work with similar challenges and opportunities faced by women regionally and globally is seen as an especially effective tool for strengthening women's individual and collective agency and for promoting women's economic interests.

Summary and Conclusion

In this chapter, we have sought to explain gender differentials in the labor market and their changes in China and Poland since socialism. An advantage of putting the two cases side by side is that it allows us to identify explanations for gender inequality – explanations that can perhaps be seen in other societies as well. A comparison between China and Poland, which shared a parallel experience of a dominant socialist state and subsequent repositioning of the state during market reforms, is also instrumental for conceptualizing how macro structural transformations impacted gender inequality. The Chinese and Polish societies are both deeply shaped by cultural and societal norms that reflect deep-rooted patriarchal ideology. At the same time, women in these two countries do have diverse labor-market experiences and outcomes, which can only be understood by contextualizing and historicizing the forces that gendered the respective economies. While we have focused more on urban situations than rural areas, the explanations we have identified have had society-wide impacts.

Similar to a number of other contributors to this volume, we have found that none of the four perspectives outlined by Logan and Fainstain can fully explain the phenomenon at hand, namely, persistent gender differentials in labor-market experiences. Tenets of modernization theory appear least convincing, as the neoclassical view that focuses on human capital and access to education explains only partially women's disadvantaged position in the labor market. Socialism, which sought to correct dependent and uneven development that is endemic to capitalism, legitimized a strong state and equalizing policies. No doubt, state socialism in both China and Poland pushed women to enter the work force, but it did not address the roots of inequality. Thus, the state equalized but it did not equalize. The results are, not surprisingly, mixed. Women's labor-force participation did increase remarkably but it was due to top-down instruments rather than efforts to empower women or to acknowledge women's identity, agency, and aspirations. Women were expected to perform and be judged both as wives/mothers and producers. Thus, mass labor-force mobilization may have had long-lasting impacts on women as economic subjects, but it has also reinforced the double burden women are expected to shoulder. The gender wage gap persisted in both China and Poland. In China, gender inequality in education and occupational attainment remained large. Polish women had made substantial gains in education and made inroads into many occupations, including professional work, but these gains were not translated into significant improvements of their position. These mixed results underscore the notion that state socialism did not bring about fundamental changes in gender inequality.

As a whole, most contributors to this volume have found that the developmentalist state (e.g., Bian and Gerber; Liang et al.; Smith and Hugo; Wu and Rosenbaum; Yusuf and Nabeshima; Zhou and Logan) and market transition (e.g., Appleton and Song; Bian and Gerber; Huang and Low; Liang et al.; Smith and Hugo; Wu and Rosenbaum; Zhou and Cai) offer more powerful explanations than modernization and dependency theories. Likewise, we have acknowledged the importance of the developmentalist state and market transition. In our analysis, we combined these two perspectives because the state's shift toward a developmentalist role meant its withdrawal and fragmentation as well as allowed for new market practices that systematically discriminated against women, as seen in sexist hiring practices and women's higher susceptibility to be unemployed. Although economic diversification meant a greater variety of work opportunities for women, they continue to concentrate in sectors that are inferior in pay and prestige. Women in the labor market are, therefore, still marginalized and are increasingly vulnerable. Even education does not protect them from unemployment, as seen especially in Poland. At the same time, more so in Poland than in China, women began to enter actively into entrepreneurship, in part to escape unemployment and the threat of unemployment. Our findings appear to be consistent with other contributions in this volume that deal explicitly with inequality. Both Appleton and Song and White et al. relate the worsening of inequality to market transition.

At the same time, they, as well as Bian and Gerber, highlight the continued role of the state and its policies in explaining inequality. Bian and Gerber, in their comparison between China and Russia, show that institutional reforms rather than economic growth explain urban class inequality. In our study, we have also found that differential economic performance (China's dramatic economic growth and Poland's much slower pace of growth) is not an important determinant of gender inequality.

While we have found Logan and Fainstain's four perspectives useful, we are not convinced that these perspectives have exhausted all prominent explanations of gender inequality. We have argued that a feminist perspective is necessary because it focuses on power hierarchy, construction of difference, and agency, all central to our explanations of women's position in the labor market, the persistent segmentation and undervaluing of their work, and their role as economic subject. We have highlighted in particular gender ideology and women's agency. Patriarchal ideology, as embodied especially in traditions and thoughts rooted in Confucianism and Catholicism, continues to be strong. It defines the sociocultural institutions such as family and church, which perpetuate societal norms for women as wives and mothers, consistently stereotype them as less productive workers, and constrain their access to and reward from the labor market. Such ideology is fundamental to explaining why state socialism did not eradicate gender inequality and why explicit gender discrimination in the economic sphere was prevalent once the state retreated from a gender-equality agenda. In this light, we concur with Huang and Low's (in this volume) view that traditions and cultural practices are deeply embedded in society and are more complex than can be explained solely by market transition or similar theories.

Most mainstream theories tend to prioritize macro forces and downplay the actors, including the marginalized and subordinate, whose life chances are assumed to be a passive function of the macro forces. A feminist perspective deviates from this masculinist approach by foregrounding agency and the processes by which the disadvantaged are empowered to challenge inequality. Women in China and Poland have responded to the retreat of the state and market discrimination by self-organizing and by cultivating a heightened sense of awareness. In both countries, women's NGOs have grown in number and activities and have sought to foster a collective agency among women. Most importantly, NGOs outside the socialist government structure have become instrumental toward advancing women's interests including labor rights and economic welfare. Perhaps because of Poland's longer and deeper engagement in labor activism, women's NGOs there are more active and visible than those in China. Nonetheless, these organizations and their advocates represent a new force, with new role models and aspirations, and they hint at the emergence of civil society where women's collective agency will play a more important role in determining their labor-market experiences.

ACKNOWLEDGMENTS

We wish to acknowledge the UCLA Academic Senate that provided partial support for our research. We would like to thank Wenfei Winnie Wang and Anna Wilkowska for research assistance and the reviewers whose comments and suggestions helped improve the chapter.

REFERENCES

All China Women's Federation (ACWF). (2004) About ACWF, online at: http://www.women.org.cn/english/english/aboutacwf/mulu.htm (accessed 10.24.2004).

All China Women's Federation (ACWF) Research Institute. (1991) *Zhongguo Funu Tongji Ziliao* [Statistics on Chinese Women] *1949–1989*. Beijing: Zhongguo Tongji Chubanshe [China Statistical Publishing House].

All China Women's Federation (ACWF) Research Institute. (1998) *Zhongguo Xingbe Tongji Nianjian* [Gender Statistics in China] *1990–1995*. Beijing: Zhongguo Tongji Chubanshe [China Statistical Publishing House].

Bauer, J., F. Wang, N.E. Riley, and X. Zhao. (1992) "Gender Inequality in Urban China." *Modern China* 18, 3, 333–70.

Beechey, V. (1986) "Studies of Women's Employment." In Feminist Review (ed.), *Waged Work: A Reader*. London: Virago Press.

Białecki, I., and B. Heynes. (1993) "Education Attainment, the Status of Women, and the Private School Movement in Poland." In V. Moghadam (ed.), *Democratic Reform and the Position of Women in Transitional Economies*. Oxford: Clarendon Press.

Centrum Promocji Kobiet (CPK). (2005) *Informator o Organizacjach i Inicjatywach Kobiecych w Polsce* [Directory of Women's Organizations and Initiatives in Poland]. Warsaw: Centrum Promocji Kobiet.

Croll, E. (1984) "The Exchange of Women and Property: Marriage in Post-Revolutionary China." In R. Hirschon (ed.), *Women and Property – Women As Property*. London: Croom Helm: pp. 44–61.

Fan, C.C. (2003) "Rural-Urban Migration and Gender Division of Labor in China." *International Journal of Urban and Regional Research* 27, 1, 24–47.

Fan, C.C. (2004) "The State, the Migrant Labor Regime, and Maiden Workers in China." *Political Geography* 23, 3, 283–305.

Fuszara, M. (2001) "Bilans na Koniec Wieku [The Balance at the End of the Century]." *Katedra* 1: 5–18.

Fuszara, M., and E. Zielińska. (2006) "Women and the Law in Poland: Towards Active Citizenship." In J. Lukic, J. Regulska and Darja Zavirsek (eds.), *Women and Citizenship in Central and East Europe*. Aldershot: Ashgate Publishing Ltd: pp. 39–60.

Gaetano, A.M. (2005) "Off the Farm: Rural Chinese Women's Experiences of Labor Mobility and Modernity in Post-Mao China, 1984–2002." PhD Dissertation, University of Southern California.

Gerber, T.P., and M. Hout. (1998) "More Shock than Therapy: Employment and Income in Russia, 1991–1995." *American Journal of Sociology* 104, 1–50.

Główny Urząd Statystyczny (GUS) [Main Statistical Office] (2004) "Women and Men on the Labor Market." Online at: http://www.stat.gov.pl/english/publikacje/kobie_mezczy_na_ryn_pracy.

Gong, Y. (2002) "Beijing dagongmei zhijia xiangmu [Migrant Women Knowing All Project in Beijing]." In X. Gao, B. Jiang, and G. Wang, (eds.), *Shehui Xingbei Yu Fazhan Zai Zhongguo: Huigu Yu Zhanwang* [Gender and Development in China: Retrospect and Prospect]. Xian: Shaanxi Renmin Chubanshe [Shaanxi People's Press]: pp. 396–403.

Gustafsson, B., and S. Li. (2000) "Economic Transformation and the Gender Earnings Gap in Urban China." *Journal of Population Economics* 13, 2, 305–29.

Haney, L., and G. Dragomir. (2002) "After the Fall: Eastern European Women since the Collapse of State Socialism." *Contexts* 1, 3, 27–36.

Hare, D. (1999) "Women's Economic Status in Rural China: Household Contributions to Male-Female Disparities in the Wage Labor Market." *World Development* 27, 6, 1011–29.

Harris, J.R., and M.P. Todaro. (1970) "Migration, Unemployment and Development: A Theoretical Analysis." *American Economic Review* 60, 126–42.

Hershatter, G. (1986) *The Workers of Tianjin, 1900–1949*. Stanford, CA: Stanford University Press.

Hershatter, G. (2004) "State of the Field: Women in China's Long Twentieth Century." *The Journal of Asian Studies* 63, 4, 991–1065.

Honig, E. (2000) "Iron Girls Revisited: Gender and the Politics of Work in the Cultural Revolution, 1966–76." In B. Entwisle and G. Henderson (eds.), *Re-Drawing Boundaries: Work, Household, and Gender in China*. Berkeley, CA: University of California Press: pp. 97–110.

Huang, X. (1999) "Divided Gender, Divided Women: State Policy and the Labour Market." In J. West, M. Zhao, X. Chang, and Y. Cheng. (eds.), *Women of China: Economic and Social Transformation*. Basingsroke: Macmillan: pp. 90–107.

Ingham, M., and H. Ingham. (2001) "Gender and Labor Market Change: What Do the Official Statistics Show?" In M. Ingham, H. Ingham, and H. Domański, (eds.), *Women on the Polish Labor Market*. Budapest, Hungary: Central University Press: pp. 41–76.

Ingham, M., H. Ingham, and H. Domański. (2001) "Women on the Labor Market: Poland's Second Great Transformation." In M. Ingham, H. Ingham, and H. Domański, (eds.), *Women on the Polish Labor Market*. Budapest, Hungary: Central University Press: pp. 1–20.

International Labor Organization (ILO). (2004) "Breaking through the Glass Ceiling: Women in Management (update)." Online at: http://www.ilo.org/dyn/gender/docs/RES/292/F267981337?Breaking%20Glass%20PDF%20English.pdf.

Jacka, T. (1997) *Women's Work in Rural China: Change and Continuity in an Era of Reform*. Cambridge: Cambridge University Press.

Jacka, T. (2006) *Rural Women in Contemporary China: Gender, Migration, and Social Change*. Armonk, NY: M.E. Sharpe.

Knight, J., and L. Song. (1995) "Towards a Labour Market in China." *Oxford Review of Economic Policy* 11, 4, 97–117.

Knothe, M., and E. Lisowska. (1999) *Women on the Labor Market. Negative Changes and Entrepreneurships Opportunities as the Consequences of Transition*. Warsaw, Poland: Center for the Advancement of Women.

Kotowska, I. (2001) "Demographic and Labor Markets Developments in 1990s." In M. Ingham, H. Ingham, and H. Domański, (eds.), *Women on the Polish Labor Market.* Budapest, Hungary: Central University Press: pp. 77–110.

Lake, R., and J. Regulska. (1990) "Political Decentralization and Capital Mobility in Planned and Market Societies: Local Autonomy in Poland and the United States." *Policy Studies Journal* 18, 3, 702–20.

Lee, C.K. (1995) "Engendering the Worlds of Labor: Women Workers, Labor Markets, and Production Politics in the South China Economic Miracle." *American Sociological Review* 60, 378–97.

Li, W. (ed.). (2002) *Shehui Xingbei Yu Gonggong Zhengce* [Gender and Public Policy]. Beijing: Dangdai zhongguo chubanshe [Contemporary China Press].

Lisowska, E. (2005) "Women's Entrepreneurship in the National Development Plan 2007–2013." Paper presented at "National Development Plan 2007–2013 – A Chance for Women?" Conference Organized by the Governmental Plenipotentiary for Equal Status and OŚKa, Warsaw, Poland (May).

Liu, B. (2003) "Zhongguo funu fei zhengfu zuzhi di fazhan [The Development of Women's NGOs in China]." In F. Du and X. Wang. (eds.), *Funu Yu Shehui Xingbei Yangjiu Zai Zhongguo* [Research on Women and Gender Studies in China] *1987–2003.* Tianjin: Tianjin renmin chubanshe [Tianjian People's Press]: pp. 409–28.

Loscocco, K.A., and Wang, X. (1992) "Gender Segregation in China." *Sociology and Social Research* 76, 3, 118–26.

Malinowska, E. (2001) "Women's Organizations in Poland." In M. Ingham, H. Ingham, and H. Domański (eds.), *Women on the Polish Labor Market.* Budapest, Hungary: Central University Press: pp. 193–220.

Mandal, E. (1998) "Sexism on the job Market: Negative Stereotypes of Women's Professional Work." *Women and Business: Journal of the International Forum for Women* 1–2, 68–71.

Mandicova, L. (2002) "Experiences in Creating a Regional Network in Central Europe." In *Women's Entrepreneurship in Eastern Europe and CIS Countries.* Series: Entrepreneurship and SMEs. Geneva, Switzerland: United Nations Economic Commission for Europe: pp. 100–1.

Maurer-Fazio, M., Rawski, T.G., and Zhang, W. (1999) "Inequality in the Rewards for Holding up Half the Sky: Gender Wage Gaps in China's Urban Labor Markets, 1988–1994." *The China Journal* 41, 55–88.

McDowell, L. (1999) *Gender, Identity and Place: Understanding Feminist Geographies.* Minneapolis, MN: University of Minnesota Press.

Meng, X., and P. Miller. (1995) "Occupational Segregation and its Impact on Gender Wage Discrimination in China's Rural Industrial Sector." *Oxford Economic Papers* 47, 136–55.

Millimet, D.L., and L. Wang. (2006) "A Distributional Analysis of the Gender Earnings Gap in Urban China." *Contributions to Economic Analysis and Policy* 5, 1, 1–50.

Mohanty, C.T. (1997) "Women Workers and capitalist Scripts: Ideologies of Domination, Common Interests, and the Politics of Solidarity." In M.J. Alexander and C.T. Mohanty (eds.), *Feminist Genealogies, Colonial Legacies, Democratic Futures.* New York: Routledge: pp. 3–29.

Nagar, R., V. Lawson, L. McDowell, and S. Hanson. (2002) "Locating Globalization: Feminist (Re)readings of the Subjects and Spaces of Globalization." *Economic Geography* 78, 3, 257–84.

National Bureau of Statistics (NBS). (2002) *Zhongguo 2000 Nian Renkou Pucha Ziliao* [Tabulation on the 2000 Population Census of the People's Republic of China]. Beijing: Zhongguo tongji chubanshe [China Statistics Press].

National Bureau of Statistics (NBS). (2003) *Zhongguo Laodong Tongji Nianjian* [China Labour Statistical Yearbook] *2003*. Beijing: Zhongguo tongji chubanshe [China Statistics Press].

National Bureau of Statistics (NBS). (2004) *Women and Men in China: Facts and Figures 2004*. Department of Population, Social Science and Technology, National Bureau of Statistics. Online at: http://www.stats.gov.cn/english/statisticaldata/otherdata/ (accessed 5.8.2006).

National Bureau of Statistics (NBS). (2005) *Zhongguo Tongji Nianjian* [China Statistical Yearbook] *2005*. Beijing: China Statistics Press.

Nowakowska, U. (ed.). (2000) *Polish Women in the 90s: The Report by the Women's Rights Center*. Warsaw, Poland: Women's Rights Center.

Pater, M. (1998) "The Employment Market for Women in Poznan." *Women and Business: Journal of the International Forum for Women* 1–2, 65–8.

Regulska, J. (1992) "Women and Power in Poland: Hopes or Reality? In J. Bystydzienski (ed.), *Women Transforming Politics: Worldwide Strategies for Empowerment*. Bloomington and Indianapolis: Indiana University Press: pp. 175–91.

Reszke, I. (2001) "Stereotypes: Opinions of Female Entrepreneurs in Poland." In M. Ingham, H. Ingham, and H. Domański (eds.), *Women on the Polish Labor Market*. Budapest, Hungary: Central University Press: pp. 177–92.

Riley, N.E. (2004) "China's Population: New Trends and Challenges." *Population Bulletin* 59(2).

Rumińska-Zimny, E. (2002) "Building Regional Networks." In *Women's Entrepreneurship in Eastern Europe and CIS Countries*. Series: Entrepreneurship and SMEs. Geneva, Switzerland: United Nations Economic Commission for Europe: pp. 89–99.

Shu, X., and Y. Bian. (2003) "Market Transition and Gender Gap in Earnings in Urban China." *Social Forces* 81, 4, 1107–45.

Środa, M. (1992) "Kobieta, wychowanie, role, tożsamość [Women, Upbringing, Roles, Identity]." In *Glos Mają Kobiety*. Kraków, Poland: Convivium: pp. 9–17.

State Statistical Bureau (SSB). (1985) *Zhongguo 1982 Nian Renkou Pucha Ziliao* [1982 Population Census of China: Results of Computer Tabulation]. Beijing: Zhongguo Tongji Chubanshe [China Statistical Publishing House].

Stiglitz, J. (1998) "Gender and Development: The Role of the State." Speech at the World Bank Gender and Development Workshop, April 2. Online at: http://www.worldbank.org/gender/events/gendev/theme1.htm.

Stockman, N. (1994) "Gender Inequality and Social Structure in Urban China." *Sociology* 28, 3, 759–77.

Szelényi, S. (1992) "Economic Subsystems and the Occupational Structure: A Comparison of Hungary and the United States." *Sociological Form* 7, 4, 563–86.

Tam, S.M. (2000) "Modernization from a Grassroots Perspective: Women Workers in Shekou Industrial Zone." In S. Li and W. Tang (eds.), *China's Regions, Polity, and Economy*. Hong Kong: The Chinese University Press: pp. 371–90.

Titkow, A. (1984) "Lets Púll Down the Bastilles before They are Built." In R. Morgan (ed.), *Sisterhood Is Global: An International Women's Movement Anthology*. Garden City, NY: Anchor Books: pp. 560–6.

Titkow, A. (1992) "Introduction." In *Głos Mają Kobiety*. Kraków, Poland: Convivium: pp. 5–8.

Titkow, A. (1998) "Polish Women in Politics: An Introduction to the Status of Women in Poland." In M. Rueschemeyer (ed.), *Women in the Politics of Postcommunist Eastern Europe*. Armonk, NY: M.E. Sharpe: pp. 24–32.

Titkow, A., B. Budrowska, and D. Duch. (eds.), (2003) *Szklany Sufit. Bariery i Ograniczenia Karier Kobiet* [Glass Ceiling: Barriers and Limitations to Women's Careers]. Warsaw: Instytut Spraw Publicznych.

United Nations. (2000) "The National Action Plan for Women." Online at: http://www.un.org/esa/gopher-data/conf/fwcw/natrep/NatActPlans/poland.txt.

Whyte, M.K. (2000) "The Perils of Assessing Trends in Gender Inequality in China." In B. Entwisle and G. Henderson (eds.), *Re-Drawing Boundaries: Work, Household, and Gender in China*. Berkeley, CA: University of California Press, pp. 157–70.

World Bank. (2004) *Gender and Economic Opportunities in Poland: Has Transition Left Women Behind?* Warsaw: The World Bank. Report no. 290205.

Part II

Changing Places

5

Urbanization, Institutional Change, and Sociospatial Inequality in China, 1990–2001

Michael J. White, Fulong Wu, and Yiu Por (Vincent) Chen

Introduction

China's recent unparalleled economic growth has brought with it growing individual economic inequality. News media and scholarly reports have examined disparities in levels of personal income and wealth, yet only limited attention has been directed to the spatial manifestation of economic growth across the landscape. Some discussion exists about the differential fortunes of individual provinces and regions. Less clear is how cities and their surrounding metropolitan regions have fared. Even less information is available about the spatial differentiation of socioeconomic groups within cities, the traditional pattern of urban ecology. In this chapter, we investigate the shifts in spatial organization and inequality accompanying China's economic restructuring and rapid development over the past two decades. In the Introduction Logan and Fainstein point to the fact that China's development path may be atypical. China's legacy of strong state authority indicates that simple Western or developmental models of urban form and interregional inequality, typically based on the experience of North America and the highly industrialized countries of Europe, cannot be accepted without examination.

Economic growth typically widens the gap between average rural and urban incomes; China is no exception. For instance, a recent report in the *Shanghai Star* (Shangyao 2004) noted that urban-rural per-capita income ratio was 1.7 to 1 in 1985 and widened to 3.1 to 1 in 2002 (see also Wei 2004).

Yet the urban-rural gap is not the only spatial differential in China. Income differences across provinces and between towns also need study.

There are several reasons for trying to understand the pattern of spatial inequality. First, the spatial organization of resources necessarily means that geographic location has implications for access to public resources. Individuals who reside in more economically dynamic cities and regions may benefit from this growth trajectory (schools, public services, etc.) and similarly, those in better-provisioned neighborhoods or districts may benefit. Second, an empirical literature in the social sciences makes the case that inequality carries measurable negative consequences for societies. Spatial inequality in particular, has been implicated in the higher level of local violence and civil strife in Latin America. Cross-national empirical evidence supports this interpretation (Messner and Rosenfeld 1997). Furthermore, higher levels of inequality at several spatial scales have been linked to inferior health outcomes. Socioeconomic inequality may erode social cohesion, which may in turn lead to disinvestment in the social capital need for investments in health (Kawachi et al. 1997). Recent work in economics argues that inequality itself may hamper economic growth and poverty reduction (Alesina and Rodrik 1994; De Ferranti et al. 2004; Persson and Tabellini 1994). Finally, spatial differentiation is a manifestation of social stratification more generally. Thus, the measurement of inequality across cities and provinces can provide a useful alternative window on changes in the society at large.

The chapter proceeds as follows. First, we introduce some of the concepts that guide our analysis, linking economic growth and inequality at multiple spatial scales. We discuss recent literature on Chinese spatial inequality. We then turn to our own empirical analysis. Our approach compares the dispersion of average income, measured as per capita gross domestic product (GDP) for various geographic units. After a brief discussion of our data and methods, we present an analysis of inequality for *provinces*, showing how it has increased in the past two decades. Second, we demonstrate that inequality *across cities* has increased since the 1980s, with some cities experiencing a relatively rapid rise in average per capita income, while other cities tended to lag behind. Finally we focus on *intra-urban spatial differentiation* and inequality in one city, Shanghai. We show that inequality across social groups is clearly manifest in one of China's largest and most economically dynamic cities.

Chinese Urban Inequality: Theory and Experience

The Kuznets paradigm and urbanization

The Kuznets paradigm (Kuznets 1955) argues that individual income inequality would initially rise with economic growth and later decline. The spatial parallel with Kuznets' paradigm is one in which spatial inequality increases with development (modernization) and then decreases with economic maturity. Inequality traces an inverted U-shaped curve with time and modernization. Whether a spatial Kuznets curve will manifest itself in contemporary China is open to debate. More recent conceptual and empirical work has challenged the generality and contemporary applicability of the Kuznets paradigm (Korzeniewicz and Moran 2005). Kuznets' original work was somewhat conjectural. Alternative patterns of labor mobility (more constrained) and institutional structures (more embedded in global networks of control) could limit its applicability. Nevertheless the Kuznets' paradigm remains provocative and persuasive. This combined with the fact that the Kuznets approach ties directly to income inequality – the outcome we examine, but across spatial units – makes it a good starting point for us. Our work looks at initial empirical evidence for China.

We can think of a simple contrast to frame the discussion of urban spatial inequality and its relationship to the Kuznets paradigm. In one scenario, we might describe economic growth as a "rising tide that lifts all boats." A simple rising tide effect might lift all boats equally; in this case all cities would benefit proportionately in the economic growth of the recent past. In this scenario, cities' proportionate economic growth means no increase seen in the inequality of economic levels across cities. In operational terms, per capita GDP rises, but the dispersion (variance in per capita income across cities) does not.

The alternative scenario is one of growing spatial disparity. Some cities and provinces do well (riding the wave of economic development) while others lag behind (and remain in the trough). Thus, even in the face of increasing economic well-being in the urban sector generally, some cities experience greater increases in per capita GDP than others. Such an outcome would be manifested through increasing values of inequality statistics across the urban system, and it would be consistent with the upswing portion of the Kuznets curve. We look for these manifestations of inequality on three spatial scales: across provinces, cities, and urban zones.

Provincial and regional experience

Jian, Sachs, and Warner (1996), using provincial-level panel data for the period 1952 to 1993, demonstrate cycles of income convergence and divergence across provinces. The initial phase of central planning, 1952–65, offers some weak evidence of income convergence. There was strong evidence of divergence during the Cultural Revolution. Jian and colleagues found a convergence of income during the subsequent rural reform period. However, divergence has recurred since 1990, which coincides with the period of economic reform of interest to us.

Kanbur and Zhang (2003) use a longer provincial-level panel to analyze regional inequality in China from 1952 to 2000. They argue that the Great Famine of the late 1950s and the Cultural Revolution during the late 1960s and 1970s brought increases in inequality. Lu and Wang (2002) attribute uneven regional development to differences in production factors and the legacy of colonial history in the coastal area, particularly the presence of industrial infrastructures.

Government policies may also contribute to regional disparity. Zhang and Fan (2000, p. 12) argue that government policy favored coastal cities and their surrounding regions (see also, Song, Chu, and Cao 2000). Yang (1999) argues further that urban-biased policies may have generated long-term regional disparity. Yang and Wei (1995) show that the average annual growth rates of GNP and industrial gross output value in inland regions have been lower than those in coastal regions since 1980. Jones, Li, and Owen (2003), using city-level data, establish a link between Special Economic Zones and "growth enhancing" foreign direct investment (FDI). Using GDP of three industrial sectors, Huang, Kuo, and Kao (2003) show that the underdevelopment of secondary industry may play an important role in increasing regional inequality.

In the Post-Mao period urban policy turned to "control the size of large cities and develop small cities" in the Third National conference on urban affairs in 1978. The policy further evolved to develop medium and small cities (Zhou 1993, p.113). Since 1990, the number of places classified as small cities grew from 291 in 1990 to 382 in 1997, and then 326 in 2002. The number of medium size cities grew from 121 in 1991 to 205 in 1997 and then to 225 in 2002 (data from China Urban Statistics Yearbook, various years). These changes were paralleled by further administrative moves, which merged other counties and county-level cities into higher-level administrative districts (Zhang and Wu 2006).

The differentiation of urban space

The classic Western (North American, European) literature on urban inequality focused on three major sources of neighborhood differentiation: social status, family status, and ethnic status. These three features of urban differentiation were generally found empirically, although not all cities manifested three simple and separable dimensions (White 1987), although this ecological approach has been criticized for its neglect of the role of political power and for seeing differentiation as a "natural" process. In the China case, we expect differentiation to be created by (1) the rapid and uneven pace of urbanization; (2) accompanying cultural changes, such as changing life styles and urbanism; and (3) the shift from a planned to a market economy.

In the West these intra-urban differences grew on top of inter-urban or regional differences. In turn, most were linked back to the changing economic fortunes of locales, as the value of access to natural resources, distance to markets, or the economic incentives for firms (and people) to relocate. Surely, the differential construction of highways, indirect or direct subsidies for extracting natural resources, and public service provision, all play a role in the differential growth of cities in the West.

There are three major scales of spatial inequalities in China: urban-rural, inter-urban, and intra-urban. First, *urban-rural inequalities* are created by the different levels of development in urban and rural areas and the policies that separate urban and rural residents (Solinger 1999 and other chapters in this volume). The socialist legacy of urban-rural division still exerts effects on spatial inequalities. For example, rural migrants mainly originate in less-developed interior regions and flow toward the coast. Within the city, various policies discriminate against migrants, since rural *hukou* may diminish access to urban resources.

Recent policy shifts have changed this landscape appreciably. The institutional barriers to urban access have been reduced, for example, the abolition of a registration card for temporary residents and compulsory return of rural sojourners to their origin. These policy shifts have probably served to encourage further migration, allowing people to move to the location of best perceived gain. The long-term effect of this migration on rural-urban disparity remains to be seen, but as we noted at the outset, the urban-rural income gap has grown recently.

Second, *inter-urban spatial inequalities* exist between the Coastal Region, Inner/Central Region and Western Region (Fan 1997). This scale of inequality should be understood through socialist economic development strategy as well as the international context: the Western Region is resource-rich but has been underdeveloped economically. National policy has tended to create a spatial division of labor among the regions: the Western Region provides cheaper resources for the Coastal Region to produce high value-added products. The state, through central planning, has supported the Western Region by expanding production capacity.

Third, *intra-urban spatial inequalities* exist between inner urban areas and suburban peripheries and between different workplace-based neighborhoods. In the socialist period, differences by neighborhood were relatively small, because workers tended to live at or near their jobs (Yeh, Xu, and Hu 1995). A basic intra-urban spatial differentiation (inner versus suburban) exists, because the socialist transformation initiated in the 1950s did not completely restructure the built environment (F. Wu, 2002). In particular, local government-housing areas have been developed in inner urban areas and managed by local housing bureaus, while state workplace housing has been developed and managed by state investment firms in alternative locations. The result is that differential housing quality exists within the public-sector; and even within workplace housing, standards vary according to the administrative rank of the workplace (see Chapter 8 in this volume; Logan, Bian, and Bian 1999; Wu 1996).

An intriguing question remains about how much internal urban differentiation should be manifest in the contemporary Chinese city. To be sure, some of the legacy of the pre-socialist and socialist past, along with market forces, will make for differences across urban neighborhoods. Yet the current situation is probably far different from the US and European cities, even given the influence of the welfare state in the latter. China also lacks the level of ethnic diversity that has characterized the American urban setting and which is now visible in European countries of immigration. What may be distinctive in China is rural migrant segregation. Migrants are still subject to the process of assimilation, although they share Chinese culture. Although there are reports of sub-cultures based on place of origin (Ma and Xiang 1998) and "sub-ethnicity" (Honig 1990), it is undeniable that cultural and linguistic barriers for migrants to the Chinese city are smaller than those in other national contexts.

Inequalities are created by more globally oriented economic sectors such as finance and services, and more traditional sectors such as light manufacturing industries. Yet, an emphasis on the global city and the demise of the national state in urban economies does not capture the continuity of the state role in urban growth, the labor market, and residential form. The state attempts to strengthen local governance by developing local forms of organization, such as neighborhood residents' committees. Market-based housing is becoming a major source of shelter for better-off residents. Social inequalities are enlarged, at least measured in terms of income differences, with new strata now identified in Chinese society (Lu 2002).

Perhaps most provocative is an assessment of the degree to which the evolution of the contemporary Chinese city will follow any other national, regional, or historical model. In one view, the Chinese case exhibits commonality with developing countries (further industrialization, weakening of traditional social ties). At the same time current changes are in keeping with industrial structure shifts in developed countries (growth of service industries, influence of globalization, and development of the civil society). Finally, Chinese societal and economic changes of recent decades have some features that are virtually unique. The particular nature of China's transition from a planned to a market-oriented economy is unparalleled, with the government retaining relatively strong state control in the midst of these changes. Most critically, perhaps, China has attempted to retain the regulation of residence and population movement (through the *hukou* system) more than most other transition economies. In no way is the die yet cast for the Chinese urban system and its form of spatial inequality, whether the intra-urban separation of social groups or the competing and differential fortunes of individual cities in China's urban system.

Data, Definitions, and Methods

Our aim is to investigate the manifestation of urban inequality in China, with much of our analysis focusing on differences across cities. We are especially interested in how the Chinese experience will map onto existing theory, mostly drawn from the West.

Urban definitions and administration in China

The urban system in China is a complex overlapping of urban and rural areas and *hukou* policy, the household registration policy operating since the 1950s (Chan and Zhang 1999). The identification of "urban" in China consists of cities (*shi* or *chengshi*) and towns (*zhen*). In China the "city" is an administrative unit, which includes a highly-built-up core, defined by "district" (*shixiaqu*), and a surrounding rural hinterland, defined by village committee. A "town" is usually a built-up area surrounded by villages. In contrast to the more strictly geographical or ecological notion of cities in the West, Chinese urban entities administer more of their surrounding landscape (including some agricultural territory in cases) than most Western urban areas. The Western notion of a highly differentiated urban core and a suburban fringe is less relevant for China. Still, the notion of a "settlement system" has utility in China. Urban agglomerations have clear boundaries, populations counted within, and a ranking based both on aggregate population size and on administrative function.

Two systems of territorial reorganization were formalized after reform in the early 1980s. The first is the "cities leading counties" (*shidaixian*) system, also known as "cities administering counties" (*shiguanxian*) system. Under this system, implemented in 1981, counties were abolished, which the central government hoped would result in the development of a regional economic network. Second, the abolition of counties and re-establishment of cities serves to classify some otherwise rural residents as "urban." There was a relaxation of the criteria for establishing cities in 1983, with many former counties and towns upgraded to cities. The criterion for prefecture-level and county-level cities was also revised in 1993. (For details of the changes, see Yeh et al. (1995) and Zhang and Zhao (1998).)

China now has a three-tier system of cities: municipal-level, prefecture-level, and county-level. A designated "city proper" at the first two levels normally consists of two kinds of city districts: urban districts (*jianchengqu*) and suburban districts (*jiaoqu*), which are mostly under the administration of street committees, although rural areas may be included (Fan 1995). A district occupies the same administrative rank as a "county." Cities usually have no more than two suburban districts (Zhang and Zhao 1998, p. 331) Some 830 districts were under the jurisdiction of cities as of 2002. The prefecture-level city (*diji shi*, literally "region-level city") is an administrative division of China that is governed directly by provinces. A prefecture-level city ranks between municipal-level city and county-level cities. In 2002, China had 275 prefecture-level cities.

A county-level city (*xianji shi*) is also a county-level administrative division. Counties and county-level cities are usually governed, in turn, by prefecture-level cities. Similar to prefecture-level cities, county-level cities also usually contain rural areas many times the size of their urban, built-up area. Towns are under the administration of counties. There were 381 county level cities and 1649 counties in China in 2002. Currently four cities have been designated with province-level autonomy: Beijing, Tianjin, Shanghai, and Chongqing.

Data

We draw on several sources of national statistical data to conduct our analysis. Data apply to provinces and various classifications of cities and urban territory. All these data are complied by the State Statistical Bureau (SSB). Chinese urban statistics are available for two levels of classification: prefecture and above, and county level. Our prefecture-level data contain information from 1990 and from 1996 to 2001. The 1996 to 2001 data are from the State Statistical Bureau electronic files [China Data Online, Machine Readable Data File, 2004], while the 1990 data are from the city statistical yearbook 2003. In order to analyze inequality at the city level during the 1990s, we select three time points at which to examine and compare inequality across cities: 1990, 1996, and 2001. We are limited to county-level data from 1997 to 2001.

For intra-urban inequality analysis, we will use Shanghai's 2000 population census data, disaggregated at the street-office level, to observe the distribution of spatial inequalities, measured by education levels, occupation sectors, and migrant status. Given the focus on a single city, we use maps to more clearly display the geographic sorting of social characteristics.

Measures

Our basic measure of province and city inequality is the dispersion of GDP per capita, measured across geographical units. Despite concerns about the accuracy of any income measure, we feel that GDP per capita will still capture overall differences in economic development and living standards across urban settings in China. We first compare the dispersion in province level average income. Then we compare the dispersion within the urban system, looking at prefecture-level cities and county-level cities.

We calculate the 75:25 ratio and the 90:10 ratio. The first of these values tells us how far apart the points are that identify the top quartile of cities and the bottom quartile. The larger this number, the more distance there is between the average income level in well-off cities from the average income level in poorer cities. The 90:10 ratio is exactly analogous representing the top and bottom decile of the urban well-being distribution. This latter measure is more sensitive to extremes, especially if a handful of cities is "pulling away" or lagging far behind. Our use is parallel to the use of percentile ratios for individual wage and household inequality; these ratios also have the advantage of being robust to price changes.

For some of our analysis we classify cities by the ratio of the nonagricultural population to the total population. We will refer to this, for ease, as "urbanization level," although we recognize that we are classifying by industrialization. Given the particular administrative definition of cities in China, the classification of population on the basis of registration rather than activity, and the close relationship between urbanization and industrialization in most empirical settings, this works acceptably well. We classify the urbanization into four levels: 0 to 25 percent; 25.1 percent to 50 percent; 50.1 percent to 75 percent; and 75.1 percent to 100 percent.

China's classification of population in official statistics creates additional complications. Migrants may be counted at current (destination) residence or not depending on the source of the statistics, their official registration, and the time frame of relevance for the data source. In general, our data consider only those who are official resident of the geographical boundary. Short-term migrants are generally underrepresented in their destinations. For us this means that the rural population can be somewhat over-counted and the urban population will be somewhat under-counted. More detail on these technical aspects is available from the authors.

For some of our intra-city comparisons we have employed the index of dissimilarity, a widely used measure of residential segregation (White 1987). The dissimilarity index compares the residential distribution of pairs of population groups. Its value varies from 0, where the group in question is evenly distributed across neighborhoods, to 1 (100 percent), where the group is isolated in specific neighborhoods, sharing no residential areas with members of the other group in the comparison.) For instance in the comparison of residential differentiation of migrants from non-migrants, the index of dissimilarity is calculated as: $D = \sum(x / X - t / T)/2$, where, x is the number of migrants in the street office district, X is the total number of migrants in the city, t is the registered population (i.e. non migrants) in the street office district, T is the total registered population in the city. The index can be interpreted as the percentage of migrants that would have to relocate in order to generate an absolutely even distribution of migrants in relation to non-migrants. This index is commonly used in residential segregation analysis of the different ethnic groups. In the US, the index between the white and black can reach as high as 80 per cent for some cities.

Results

Provincial differences

The province is the most general geographic scale at which we might be expected to observe growing spatial inequality. Provinces of course have varying access to markets and resources, and they have varying institutional actors and historical legacies. There has been much attention given to the varying fate of provinces within China. In the earliest days of reform, much attention was focused on the rapid economic growth of Guangdong Province. This growth was accompanied by in-migration. Thus, while the overall level of economic activity grew, less is known about the trajectory of per capita income. As economic growth spread and persisted through the 1980s and 1990s, other provinces, particularly those along the coast, began to exhibit differential economic growth or "success."

The question then, is this: What has been the net shift in the disparity across provinces during the period? Table 5.1 presents the 75:25 ratio and the 90:10 ratio for 1990, 1996, and 1999. The larger the ratio, the more unequal the average income level is across the several provinces. Under the hypothesis of differential spatial manifestation of economic growth (the spatial Kuznets hypothesis) we expect these ratios to rise over time during the period we analyze. *The fortune of China's provinces does become more unequal over the 1990s.* In 1990 the per capita income level that marked the top 25 percent of provinces was 1.81 times that of the income level that marked the bottom 25 percent. This ratio crept up to nearly 2 in 1996 and grew to exceed 2 (at 2.07) by 1999. The trend of increasing disparity is confirmed and extended by looking at the 90:10

Table 5.1 Spatial inequality at provincial level 1990–99: per capita GDP ratios

Year	*Provincial level per capita GDP ratio*		*N*
	Ratio 75:25	*Ratio 90:10*	
1990	1.81	4.07	28
1996	1.97	4.76	28
1999	2.07	4.91	28

For data consistency, Hainan is included in Guangdong, Chongching is included in Sichuan, and Tibet is not included in provincial level data.

ratio. Whereas the 10 percent of most wealthy provinces exhibited incomes about four times that of the bottom 10 percent of provinces in 1990, the disparity was five-fold by 1999.

It is important to recall that in these provincial comparisons (and in all city comparisons below) we are comparing aggregate snapshots at successive points in time. The values do not represent directly inequality among individual workers or household residents, and the values represent the income (regional GDP) of the population's residents in a particular place in a particular year. Still, sub-provincial level data are considered to be reliable and consistent due to the same method of collecting and accounting system across cities (Fujita et. al 2003, p. 74). Various hypotheses of increasing spatial inequality argue that the disparity should grow. An examination of the actual trend in mean income by percentile cut-point (detail not shown) indicates that, indeed, urban areas at all levels grew in per capita income. For example in the case of counties administered by cities, the ninetieth percentile income level grew by 5.8 percent in 1990–2001, while the tenth percentile income level grew 4.0 percent. Growth at the seventy-fifth and twenty-fifth percentile was intermediate.

Inter-city inequality

We now examine the manifestation of inequality across the urban settlement system. Again, larger values of the 75:25 and 90:10 ratios of GDP per capita income point to more inequality. We make these comparisons across the 1990s and for different parts of the settlement classification in China.

Table 5.2 presents our analysis. First we begin with an analysis of the prefecture level (Panel A). In 1990 the 75:25 ratio for the 215 cities-proper (the most restrictive definition) in the sample was 2.12. This means that the average per capita GDP in the city that marked the beginning of the upper quartile of urban well-being (or productivity) was about twice that of the city that marked the lower quartile. In the interval 1990–2001, the 75:25 ratio increased modestly, pointing to somewhat greater income inequality among cities. Additional information is available from the 90:10 ratio. This value increases from about 4 to 4.8 over the decade. The 90:10 ratio is necessarily larger, since it expressed the ratio of the income distribution closer to its "tails." What is noteworthy, however, is that this ratio increases noticeably more than the 75:25 ratio. This means that the tails have spread out between 1990 and 2001. In more ordinary terms, one can say that the rich (cities) are getting richer and the poor (cities) poorer, or perhaps more accurately that economic growth has been experienced by a smaller segment of urban areas. It is likely that selected urban areas have been able to capture foreign direct investment or industrial expansion, so as to have a growth differential from China overall and even more rapidly than some better-than-average cities.

Table 5.2 Spatial inequality in the urban system 1990–99: per capita GDP ratios

	Ratio 75:25	*Ratio 90:10*	*N*
A. Prefecture level cities proper			
City proper in 1990	2.12	3.97	215
City proper in 1996	2.14	4.54	232
City proper in 2001	2.23	4.80	233
B. Prefecture level counties managed by cities			
Counties managed by cities in 1990	1.58	2.71	167
Counties managed by cities in 1996	1.92	3.64	202
Counties managed by cities in 2001	1.96	3.94	220
C. County level cities and surrounding counties			
Rural county 1997	2.15	4.26	1541
Rural county 2001	2.20	4.52	1605
County level city 1997	2.09	4.03	345
County level city 2001	2.22	4.71	341
Total (county + city) 1997	2.28	5.07	1886
Total (county + city) 2001	2.40	5.41	1946

We next turn to the next-higher level of administrative organization for urban areas: counties managed by prefecture-level cities (Panel B). Here we find that the 75:25 ratio is lower than for the city-proper group. Thus, these settlements are more compact in their income distribution (in every year) than the city-proper group. Here again, income inequality increases steadily over the 1990s. As with the city-proper, the county level shows a more dramatic increase in disparity for the 90:10 ratio. Thus, the leading 25 or so (top 10 percent) city-counties pulled away from the others. While the set of cities at the top or bottom of the distribution may shift (although we doubt it changes very much) the fact of changing disparity holds. Our examination of the underlying income data indicates that this is indeed driven by superior economic growth from the "successful" or advantaged cities, rather then attributable to outright decline of cities at the bottom of the distribution.

Panel C of Table 5.2 shows the county-level data. Due to data limitations, we can only compare data for 1997 and 2001. Although the time frame is short, we are able to identify an increase in overall inequality. While the inequality of county-level cities increased 10 percent when considering the 75:25 ratio, the

90:10 ratio shows an 18 percent increment and a substantial increase in inequality between the richest and poorest county-level cities. While strict comparisons are difficult, we note that this growth of inequality among county-level cities is greater than the growth in inequality we identified above for larger geographical units.

Urbanization and economic inequality

In Figure 5.1 we consider whether urbanization level (here the ratio of the non-agricultural population to the agricultural population) is broadly associated with level of per capita income and increases in income dispersion. Figure 5.1 shows the box plot of the urbanization level and nominal GDP per capita at

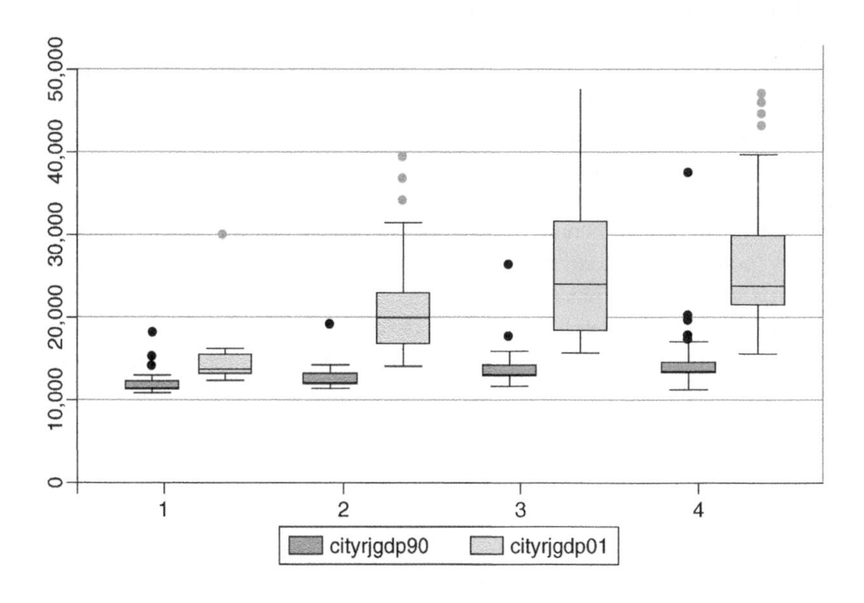

Key:

cityrjgdp90 = prefecture level cities' GDP per capita in 1990
cityrjgdp2001 = prefecture level cities' GDP per capita in 2001

X-axis: Urbanization level:
Urbanization level = non-agri. Pop/ total population
where
1 = if urbanization < .25
2 = if urbanization >= .25 & urbanization < .5
3 = if urbanization >= .5 & urbanization < .75
4 = if urbanization > .75

Figure 5.1 Box & whisker plots GDP per capita at prefecture level city proper at 1990 (left box of pair) and 2001(right box of pair), by level of urbanization [1, 2, 3, 4] of the prefecture

the prefectural city level at 1990 and 2001. While it may seem counterintuitive, Chinese cities, since they are administrative units that can include their hinterland, may incorporate agricultural as well as non-agricultural populations. In Figure 5.1 we represent the level of urbanization in these prefecture-level cities in four levels of increasing urbanization (urbanity). For 1990 one sees a relatively compact distribution at each level of urbanization. Prefecture-level cities at the lowest level of urbanization – effectively small cities and towns in the hinterland – show very little dispersion. As the urbanization level rises, so, too, does the spread of income across cities in the groups. Stated another way, the more "urban" (and less agricultural) places exhibit greater inequality. This may replicate the diversity of their economic and industrial structure. The comparison shows that the inequality levels are higher during 2001, and moreover, inequality ratios increase with urbanization level.

As we shift focus to data from 2001, we see the rather dramatic impact of a decade of sustained growth. First we see that median income levels (the middle line of each "box" in the box-and-whisker plot) is higher for 2001 than 1990 in the corresponding urbanization group. More notable, however, is the change in relative position across urbanization level. The more urban locations exhibit much higher per capita income levels and a much wider distribution. There is appreciable skew in these city income distributions. The fact that a handful of cities are outliers (points out beyond the upper portion) indicates appreciable economic shifts in a smaller set of places. Such a widening distribution is reminiscent of the "take-off" notions circulated in economic development theory some time ago. To be sure, the overall economic growth of China in the 1990s (and doubtless extending back through the 1980s) was realized more fully in some cities than others.

Intra-urban analysis

The intra-urban analysis requires detailed small area statistics, such as census tracts. For much of recent time, there has been a dearth of data at this level of geographic detail for China. Recently, however, census data began to become available at a very crude resolution (finest being the "street office district" or *jiedao)*. The scale is still far too large to yield the same analytical information as a census tract. Such a scale is equivalent to a ward in the UK census or perhaps a community area in the US. Moreover, the attributes collected in the census are limited. The definition of "migrants" in the most recent census is more problematic: it only recognizes residents who have been living in the current location for over 6 months but without local household registration. While this would mainly include rural migrants, it also includes the households moving toward the urban periphery who prefer to keep their registration in the inner area. Still, despite the disadvantages of scale and data content, these sub-city statistics do allow a window on the sorting of population into urban sub-areas. What is more, they offer some opportunity, however approximate, to compare contemporary Chinese urban ecology to expectations based on Western theory or recent trends in Chinese economic development.

Figure 5.2 shows the sub-area distribution of the population with above-college-level education in Shanghai. Education is important for its link to income and occupation (Bian and Logan 1996). Residents with university degrees have a much better chance of accessing the formal work-unit system in the labor market than those who do not have such education. People with higher education are guaranteed urban *hukou*, either through state job allocation, enrollment in the formal workplace, or preferential treatment to attract their skills. Since education has strong implications for overall social status, educational differentiation in space is, therefore, indicative of the pattern of sorting by socioeconomic status.

Figure 5.2 also shows a very clear pattern of concentration of more highly educated individuals southwest and northeast of the central districts. These areas are close to education and research institutions. The level of concentration, though we lack a strict comparable baseline, seems to be increasing (Wu and Li 2005). The development of several high-tech industrial development parks in Chaohejin and Zhangjiang has attracted highly skilled laborers.

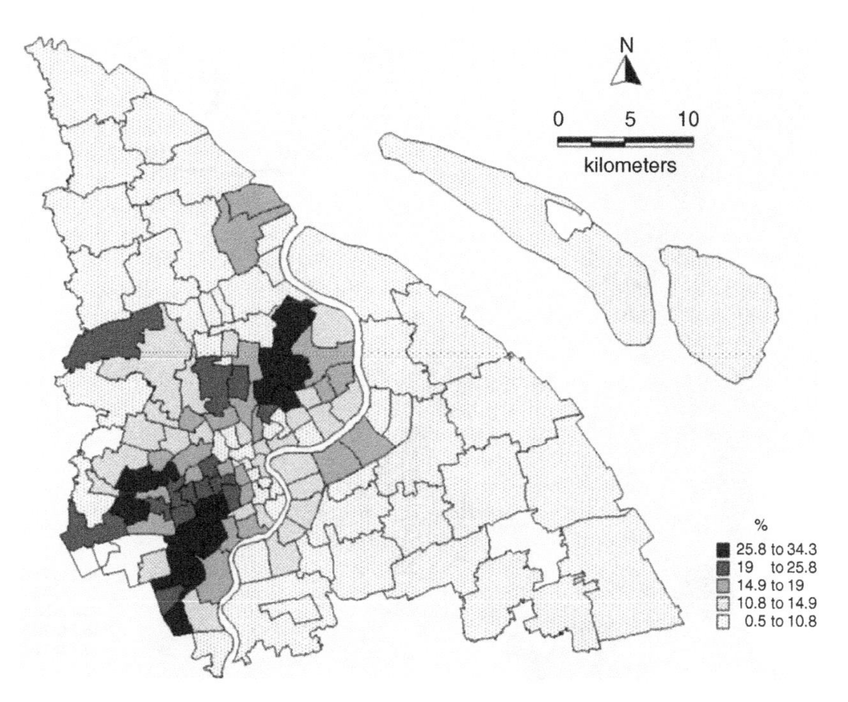

Figure 5.2 Distribution of the population with above college-level education in Shanghai, derived from census 2000

In some inner areas such as Xiaodongmeng (literally, "little east gate" of the old Chinese walled city) and Pingliang Road in Yangpu (industrial areas), we have seen a concentration of low-educated residents.

Figure 5.3, also using data from the 2000 census, shows that the areas immediately adjacent to the central districts are the places favored by migrants. (As we described earlier, the household registration system complicates the identification of urban and rural populations (Chan and Zhang 1999). Fan (2002) argues that according to registration status, laborers in China are differentiated into three types: (urban) *hukou* holders as native residents, migrants relocated through formal channels (and hence possessing the urban *hukou* entitlement) as elite residents and non-urban-*hukou* migrants as "outsiders.") Discriminatory housing policy, the location of less expensive rental units, and job sites in peripheral industrial areas probably all contribute to this phenomenon in Shanghai. Our finding here is consistent with other research, which also reveals a concentration of migrants at the periphery of the city (Ma and Xiang 1998; W. Wu 2002; see also the Introduction to this volume). The index of dissimilarity (White 1987) between migrants and the total registered population is 0.24. This would be much lower than the level of segregation exhibited by many ethnic groups in the US, probably even after accounting for the larger neighborhood units in Shanghai.

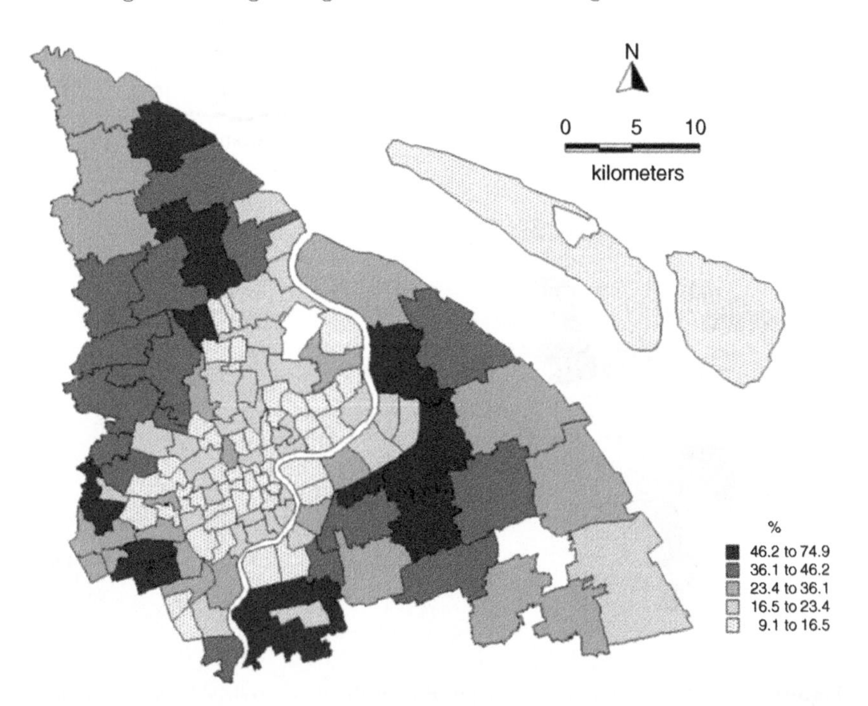

Figure 5.3 Distribution of non-locally registered residents (mainly migrants) in Shanghai, derived from census 2000

We draw on the index of dissimilarity (discussed above among methods) to asses the residential intermingling of social status groups, as indicated by educational attainment. Comparisons of the index of dissimilarity across the three educational attainment categories further reveal social sorting. The differentiation between the population with the below primary-school education attainment and the population with the junior and high school-level attainment gives an index of dissimilarity of 0.13. The index is 0.34 between the population below primary school education and above college education; and 0.24 between the population below primary education and the population above college education.

While these numbers are fairly modest, they are calculated for relatively large geographic units (districts of 50,000) in Shanghai. These neighborhoods would be about ten times the population of a US census tract. To the extent that there is further segregation within each of these districts, which is likely, city-wide educational segregation would be much larger if measured at a comparable neighborhood scale. Low-income groups are more likely concentrated in particular residences (literally, micro-residential region, *xiaoqu*) or neighborhoods in accordance with different types of building provision. If this is the case, and if one had a finer geographic resolution in statistics, the index of dissimilarity or any summary segregation statistic would be larger.

Logan (2004) reported a high index of dissimilarity based on the 1990 population census data of Tianjin, one of the four autonomous cities that report directly to the central government. It is found that the index of dissimilarity between those with postsecondary education and those with less education has a value of 0.40. This level of segregation is about the same as the segregation of the college-educated and non-college educated population in New York and Los Angeles, where entrenched spatial inequalities are well-known.

At first pass, Chinese urban social space may appear to be less differentiated. To be sure, the larger scale at which data are gathered and calculations made would suggest that these segregation statistics underestimate true neighborhood segregation by socioeconomic status. In addition, and more importantly substantively, social differentiation may be just beginning in urban China. Undoubtedly it takes time to translate social differentiation into the urban settlement patterns. The built environment is more stable, so physical effects accompanying social changes take somewhat longer to emerge.

Discussion

Inevitable inequality?

The analysis of Chinese cities seems to suggest that increasing inequality is inevitable under market transition, although the exact form of inequality is dependent upon social and cultural contexts as well as the particular mechanisms driving economic development. Echoing the thesis of Szelényi (1983) that public ownership does not totally eliminate inequality but merely transforms it from one form (market-based) to another (redistributive state or institutionally based), we can suggest that the Chinese case might point to a broader spectrum of inequalities: the difference in residential distribution migrants should also be related to modernization and urbanization processes This form is parallel to ethnic segregation in that it is related to how "outsiders" become or resist becoming assimilated (see Chapters 10 and 11 in this volume), but the exact institutional mechanism that plays a role to separate social groups with and without registered *hukou* is different from that of transnational immigrants. This is a topic worthy of more analysis, given the urban economic transformation in China and the particular place administratively and spatially occupied by temporary migrants.

While data at the moment do not allow us to judge for sure whether educational attainment and *hukou* status are the key traits associated with residential segregation in China, we have seen that these attributes are clearly important. They are very meaningful in the construction of social space: education is related to human capital, but more importantly in the context of Chinese cities, it points to different institutional routes to acquiring residential space; *hukou* as a legacy of state socialism defines the boundaries of urban and rural regimes.

China and the Western city

Besides the replication of spatial inequalities in the Western city associated with a market regime, we witness perhaps a different category of inequalities. Pickvance's (2002) review of Central and Eastern European residential patterns argues that there is no unitary model for urban social space. It might be difficult to develop a model for Chinese spatial inequalities since these are manifest more and more through enclave-like residences. In the suburbs, for example, the best-quality commodity housing estates are mixed with migrant enclaves in spatial proximity but segregated by walls and gates and different territorially based governance. Individual households are relocated and re-sorted in space (Li and Wu 2004).

Some of these relationships are broadly consistent with evidence for the Western city. Haworth, Long, and Rasmussen (1978) found that the Gini coefficient of total income inequality rises with city size and growth in the US's cities during the 1970s. A considerable accumulation of work has often shown that urban size and growth is associated with higher levels of intra-urban inequality, but the evidence for cross-city variation and dynamics is relatively sparse. While ethnic separation in the Western city has been widely documented, there is some evolving evidence of ethnic (and origin region) clustering in Chinese cities (Ma and Xian 1998; W. Wu 2005). In urban China, the absence of ethnic segregation in large cities is obvious. In Shanghai the index of dissimilarity is only 0.473 for the migrants arriving within 6 months and 0.188 for over 6 months (Li and Wu 2006).

The segregation of the rich and poor is becoming one of the "hottest" issues in China at the moment. In the beginning of 2006, the CEO of Huayan, Mr Ren Zhiqiang, openly argued that the poor and the rich should live separately, leading to over 3,000 replies in one of the Chinese largest news websites (sina.com). This event suggests that residential segregation is new to most urban residents. Yet it is increasingly visible through such manifestations as exclusive suburban luxury estates (F. Wu 2005).

While a detailed inquiry is beyond our present scope, we suggest that our analysis helps point to the value of linking inter-regional and intra-urban spatial inequalities. Further thinking could grapple with the need of an overarching analytical framework, applied to China, that uses social science inference to understand spatial inequality at multiple geographic scales. China's unique settlement legacy, the residual importance of the *hukou* system, and the spatial dynamics of the Chinese economy all suggest that such analysis will help us understand both Chinese patterns, and by reflection, those in other parts of the world.

Conclusion

Economic growth is accompanied almost inevitably by increasing inequality, as some sectors and workers benefit from the upswing in economic development. So it is the case in China. China's appreciable – some would say astounding – economic growth over the past two decades has brought with it increasing inequality. Many reports have called attention to increases in overall income inequality and the growing urban-rural gap. But that is not the entire picture. Our objective has been to examine the broader spatial manifestation of inequality. We look at dispersion in the distribution of per capita gross domestic product (GDP) across provinces and across cities in China. We also look further at differentials within the city, examining socioeconomic segregation across Shanghai's residential districts.

We look at this dispersion for several reasons. One the one hand it is simply another useful window on the evolving social and institutional structure of China in the wake of reform. What is more, spatial inequality both mirrors social inequality and has implications for social opportunity. A widening gap among cities may mean that some urban residents have more access to emerging opportunities and/or public resources that accompany economic growth, while others may be left behind.

In the end we find significant evidence for growing spatial inequality in the 1990s in China. The urban-rural income gap widened. The economic distance between rich and poor provinces grew. The income gap that separates the top 10 percent of administrative counties from the bottom 10 percent grew by about 50 percent in the 1990s, clear evidence that some locales are "winners" in the economic bonanza. Other geographic units exhibited more modest increases in disparity, but in all cases the gap between the top and the bottom places widened. All of this is broadly consistent with the spatial manifestation of the Kuznets paradigm. Most likely we are observing the initial upswing in inequality (personal and spatial) with the initial phases of economic growth and modernization. It remains to be seen whether further economic growth will reduce disparities, as some predict.

The Chinese case might not be unique, when compared with other newly industrializing economies in East Asia. What is interesting is that China has been operating an explicit regional policy to reduce regional inequalities through state intervention. Nevertheless, the strategy to develop major globalizing Chinese cities may have counteracted this regional policy, by favoring the coastal region and established metropolitan areas such as Beijing, Shanghai, and Guangzhou, and by further favoring certain smaller cities in the urban hierarchy.

Finally, there is evidence from our work that *intra*-urban differentials are manifest in the contemporary Chinese city. Spatial inequalities might not be entirely new, but there are two distinctive features: first, the clustering of non-local residents in the metropolitan periphery; second, significant differentiation based on human capital (education attainment). This pattern is built upon the separation of occupation in the pre-reform period plus new housing and labor market activity in the past decade or two; such is the evidence from Shanghai. So far segregation within cities has not been identified as a distinct policy problem. This is partially because segregation is still modest and partially because policy-makers still promote aggregate growth and strive for revenues from land development. However, there is every suggestion that market transformation will continue to promote increasing spatial differentiation in the twenty-first-century Chinese city.

All evidence points to some replication in China of the patterns of urban spatial inequality that have been seen during the industrial and post-industrial growth of Western cities. Our interpretation of theory and our empirical results at several geographic scales point to this prognosis. Yet, in many ways the case of contemporary China's cities is unique: decades of central planning followed by the liberalization into the socialist market economy, maintenance of significant overarching administrative control of urban development, the residential strictures (albeit now nominal) of the *hukou* system, and the rapid pace of economic development in a economic environment of globalization and service sector activity. These shifts in relative spatial inequality take place against a backdrop of declining absolute poverty during economic development (see Chapter 2, this volume). While our analysis does not point to an end point for the sociospatial evolution of the Chinese urban system, it does strongly suggest that there will remain a role for urban planning and social policy to examine and perhaps redress an urban development trajectory that is pointed toward increasing inequality over space.

Logan and Fainstein in the Introduction to this volume offer four possible explanations of Chinese urban development: modernization; dependency theory; the developmental state, and the transition from socialism. Our research on spatial inequalities shows that these approaches are relevant but no single approach can fully explain the pattern of inequalities. The modernization thesis duly invokes changing economic structure. The Kuznets-inspired paradigm we discuss can be viewed as a model using modernity or economic restructuring to explain spatial inequality. However, the thesis of modernization is problematic in the sense it "naturalizes" economic structure changes, whereas reindustrialization and consequent urbanization are closely related to state policy.

Dependency theory explains the historical root of regional imbalance in China and seems to continue to offer a possible explanation of under-development of inner regions in China's re-orienting its policy towards the coastal region. For understanding urban inequalities, dependency theory goes beyond "local" inequalities and positions the marginalized in the context of global economy. Despite flowing foreign investment in the coastal region bringing economic prosperity to the place, much profit is controlled by multinationals and their agents and does not spread out to the 'direct' producers – those who are actually providing the labor to the production of global commodities. Migrants are engaged in market-based production but receive limited rewards. They cannot afford so-called commodity housing and tend to live in private rentals in 'urban villages,' a likely contributor to intra-urban stratification.

The thesis of developmental state would appear to have strong appeal in that the Communist Party and the state maintain rather strong control over the society. Indeed, the remaking of large Chinese cities such as Shanghai and Beijing is more like a 'state-project' (Wu 2003) than an outcome of global trade flows that is identified in the classical global city thesis.

Finally, transition from socialism perhaps not only offers an explanation for the post-socialist urban pattern but also links current urban forms to their historical roots in state socialism. Many present divisions – urban versus rural institutions for instance – are often legacies of socialism. The mobility of rural laborers into the city brings this division to the intra-urban level. When the creation of new urban poverty is analyzed, one finds a role for the state (Liu and Wu 2006).

These four paradigms start from very different theoretical stances. Modernization theory views social structure essentially being determined by economic structure (including the level of urbanization), which assumes that the change in the former correspond to the stage of the latter, Dependency theory views the social structure as conditioned by relationships across nation states in a world system. The theory of the developmental state emphasizes the active role of the nation state, and argues that economic growth can be boosted by an interventionist state. This orientation runs against the "determinist" view of dependency theory by emphasizing national sovereignty.

Finally, post-socialist transition theory argues social structure is determined and continues to be determined by internal political structure.

These paradigms are theorized from particular contexts and do not necessarily fit into every national experience. Modernization theory is oriented towards neoclassical economic explanation, and its strength lies in its view of temporal change in a national economy, especially economic structure and urbanization. Its problem, in contrast to a structuralist view, is the alleged very direct connection between economic stage and social structure, leaving less room for a whole range of national and international policies. Dependency theory goes beyond local and national politics to view the nation in the world context. Dependency theory tends to overlook how globalization of productive forces is a process rather than an end result. Also, there is no "united" class of capitalists in the world; rather they are embedded into local and national politics. The developmental state duly points out this local dimension but its explanation of this state role suffers from lack of a detailed account of the agents behind the development state, albeit the original version does implicitly relate the development mentality to elite bureaucrats. That is, the development state emphasizes the role of the state but from a more economic stage of view. In this aspect, it is closer to modernization theory. Post-socialist transition theory probes deeply into this power structure of socialist and post-socialist state. But the key problem is that it views the socialist state narrowly without situating it in the particular stage of economic development (late development) and its position in the world system (semi-periphery).

Although the differences in theoretical stances mean these theories are not completely compatible, the elements upon which these theories build can be re-combined to form an improved explanation. We disagree with modernization theory in its determinism of economic stages, disagree with world system theory in its determinism of world politics, disagree with the developmental state approach in its rosy picture of state capacity, and disagree with post-socialist transition in its notion of "socialist legacies." Building on our insights from analyzing urbanization, institutional change, and socio-spatial inequalities in this chapter, we lean toward a hybrid theory.

Our approach places state action in context. In this particular historical stage of economic development (under-urbanization) and the world system (globalization furthering a new international division of labor for socialist countries), the socialist state, to further its own legitimacy uses economic instruments to boost economic growth. Thus this contemporary state entity acts like a developmental state but without the coherence of the latter. In turn, the new state subsequently turns the particular forms of social structure existing in history into the conditions for capitalist production of the global economy. The forces of population redistribution (migrant labor) and differential regional growth (tied to the world economy) further shape inter-and intra-urban inequalities as new forms of "uneven spatial development."

Reading from this picture of sociospatial inequality, we believe that China's transition is moving towards a very "advanced" stage of capitalist development. Its newness has so far been neglected in the literature. Following Wu, Xu, and Yeh (2007), we see a picture of "advanced" market-oriented urban growth in transitional China. We see a "profound shift from the 'developmental' state, which emphasizes the use of industrial policies to guide national economic growth in late industrializing countries, to the 'entrepreneurial' city" (Wu et al. 2007, p. 309). The state is surely a market actor. The legacy of state socialism and its attendant political resource mobilization help direct the trajectory to a market setting. As Wu et al. (2007, p. 309) argue: "It is the distinctive combination of 'path-dependent' politics and a vibrant market economy that demands future scholarship." Our chapter highlights how this process generates sociospatial inequality at this particular stage and in this particular context of urbanization and institutional change.

REFERENCES

Alesina, A., and D. Rodrik. (1994) "Distributive Politics and Economic Growth." *Quarterly Journal of Economics* 109: 465–89.

Bian, Y., and J.R. Logan. (1996) "Market Transition and the Persistence of Power: The Changing Stratification System in Urban China." *American Sociological Review* 61: 739–58.

Chan, K.W., and L. Zhang. (1999) "The Hukou System and Rural-Urban Migration in China: Processes and Changes." *The China Quarterly* 160: 818–55.

De Ferranti, D., G. Perry, F. Ferreira, and M. Walton. (2004) *Inequality in Latin America: Breaking with History*. Washington, DC: World Bank.

Fan, C.C. (1995) "Developments from Above, Below and Outside: Spatial Impacts of China's Economic Reforms in Jiangsu and Guangdong Provinces." *Chinese Environment and Development* 6, 1–2: 85–116.

Fan, C.C. (1997) "Uneven Development and Beyond: Regional Development Theory in Post Mao China." *International Journal of Urban and Regional Research* 21, 4: 620–41.

Fan, C.C. (2002) "The Elite, the Natives, and the Outsiders: Migration and Labor Market Segmentation in Urban China." *Annals of the Association of American Geographers* 92, 1: 103–24.

Fujita, M., J.V. Henderson, Y. Kanemoto, and T. Mori. (2003) "Spatial Distribution of Economic Activities in Japan and China." Unpublished manuscript.

Haworth, C.T., J.E. Long, and D.W. Rasmussen. (1978) "Income Distribution, City Size, and Urban Growth," *Urban Studies* 15: 1–7.

Honig, E. (1990) "Invisible Inequalities – The Status of Subei People in Contemporary Shanghai." *China Quarterly* 122: 273–92.

Huang, J.-T, C.-C. Kuo, and A.-P. Kao (2003) "The Inequality of Regional Economic Development in China between 1991 and 2001." *Journal of Chinese Economic and Business Studies* 1, 3: 372–85.

Jian, T., J.D. Sachs, and A.M. Warner. (1996) "Trends in Regional Inequality in China." NBER Working Paper no. 5412.

Jones, D., C. Li, and A.L. Owen. (2003) "Growth and Regional Inequity in China During the Reform Era." William Davidson Working Paper no. 561.

Kanbur, R., and Z. Zhang. (2003) "Fifty Years of Regional Inequality in China: A Journey Through Central Planning, Reform and Openness." Paper presented for the UNU/WIDER Project Conference on *Spatial Inequality in Asia* United Nations University Centre, Tokyo, March 28–29.

Kawachi, I., B. Kennedy, et al. (1997) "Social Capital, Income Inequality, and Mortality."*American Journal of Public Health* 87: 491–8.

Korzeniewicz, R.P., and T.P. Moran. (2005) "Theorizing the Relationship between Inequality and Economic Growth." *Theory and Society* 34, 3: 277–316.

Kuznets, S. (1955) "Economic Growth and Income Inequality." *American Economic Review* 65: 1–28.

Li, S.-M., and F. Wu. (2004) "Contextualizing Residential Mobility and Housing Choice: Evidence from Urban China." *Environment and Planning A* 36, 1: 1–6.

Li, Z., and F. Wu. (2006) "Socioeconomic Transformations in Shanghai (1990–2000): Policy Impacts in Global-National-Local Contexts. *Cities*.

Liu, Y., and F. Wu, (2006) "The State, Institutional Transition and the Creation of New Urban Poverty in China." *Social Policy and Administration* 40, 2: 121–37.

Logan, J.R. (2004) "Socialism, Market Reform, and Neighborhood Inequality in Urban China." In G. Knaap and D. Chen (eds.), *Emerging Land and Housing Markets in China*. Cambridge: MA: Lincoln Institute of Land Policy.

Logan, J.R., Y.J. Bian, and F.Q. Bian. (1999) "Housing Inequality in Urban China in the 1990s." *International Journal of Urban and Regional Research* 23, 1: 7–25.

Lu, M., and E. Wang. (2002) "Forging Ahead and Falling Behind: Changing Regional Inequalities in Post-Reform China." *Growth and Change* 33, 1: 42–71.

Lu, X. (ed.). (2002) *China's Social Stratum Research Report* [*dangdai zhongguo shehui jiecheng yanjiu baogao*]. Beijing: Social Science Documentary Press.

Ma, L.J.C., and B. Xiang. (1998) "Native Place, Migration and the Emergence of Peasant Enclaves in Beijing." *The China Quarterly* 155: 546–81.

Messner, S.F., and R. Rosenfeld. (1997) "Political Restraint of the Market and Levels of Criminal Homicide: A Cross-National Application of Institutional-Anomie Theory." *Social Forces* 15: 1393–416.

Persson, T., and G. Tabellini. (1994) "Is Inequality Harmful for Growth? Theory and Evidence." *American Economic Review* 84: 600–21.

Pickvance, C. (2002) "State Socialism, Post-Socialism and their Urban Patterns: Theorizing the Central and Eastern European Experience." In J. Eade, and C. Mele (eds.), *Understanding the City: Contemporary and Future Perspectives*. Oxford: Blackwell: pp. 183–203.

Shangyao, C. (2004) "Urban-Rural Income Gap." *Shanghai Star*. (English Online Edition, accessed October 12, .2004.

Solinger, D.J. (1999) Co*ntesting Citizenship in Urban China: Peasant Migrants, the State, and The logic of the Market*. Berkeley: University of California Press.

Song, S., G.S.-F. Chu, and Cao, R. (2000) "Intercity Regional Disparity in China." *China Economic Review* 11: 246–61.

State Statistics Bureau (SSB). (1990–2003) *China Urban Statistics Yearbook* (Zhongguo Chengshi tongji nainajian). Beijing: China Statistics Press.

Szelényi, I. (1983) *Urban Inequality under State Socialism*. New York: Oxford University Press.

Wei, L. (2004) "Survey shows Wider Rural-Urban Per-capita Income Gap." *People's Daily* (English Online edition) accessed 12.10.2004.
Wei, S.J., and Y. Wu. (2001) "Globalization and Inequality: Evidence from Within China." NBER Working Paper no 8611, National Bureau of Economic Research, Inc.
White, M.J. (1987) *American Neighborhoods and Residential Differentiation.* New York: Russell Sage.
Wu, F. (1996) "Changes in the Structure of Public Housing Provision in Urban China." *Urban Studies* 33, 9: 1601–27.
Wu, F. (2002) "Sociospatial Differentiation in Urban China: Evidence from Shanghai's Real Estate Markets." *Environment and Planning A* 34: 1591–615.
Wu, F. (2003) "The (Post-)Socialist Entrepreneurial City as a State Project: Shanghai's Reglobalisation in Question." *Urban Studies* 40, 9: 1673–98.
Wu, F. (2005) "Rediscovering the 'Gate' under Market Transition: From Work-Unit Compounds to commodity Housing Enclaves," *Housing Studies* 20, 2: 235–54.
Wu, F., and Z.G. Li. (2005) "Sociospatial Differentiation: Process and Spaces in Subdistricts of Shanghai." *Urban Geography* 26, 2: 137–66.
Wu, F., J. Xu, and A.G.O. Yeh, (2007) *Urban Development in Post-Reform China: State, Market, and Space.* London: Routledge.
Wu, W. (2005) "Migrant Residential Distribution and Metropolitan Spatial Development in Shanghai." In L.J.C. Ma and F. Wu (eds.), *Restructuring the Chinese City: Changing Society, Economy and Space.* Routledge: p. 222–42.
Wu, W. (2002) "Temporary Migrants in Shanghai: Housing and Settlement Patterns." In J. R. Logan (ed.), *The New Chinese City: Globalization and Market Reform.* Oxford: Blackwell Publishers: pp. 212–26.
Yang, D.L., and H. Wei. (1995) "Rural Enterprise Development and Regional Policy in China." Paper presented at the annual meeting of the Association of Asian Studies, April 6–9, Washington DC.
Yang, D.T. (1999) "Urban-Biased Policies and Rising Income Inequality in China," *American Economic Review* 89, 2: 306–10.
Yeh, A.G O., X.Q. Xu, and H.Y. Hu. (1995) "The Social Space of Guangzhou City, China." *Urban Geography* 16, 7: 595–621.
Zhang, J., and F. Wu, (2006) "China's Changing Economic Governance: Administrative Annexation and the Reorganization of Local Governments in the Yangtze River Delta." *Regional Studies* 40, 1: 3–21.
Zhang, L., and S.X.B. Zhao. (1998) "Re-Examining China's 'Urban' Concept and the Level of Urbanization." *The China Quarterly* 154: 330–81.
Zhang, X., and S. Fan. (2000) "Public Investment and Regional Inequality in Rural China," Environment and Production Technology Division Discussion Paper no.71, International Food Policy Research Institute.
Zhou, M., and G. Cai, "Trapped in Neglected Corners of a Booming Metropolis: Residential Patterns and Marginalization of Migrant Workers in Guangzhou." In this book volume.
Zhou, Y. (1993) "New Trends of Chinese Urbanization in 1980s." In Y. Yeung (ed.), Zhong guo Chenshi Yu Quyu Fanzhan: Zhanwang Ershiyi Shiji [Urban and Regional Development in China: Prospects in the 21st Century], Hong Kong: Hong Kong Institute of Asia Pacific Studies, The Chinese University of Hong Kong: pp. 105–31.

6

Growth on the Edge: The New Chinese Metropolis

Yixing Zhou and John R. Logan

Introduction

The rapid outward expansion of urban space is one of the most visible changes in Chinese cities in the last two decades. City centers are beginning to hollow out at their core and shift both people and economic activity toward outer zones (Wang and Zhou 1999; Zhou and Ma 2000). This phenomenon of growth at the edge of the metropolis is found in many countries, and it is sometimes regarded as a natural process. It is "natural" from the perspective of central place theory and density gradient laws for development to be concentrated in zones where land is more available, less expensive, and newly accessible to public and private transportation. Although these factors are important in the Chinese case, the pattern of suburban growth in this country is surprising in two ways.

First, it is new. Until the 1980s there were well defined and long-standing distinctions between urban and rural zones, and indeed the term "suburb" typically referred to agricultural towns outside of cities. It is only in the past 20 years that cities have spilled over their traditional boundaries and created a large transitional space whose residents have predominantly urban occupations. The city is becoming a metropolis. One goal of this chapter is to explain why and how this has occurred.

Second, the pattern of settlement in these zones does not match other international models of suburban development. In North America suburbanization in the twentieth century was mainly associated with relatively high-income commuters. In many European cases the central city has retained its more affluent residents, and suburbs have developed mainly to house the working classes. In much of the Third World, suburbs have grown up as marginal and weakly regulated zones where migrants from the countryside were concentrated in shanty towns and squatter settlements. In contemporary China these three functions are found together in nearly equal proportions. Large scale urban renewal and reconstruction in the central cities since the 1980s have forced the relocation of many city residents to suburbs. Luxury villas and gated communities have been built for a growing class of car-driving homeowners. And at the same time massive waves of migrants from other regions have been settled in peripheral areas. Although there may be considerable social differentiation among communities in the suburban zone around cities elsewhere in the world, the Chinese case stands out for the intensity of spatial inequalities.

China is important also as an example of urban patterns under socialism and the changes experienced in the post-socialist period. A considerable literature has emerged to evaluate whether there is a distinctively socialist urban form and how the introduction of market mechanisms affects it. Scholars who have specifically studied suburbanization in East Europe point to two kinds of patterns. Sjoberg's (1992) research on Albania emphasizes the state's strong policies of retaining the rural population, limiting urban growth, and preventing migrants from finding housing in cities. This only partly succeeded, as persons attracted to industrial jobs in larger towns and cities were simply diverted to peripheral settlements that mixed peasants with working class migrants. Tammaru (2001) suggests that such "diverted migration" was typical of Eastern Europe. In contrast, the Soviet Union (and Soviet-influenced nations like Estonia) accommodated internal migration through investments in industrial satellites around major cities, more similar to new town development in the West.

Despite the evidence of such particularities, Enyedi (1992) argued that the main lines of urban development, including suburbanization, have been universal, not substantially different in socialist and market societies. Wang and Zhou (1999) echo this view, stating that many of the sources of suburbanization in China are the same as those found in the West. "Economic forces are universal," they believe, "so even socialist countries cannot escape" (1999, p. 279). At the same time, however, these researchers point to institutional changes and infrastructure investments associated with market reform in China as factors that unleashed suburbanization in its current form.

We will point out the importance of state economic policy and the absence of market mechanisms as determinants of China's limited suburban growth prior to 1980, and identify specific changes that have made a difference in the post-socialist era. But neither the market model nor market transition can offer more than a reference point against which to assess this case. Suburbanization responds broadly to the rapid economic development of

China since 1980, a process carried out through explicit public policies of re-engagement with the global economy and removal of barriers to urban growth. Indeed the economic and spatial restructuring of the metropolis was an intentional goal of the Chinese government as it sought to secure a new and stronger international position. Thus in terms of the theoretical perspectives set forth in the Introduction to this volume, our emphasis is on the developmentalist state. At every level the economic processes that facilitate and shape suburban growth are grounded in policy choices. Theoretical models that have been relied on in the West, such as distance-density gradients, the natural operation of the property market, or the impact of economic expansion play their part, as do models of socialist urbanism, but in a very specific historical and political context. From a theoretical perspective our contribution is to demonstrate the importance of contextualizing this case.

In the following sections we begin by describing the timing and pace of suburban growth, emphasizing the case of Beijing that has been intensively studied by Zhou and his colleagues (Feng and Zhou 2003a; Zhou and Ma 2000; Zhou and Meng 1998). We then analyze the processes involved in this transformation and identify the pattern of metropolitan spatial inequality that it is creating.

The Timing and Pace of Suburbanization

Before the 1980s, Chinese planners created blueprints to decentralize the largest cities but with little effect. Certainly there was a potential for suburbanization, especially to meet the needs for housing. The population density in central cities reached very high levels. Per capita living space was only 3 to 4 m^2, so that typically three persons would share a bedroom that was only 10 ft by 10 ft. Population growth, even after fertility was limited by the one-child family policy and despite stringent restrictions on rural-urban migration, was greater than the rate of housing construction. Environmental quality was poor, especially where industrial plants were located in the inner city. In the 1960s and 1970s, some new industrial satellite towns were built around cities like Beijing and placed downwind of the city. The northwestern sector of the city (upwind) had already been designated for development of scientific research and education since the 1950s. But although many plants were constructed, their employees mostly remained in the city and commuted to the suburbs to work. Hence there was a potential demand for expansion on the periphery of the city, but it was not satisfied. Why not?

There is broad agreement that the poor performance of the national economy is one reason. Relatively underdeveloped anyway, China emerged from the 1940s with a decades-long legacy of foreign occupation and civil war. Central planning directed investments especially toward the heavy industries believed to be essential for national defense. Officials chose to repeat the Soviet

Union's strategy of redistributing resources to meet very basic needs for food and shelter, investing heavily in industry but postponing "unproductive" investments in property and services. Failed experiments during the "Great Leap Forward" from 1958 to 1960, the "Black Years" from 1961 to 1963, and the Cultural Revolution further undermined the economy.

Beyond the overall lack of resources, the organization of the planned economy favored a continuing buildup of central cities. One factor is the refusal to treat land as a scarce resource. Urban land was allocated by the government at no cost to end users and without regard to the relative value of different locations. As a result, for example, large industrial zones continued to be built within the central cities in spite of environmental factors that were well known at the time. In Beijing, industrial plants were disproportionately located in the four central districts through the 1970s. Another is the centralization of fiscal resources in the hands of national ministries. Municipal officials did not have the means or authority to support large-scale construction of new residential districts, or of high-grade roads to connect cities and satellite towns, or to create infrastructure that would bring standard public services to outlying areas. Consequently neither work unit leaders nor their employees perceived any benefit to relocating to suburbs and pressed instead to remain in central locations. A third factor is the treatment of housing as a welfare service. In urban areas most housing – aside from a leftover stock of private residences of generally low quality – was provided by work units or city housing authorities at rents that did not even cover basic maintenance costs. Consequently no investment in new housing could be self-supporting, and public housing in suburbs (because it required additional infrastructure investments and offered no savings in the cost of land) was especially costly.

Nevertheless suburbanization took hold in China's major cities in the 1980s. The earliest studies identified this trend in Beijing (Zhou, 1992; 1995). From 1964 to 1982 the average annual population growth rate of Beijing was 1.1 percent, but only 0.2 percent for four central districts (Dongcheng, Xicheng, Xuanwu, and Chongwen). From 1982 to 1990, the core lost 82,000 population (–3.4 percent), the inner suburbs gained 1,149,000 (up 40.5 percent), and the outer suburbs gained 521,000 (up 13.1 percent). The trend intensified in the period from 1990 to 2000, as the core lost 222,000 (–9.5 percent), the inner suburbs grew by 2,400,000 (60.2 percent), and outer suburbs grew by 572,000 (up 12.7 percent). The inner suburbs became the main destination for movers from the central city, as well as the area where temporary residents from other regions were mostly concentrated. Both centripetal migrants and centrifugal migrants met in this zone. The shifts in population density between 1990 and 2000 are mapped in Figure 6.1.

Similar results have been found other big cities, such as Guangzhou (Xie and Ning 2002; Zhou and Xu 1996), Shanghai (Gao and Jiang 2002; Ning and Deng 1996), Shenyang (Zhou and Meng 1997), Dalian (Chai and Zhou

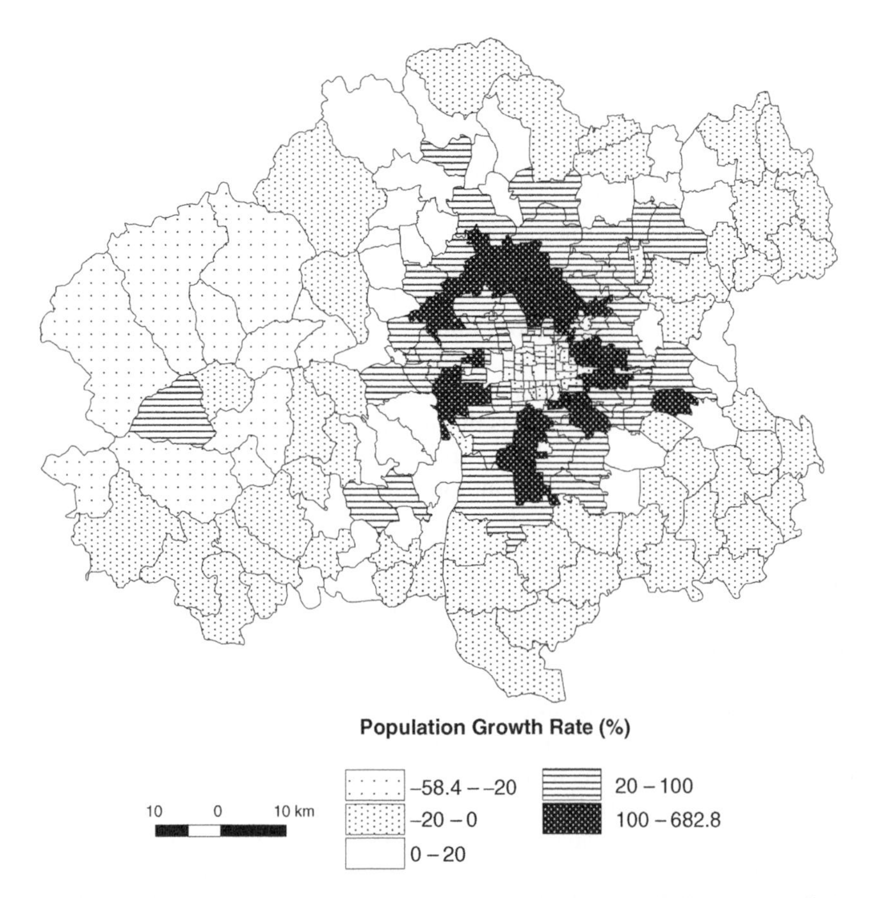

Figure 6.1 The spatial distribution of population growth rate in Beijing metropolis by sub-district, 1990 to 2000

2000), Hangzhou (Feng and Zhou 2002; Zhou1997), Suzhou, Wuxi, and Changzhou (Zhang 1998). Our task is to understand why this happened.

Figure 6.2 reproduces a conceptual model that identifies several aspects of the process (drawn from Feng et al. 2004). Feng depicts all of these as consequences of development of a market economy, but distinguishes processes "at the level of the economy" from processes "at the level of society." Suburbanization itself is shown to include shifts in economic activity (industrial decentralization and development of large supermarkets, which represents the wider shift in retail services) as well as changes in residential locations (including both permanent and seasonal housing). The more "economic" sources of change involve urban land use transformation and the renovation of the inner city, on the one hand, and real estate investments (both domestic

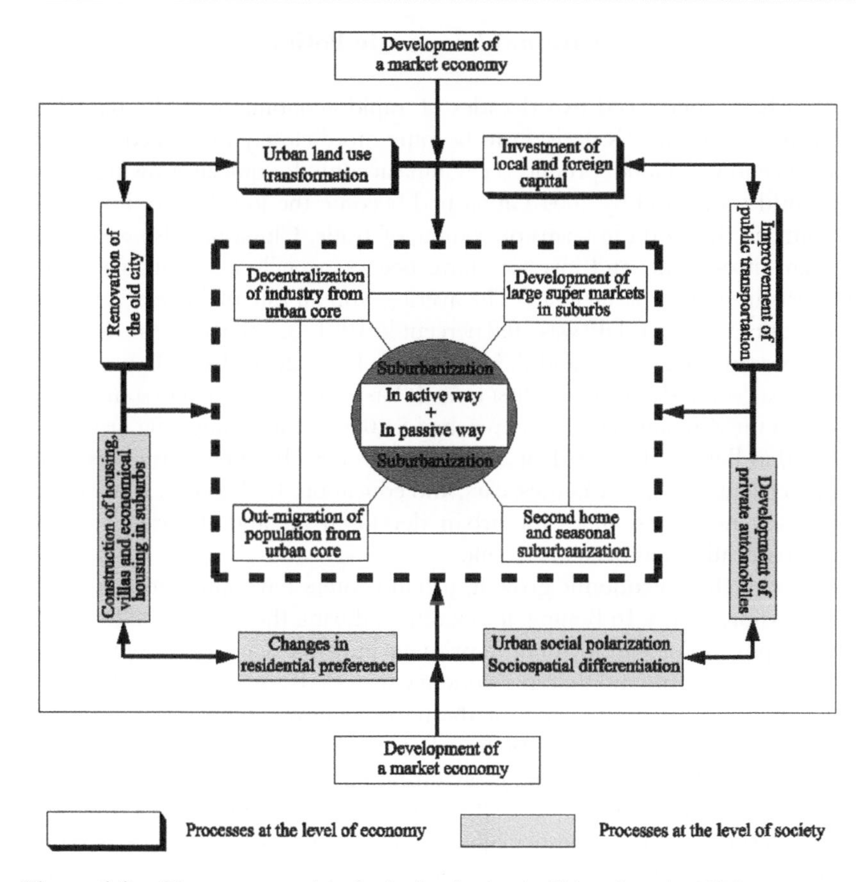

Figure 6.2 The process model of suburbanization in China since the 1990s

and foreign) and improvement of public transportation, on the other. Feng also notes two types of social changes. One is in the class structure – the growing class inequalities that are manifested in spatial inequalities and represented in both practical and symbolic terms by growth in the use of private automobiles. The other is in residents' attitudes toward the use of space, which provide support for a mix of high-class and moderate-income housing in suburbs.

This model calls attention to the many components of suburbanization. We focus on only a few of these: national changes oriented to economic development; the more specific reforms directed at the system of housing and real estate development; and the decentralization of development controls.

National Economic Policy

China has experienced two decades of rapid economic growth coinciding with the government's decision to become more closely integrated with the global economy. Goods produced in China account for quickly growing shares of world trade, and by 2004 China had become the fourth most important country in the world in terms of volume of trade. China has also welcomed foreign investment, and investors have been especially active in real estate projects in major coastal cities. The average annual rate of increase in Gross Domestic Product (GDP) was 10.8 percent in 1981–5, 7.9 percent in 1986–90, 11.6 percent in 1991–5, and 8.3 percent in 1996–2000. In 2003 per capita GDP exceeded $1,000 for the first time. It has exceeded $5,000 in some major cities in the Zhujiang Delta, and reached $4,000 in the Changjiang Delta and $3,000 in Beijing. The residents of these cities have become consumers at an unprecedented scale for houses, cars, and education. In this context, there are obviously new possibilities for urban development, and the trend toward suburbanization has been one result.

Besides sheer economic growth, urban centers have undergone a sectoral transformation. In Beijing, for example, during the period 1982 to 2000 the proportion of employment in primary industry decreased from 29 percent to 13 percent, and that in secondary sectors decreased from 39 percent to 29 percent. As the location of the national government, service sectors were traditionally stronger in Beijing than in most cities, but service employment is now dominant, having grown from 32 percent to 58 percent of the workforce. Among service activities, banking, insurance, real estate and social services increased especially rapidly from 2 percent to 14 percent, and transportation, storage, postal and telecommunication services, and wholesale, retail trade and catering services more than doubled from 11 percent to 23 percent.

The impact of these changes on spatial patterns depended on policy choices by the national and local governments. Older heavy industries were good prospects for relocation to suburbs on environmental grounds, but since industry was not the main component of economic expansion this might not have contributed much to suburban growth. The new service activities – especially business services – could have been concentrated in central cities, restoring to cities the "central place" functions that they would naturally have had earlier in the twentieth century. But they could as well have been shifted to outer zones.

In the era of the socialist planned economy the actual direction of urban development was governed by urban planning. The guiding ideology in the planned economy was strong controls by government and limitation of population growth, although specific sorts of development were expected and carried out. In the late 1950s, it was decided in the city's master plan that

Beijing should be developed into a modernized manufacturing center. The city's built-up area increased from 109 km^2 in 1949 to 221 km^2 in 1959. Emphasis was placed on construction in the industrial quarter in the northwestern and eastern suburbs. Later, the industrial base of the Capital Iron and Steel (*Shougang*), the culture-education quarter in the northwest and the embassy quarter in the central core were constructed one after another. Some satellite towns in the outer suburbs were also listed in the development plan at the time, but this aspect of the plan was not fully implemented.

Although the national government has maintained weaker planning controls in the period of market transition, formal plans have continued to affect development. In the master planning completed in 1982, officials decided to reinterpret Beijing's economic function from an industrial center to a center of politics and culture (Beijing's Planning Committee 1982). This anticipated the restructuring of the labor market. In the 1993 master plan, more specific steps were taken to promote spatial restructuring: it was decided to promote dispersed clusters of growth (*Fensan Jituanshi Buju*) throughout the region, to invest in transportation systems to link the city with towns in the outer suburbs, to coordinate the development of new districts and the renewal of the old city, to protect and improve the urban physical environment, and to shift from heavy industry to new high-tech industry (Beijing's Planning Committee 1993). Without doubt these policies have facilitated suburbanization.

Industrial decentralization was not limited to Beijing. Gaubatz (1999) documents the same process in Shanghai and Guangzhou. In all of these cities manufacturing is increasingly found in the suburbs and rural counties. Some of the change is due to relocation. A larger share results from the establishment of new businesses, particularly in special development zones placed on the edge of the city.

Decentralization of employment is a familiar aspect of suburbanization in other parts of the world. For example since the late 1950s, the bulk of new manufacturing and trade employment in the US metropolis has been located in small and middle-sized cities in the suburban ring (Berry and Kasarda 1977, ch. 13). Downtown department stores compete with new suburban shopping malls. The highly developed expressway network around central cities frees manufacturing plants to take advantage of the lower land prices and taxes and the superior access to the skilled workforce offered by the suburbs. In the latter half of the twentieth century total employment growth in the suburbs outpaced the growth of population (Logan and Golden 1986). This is the heart of the phenomenon popularized by Garreau (1991) as the creation of "Edge City." Perhaps the difference in the Chinese case is that housing and job development have generally taken place simultaneously. China has only recently experienced the creation of dormitory suburbs whose residents mostly commute to central city jobs.

Reform of the Housing and Urban Land Use System

The emergence of real estate markets, including a partial privatization of housing, has significantly altered the urban development process (Wu 1995). The first wave of suburbanization in the 1980s was driven by relocation of persons displaced by urban redevelopment projects, in housing built and financed by government, and by relocation of industry. Since then, as Zhou and Ma (2000) point out, the most powerful force for change has been the creation of a system requiring land users to pay for development rights based on land value and other market factors. In the mid-1980s, payment for use rights of urban land was established, and this caused a fundamental shift of land use in the central city core – from industrial and administrative to commercial and other tertiary uses. Survey data shows that in Beijing's central section the profit per m^2 of space for industry is only one tenth – in some cases even one fiftieth – as high as that for business. This disparity has begun to force industry out of the central city, starting with enterprises that pollute the environment, or needed space to develop, or needed money to improve technology. Similar trends have taken place in Eastern Europe, where location formally had no value in a socialist economy, but market forces have revalorized central locations (Keivani, Parsa, and McGreal 2001).

For the half century from 1949 to 1998, housing provision and allocation had been considered to be a kind of welfare, with the majority of housing owned by work units. Housing was constructed or purchased by work units near their premises and rented at a very low rate to their workers, thus, residential location was governed by the location of housing provided by the work units, often within or near them (Yeh, Xu, and Hu 1995). Some powerful work units, such as government departments and large enterprises, had the ability to construct large concentrative residential quarters for their cadres and workers, creating whole neighborhoods of relative privilege. Residential mobility before the mid-1980s was low and had little impact on urban social spatial structure, because of a housing shortage and a housing allocation system based on the construction and allocation of housing by work units.

In the early years of market reform, the most significant changes in the structure of housing provision included the movement toward market rents, experiments with ownership schemes and home mortgages, and creation of partially autonomous real estate development and construction companies. These changes accelerated in the latter half of the 1990s. The share of housing units built by the central government agencies or local units has dropped sharply, while the share built by real estate enterprises has risen. In 2000, the proportion of the housing built by the central units was only 18.6 percent, while in 1997 it had been 25.4 percent . The proportion of the housing built by work units was 13.8 percent in 2000, compared to 26.6 percent in 1997. The proportion of housing built by real estate enterprises increased from 48.0 percent in 1997 to 67.7 percent in 2000.

After 1998 the role of work units in allocating housing changed. At that time the traditional system of municipal and work unit housing allocation was cancelled, and a new system based on subsidized homeownership was established. Work units now encourage employees to buy private housing through a housing accumulation fund, loans for private housing, and national and work unit housing subsidies. More powerful work units (such as state-owned enterprises) were reluctant to lose control of this important aspect of their employees' lives, which had long been more consequential than salaries, but they have accommodated to the new situation by manipulation of housing subsidies. As a result, people have greater freedom and flexibility in selecting housing location. It is no longer necessary for employees in the same working units to live together or for the new housing purchasers to live near their working units. Thus, there is a new spatial separation between residence and employment that facilitates relocation to suburbs.

As housing reform has progressed, a higher share of suburban housing has been developed through market mechanisms. A growing proportion of housing purchases are made by individuals rather than by work units. A survey of residents in new dwelling units from 1990 to 1995 in Beijing found that at that time only 14 percent purchased their own unit, and for those moving from the central city to inner suburbs the rate was only 8 percent. At that time Guangzhou had a more advanced private housing market, but still only 38 percent of movers had purchased homes as individuals (Zhou et al. 2000). Individual purchases now account for above 70 percent of transactions in most cities and up to 90 percent in some coastal cities.

Wu and Yeh (1997) show that suburbanization and urban sprawl around Guangzhou accelerated after the institution of the land-leasing system. In a market setting, private real estate developers extensively promote suburban locations, advertising the higher living quality and more comfortable space in suburbs. They design and sell housing for all segments, ranging from apartments for people with modest incomes, to townhouses and single-family villas. These tend to be located in different parts of the metropolis, adding to spatial differentiation. In Beijing, for example, 70 percent of new housing affordable to low income people is found in inner suburbs, 20 percent in outer suburbs, and only 10 percent in the central city. In contrast 60 percent of villas for the wealthy are located in outer suburbs, more than 30 percent in inner suburbs, and almost none in the city itself (Feng et al. 2004). Gaubatz (1999) documents the construction of suburban villa developments around Beijing, Shanghai, and Guangzhou, often in the form of gated communities. Higher density apartments serve a wide range of consumers, from average workers relocated from the central city to those who can afford luxury condominiums.

Another important component of the suburban housing market is the growing population of rural-urban migrants. At one time the full array of state powers had been directed against migration, with strict enforcement of a household registration system and procedures for food rationing that made it difficult for unauthorized migrants to live or work in cities. The registration system has been in flux in China since the 1980s. For example, the external population (*wailai renkou*) in Beijing increased from 170,000 in 1982 to 600,000 in 1990, a growth rate of 255 percent. It then boomed in the 1990s, rising to 2,570,000, for a growth rate of 326 percent. As a result, concentrated areas of external population have become a specific form of urban settlement. And these tend to be located in peripheral areas for some of the same reasons that "diverted migrants" settled in suburbs of Eastern European cities and marginal migrant populations created shantytowns around Latin American cities. Migrants generally have neither the financial resources nor the legal right to compete for space in city neighborhoods.

The New Politics of Suburban Growth

One of the defining characteristics of suburbs in the US is their political autonomy – suburbia in many respects starts at the city limits, and some of the most important issues for American suburbs stem from this fact. To be "suburban" means to have an independent school system, to be reliant on your own tax base, and to control local development. Suburbs often use these tools to compete with one another for favorable growth patterns (Logan and Molotch 1987). Some of the same processes are now emerging in Chinese metropolitan regions. In the reform period local governments gained increased autonomy from the provincial and central government, and they have also absorbed greater responsibility for local services, to be paid from local revenues.

This administrative change mirrors events in Eastern Europe, where local authorities have emerged as the main decision-makers about urban development and public services. At the same time, local governments, even districts within the same city, are now rivals for private investment (Keivani et al. 2001). The largest Chinese cities cover extensive geographic areas that include much undeveloped rural territory. But even when a territory is nominally within the borders of the central city, the municipal government may not be responsible for planning or services. Districts within cities have become an increasingly important level of local government, and they now compete with one another for the revenues that can be raised by selling development rights within their jurisdiction. In many districts revenues from land leases are the primary source of financing for local services.

Some rural villages that have been enveloped by leap-frog development may retain control of their local services, development controls, and collectively held land. They continue formally to have the status of "village" and their residents may still have a rural registration despite living well within what appears to be the city. Other peripheral areas that have potential for suburban development are still classified as rural counties or towns. Historically these governmental units were provided with lower standards of public services than cities, and they had less political independence from higher authorities to manage their own affairs. For officials in these places rapid growth is a route to new resources and possibly to a formal upgrading of political status. As a result local officials are often eager to make deals for residential and commercial projects. It is common now for town and county governments to seek the gains from land developments without accepting responsibility to serve the new population. Residents of more affluent gated communities often depend on the real estate developer for services, including private schools that may charge high tuitions. Migrants without a local registration often have nowhere to turn for services.

Reorganization of the metropolis

Changes in the context and processes of development are reflected in the kinds of neighborhoods that are being produced in the reform period and more specifically the characteristics of suburban neighborhoods.

As noted earlier, "suburbs" have a different social connotation in different parts of the world. A common pattern in the US is for central cities in most metropolitan regions to have a less affluent residential population than their surrounding suburbs. Suburban housing stock is newer, suburban land is less expensive, and suburbs are more accessible to automobile and truck traffic. In the American context these differences draw more affluent residents to suburbs where they can buy larger houses as well as enjoy superior public services. There is much debate, however, whether this class segregation between central cities and their suburbs is a natural sorting out of social classes through the private market or whether its causes are political and institutional (Danielson 1976).

The suburbanization process in the US also increasingly involves minorities and immigrants, and the incorporation of these groups into suburban areas has become an important issue. As Massey and Denton (1987) document, the rate of growth of non-whites and Hispanics in metropolitan areas is far outstripping the rate of growth of non-Hispanic whites. Much of this growth is occurring in suburbs. During the 1970s, for example, the number of blacks in the non-central-city parts of metropolitan areas increased by 70 percent, compared to just 16 percent in central cities; and the number of other non-whites in them shot up by 150 percent, compared to approximately 70 percent in central cities. One reason for the rapidly increasing racial and ethnic diversity of suburbs may be that some new immigrant groups are bypassing central cities and settling directly in suburbs. Equally important is the increasing suburbanization of older racial and ethnic minorities, such as blacks (Frey and Speare 1988).

This phenomenon has encouraged researchers to study suburbanization as a mirror on the social mobility of minorities. Consistent with classical ecological theory, suburbanization has often been portrayed broadly as a step toward assimilation into the mainstream society and as a sign of the erosion of social boundaries. For European immigrant groups after the turn of the century, residential decentralization appears to have been part of the general process of assimilation (Guest 1980). In contrast the suburbanization process for blacks appears largely to be one of continued ghettoization (Farley 1970), as indicated by high and in some regions increasing levels of segregation and by the concentration of suburban blacks in communities with a high incidence of social problems (e.g., high crime rates), high taxes, and underfunded social services (Alba, Logan, and Bellair 1994; Logan and Schneider 1984, Reardon 1997). These findings regarding black suburbanization have been interpreted in terms of processes that impede the free mobility of racial minorities: steering by real estate agents; unequal access to mortgage credit; exclusionary zoning; and neighbor hostility (Foley 1973).

Although race has an unusually strong impact on neighborhoods in the US, similar cleavages arise in other societies in relation to minority ethnic groups, rural-urban migrants, and other visible categories with low socioeconomic status. Preteceille (2003) points out the importance of the traditional Red Belt of working class suburbs around Paris, now increasingly also a settlement zone for North African immigrants (see also Wacquant 1993). These towns are predominantly twentieth century creations of the French government, whose policy was to preserve Paris itself as a middle-class city. A typical Latin American pattern is associated with what scholars term over-urbanization (see the discussion in the Introduction to this volume). Rural-urban migrants, often mainly of native American peasant background, tend to concentrate in marginal, often self-constructed settlements on the periphery of the city. This is the case in such cities as Sao Paolo (Caldeira 1996; Taschner and Bogus 2001) and Mexico City (Giusti 1968). The general tendency does not imply a uniform pattern, however. Especially in the 1990s there was a tendency for some elite groups in these places to relocate to outer suburbs, and Castells (1983) has emphasized the diversity of composition and location of marginal settlements, which can also be found in inner city slum areas.

In the socialist bloc Ruoppila (2004) reports that the historic pattern in Central European cities as of the 1930s placed higher status populations in central neighborhoods, with working class districts on the edges of the city near manufacturing zones. There were exceptional cases, however, of exclusive residential zones and garden suburbs on the fringe of most cities. Housing redistribution and limited new construction the early 1950s socialist period tended to reproduce this pattern. Housing estates developed in the 1960s and 1970s on the outskirts of cities represented the first large-scale suburban development in several countries. Ruoppila argues that these apartments were initially assigned with priority to higher status persons, but access was quickly equalized. Meantime, the introduction of market processes in the 1980s (promotion of cooperative apartments and single family houses) provided a new source of advantage to persons with political influence or high occupational status. Many larger apartments of higher quality in this period were located in the central districts. Dangschat and Blasius (1987) show that in 1987 Warsaw's lower-status residents tended to live either in privately owned single family dwellings in the rural outskirts of the city or in aging apartments in the center, adjacent to high-status neighborhoods that were rebuilt after World War II.

There is some evidence of changes that result from market reforms in these countries. For example Brown and Schafft (2002) conclude that industrial suburbs were developing outside Budapest and other Hungarian cities in the 1960s and 1970s, intermixing an industrial working class population with rural households. Between 1990 and 1997, suburbs on the fringes of larger cities were the only category of places with net population gains, while cities and outlying rural villages both lost population. In this period, however, suburban development is more oriented to commuters with jobs in the central cities. This shift appears to bear out Musil's (1993) predictions about post-socialism. He argued that the commodification and rising cost of housing, increasing income inequality and the reintroduction of the private housing market would naturally change the socio-spatial organization of cities. Low-income groups and retired people, he expected, would be squeezed out of inner cities by rising land prices. At the same time suburbs would expand in part to satisfy the demand by more affluent people for high environmental quality and the demand by working class people for low-cost housing.

Against this comparative backdrop, what is the spatial form of the Chinese metropolis? The best evidence is from extensive analyses of changes in the sociospatial differentiation of the Beijing metropolis between 1982 and 2000 carried out by Feng and Zhou (2003b). Their research is based on the methods of factorial ecology, which are used to identify the main sources of variation in the social composition of neighborhoods and to show where different kinds of neighborhoods are located. In most market societies social areas mainly reflect differences in socioeconomic status, family life cycle (young families live in different areas than single adults or the elderly), and race or ethnicity. The Beijing data reflect a very different but rapidly changing society.

The 1982 social areas map was relatively simple. The majority of territory in the Beijing metropolis was still primarily agricultural, and there was a very strong distinction between low density agricultural villages and other zones.

There were also mining areas located mainly in the western suburbs. The typical neighborhoods in the urban core were distinguished from these agricultural and mining zones by their high population density and urban working class occupations. To the extent that there was a class gradient at this time, it represented the inequalities between city and countryside rather than an urban-suburban dichotomy. Early "suburbanization" had made little impact on the composition of the periphery, which appears as a vast and undifferentiated zone.

An interesting feature of the 1982 map is that it also identifies spatial inequalities within the city. "High density, worker areas" dominated the four oldest districts in the heart of the city as well as newer industrial zones in the northwest and to the east. Intellectual areas (standing out for high average education levels) are found in the northwest adjacent to the urban core, where planners had located Beijing's most important universities and research centers. Cadre areas, the site of housing constructed for government employees, are found along the inner city's eastern edge. Hence in the era of the socialist planned economy, the types of social areas mainly reflected the location of different types of jobs.

The 2000 map (shown in Figure 6.3) shows much greater spatial diversity, and it also marks the emergence of some new kinds of neighborhoods. Let us review the changes one at a time, starting with the center and moving out to the periphery:

1 First, the districts in the very center of the region continue to be identified by their very high density, and there are now some similar areas not only in the northwest inner suburbs, but also farther to the west.
2 Cadre areas have disappeared, suggesting that there has been a dispersal of government employees to a wider range of neighborhoods. Intellectual areas are still found in a band around the west and north of the inner core. But they have added an ethnic component to their makeup. Noticeable shares of residents are members of Chinese minority groups, many of whom have come to Beijing to study or drawn to jobs requiring a higher level of education. There are also some migrant villages mixed into this part of the city, such as the Hui nationality settlement in Xinjiang Village.
3 Turning to the suburbs, the territory identified by the predominance of agricultural employment – still important at the edges of the metropolis – has diminished, and no areas are typified by mining employment. It is not that mining has disappeared, but rather that population growth with other characteristics has become more important in those areas.
4 A considerable portion of former agricultural zones are now characterized by their low population density and larger housing units, but now predominantly settled by people with urban occupations who previously lived in the

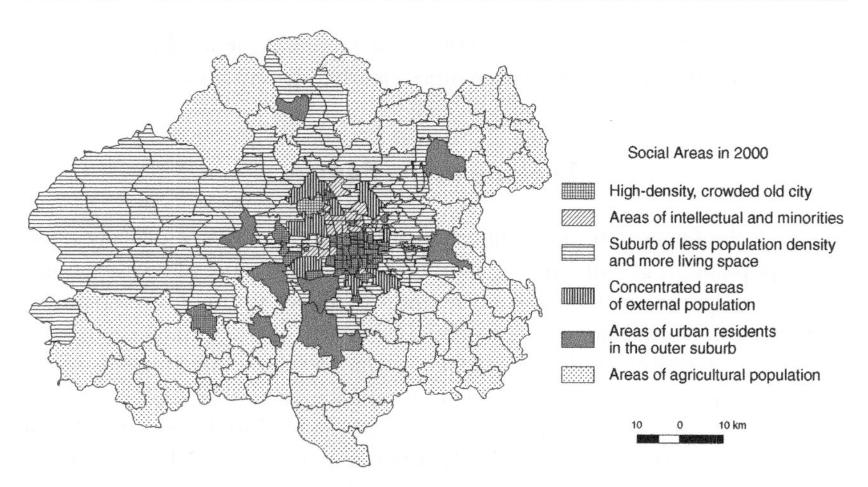

Figure 6.3 Social areas in Beijing metropolis in 2000

central city. This stratum includes many inner suburbs on the south, east and north, as well as an extended suburban belt on the west of the city.

5 Another new and critically important kind of suburban neighborhood houses migrants from outside the region. These are Beijing's floating population, people who do not have the same rights to rent public housing or purchase on favorable terms from their employers. They often rent in the private housing market provided by villagers on the edge of the city and constitute enclaves of persons who congregate based on their province of origin and the sector of the economy in which they work.

6 Finally some suburban neighborhoods are described as "areas of urban residents in the outer suburb." These are defined in part by the low occupational and educational level of their residents, who nonetheless do not tend to have agricultural jobs. Several of these neighborhoods are county seats in outlying sub-districts.

Research on spatial inequalities in China is handicapped by the lack of data on household incomes for small areas. It is widely expected that the increasing inequality in earnings that characterizes the class structure will be mapped into space, but the tools to measure this phenomenon are not yet available. In addition, the geographic units on which Feng and Zhou based their work are too large to reveal the distinctions between villa developments, gated communities, and standard housing built for persons of different income levels – housing projects that are often in close proximity to one another despite serving very different populations. Still, we see much evidence that as suburbs have grown, they have also become more clearly differentiated.

Looking to the Future

Many of the forces that drove suburbanization in the 1980s, especially rapid urban population growth and the pent up demand for more and better housing, are still operative. China is in the midst of the transition from a planned economy to a socialistic market-oriented economy. It is also undergoing this change at a relatively early phase of economic development. If we compare China's economy at the starting point of suburbanization (a 1985 per capita GNP of $289 and an urbanization level of 24 percent) to that of the US just prior to its post-World War II suburban boom (per capita GNP of $934 and urbanization at 57 percent in 1941), the gap is large indeed. One consequence is that the Chinese middle class is much weaker than was that of the US, and suburbanization in China could not be based primarily on lifestyle choices by more affluent people. Another is that the role of the private car has been very different. The automobile was a necessary structural support for large-scale suburbanization in Western Europe and the US (Muller 1981), but until very recently was nearly invisible in China (Zhou and Ma 2000). Even in Beijing in 1985 only 1,107 cars were owned by households in Beijing; most people traveled on bicycles and buses.

Seen in this light we could explain many facets of Chinese suburbanization in terms of lagged development – beginning slowly due to obstacles placed by state central economic planning and disproportionately controlled by decisions of work unit managers and urban redevelopers in its early phase, but destined to catch up in the future. Indeed there are some signs that this is happening. In many respects – rapid economic development and improvement of standard of living, commodification of space and housing, and residents' perceptions of their housing choices – the Chinese situation is becoming more similar to Western countries.

One key to the future is a change in people's preferences about where to live. In the past a common saying was "preferring a bed in the central city to a house in suburbs" (Wang and Zhou 1999), reflecting a sense that the city offered a better location even at the cost of less space. Early suburban development was largely decided by the state and by work units rather than by the residents. A survey in Shenyang found that forced moves accounted for 64 percent of all city-to-suburb mobility (Zhou and Meng 1997). In Dalian the share was 67 percent (Chai and Zhou 2000). In every city many people move out of the central city while still keeping their household registration there in order to retain urban residence rights (Zhou and Meng 1997). Yet this sentiment is changing, and people are giving more weight to housing prices and the living environment. The pursuit of a comfortable environment for a large share of families is becoming more important than simple demand for living space.

In China there is no tradition of suburban single family homes (the current preference seems to be gated developments), and no organized local political voice to object to high-density development. Low land prices, then, can facilitate apartment developments for working class people. In addition the municipal government itself is a major land developer, under increasing pressure to re-house people who have been displaced by redevelopment projects in the central city. Therefore, market forces are reinforced by local development policy. A sort of Chinese suburban growth machine (Logan and Molotch 1987) is in place for the foreseeable future.

Further, the automobile has arrived in China. By 2000, the total number of private automobiles in Beijing had amounted to over a million, with 12 percent of all households owning cars (Feng et al. 2004). And there are other ways that prosperity is promoting suburban growth. There are signs of development of a market for second houses in suburbia, and suburban shopping malls and supermarkets are transforming the spatial distribution of urban retail trade – increasing at an average annual rate of 68 percent between 1998 and 2001 (Wang and Zhou 2002).

In these respects Chinese suburbanization does seem to support Enyedi's view of the universality of major trends in urban development. It is nevertheless distinctive in its social composition. As seen in the US, there is a rapidly growing middle class and a segment of the very rich who move to suburbs as a lifestyle choice. China's economic expansion makes this a powerful current especially in the coastal cities. As in Europe, another large share of growth represents the relocation of working class people from the central city to industrial satellites, and relatively low-cost "new town" developments, sometimes uprooted by inner city gentrification and commercialization. And somewhat like Third World cities in Latin America and elsewhere an important segment of the suburban is a marginalized migrant population. In metropolitan regions like Beijing, the "floating population" is as much as 25 percent of all residents and a much larger share outside the city. The difference is that the labor market is fully capable of absorbing migrants in construction and in trade and services in the growing private sector. These people's marginality is mainly due not to unemployment but to the unequal formal legal status imposed on them by the state. Unlike most formerly socialist countries, although China relaxed its migration policies sufficiently to bring these people into the metropolitan sphere, it has only recently begun to experiment with reforms that would equalize their access to housing and public services.

All of these currents of change converge to shape the new Chinese metropolis. If one wants to find signs of suburbanization in China that are reminiscent of other models, they can be found. But it would be risky and misleading to make too much of these parallels, because they have to be grounded in a specific historical and political context. If we begin to see that moving to the suburbs reflects people's hopes for better housing, more space, and a cleaner environment, this is not yet the reality encountered by many of China's new suburbanites. More generally, though we have emphasized ways in which China is an economy in transition, it is far from a society dominated by private capital and a free market. The future depends as much on the continuing role of central government policy and direction as on the market trends made possible and stimulated by public policy.

ACKNOWLEDGMENTS

The authors thank Dr Jian Feng, Peking University, and Mr David Chunyu, University at Albany, for their assistance in the preparation of this chapter. Yixing Zhou acknowledges the financial support of the National Natural Science Foundation of China (project number 40335051).

REFERENCES

Alba, R., J.R. Logan and P. Bellair. (1994) "Living with Crime: The Implications of Racial and Ethnic Differences in Suburban Location," *Social Forces* 73: 395–434.

Beijing's Planning Committee. (1982) *City Comprehensive Planning of Beijing (1981–2000)* [in Chinese].

Beijing's Planning Committee. (1993) *City Comprehensive Planning of Beijing (1992–2010)* [in Chinese].

Berry, B., and J. Kasarda. (1977) *Contemporary Urban Ecology*. New York: Macmillan.

Brown, D.L., and K.A. Schafft. (2002) "Population Deconcentration in Hungary during the Post-Socialist Transformation." *Journal of Rural Studies* 18: 233–44.

Caldeira, T.P.R. (1996) "Building Up Walls: The New Pattern of Spatial *Segregation* in *Sao Paulo*." *International Social Science Journal* 48, 1(147), March: 55–66.

Castells, M. (1983) *The City and the Grassroots*. Berkeley: University of California Press.

Chai, Y.W., and Y.X. Zhou. (2000) "The Characteristics, Mechanisms and Tendency of Suburbanization in Dalian City." *Scientia Geographica Sinica* 20, 2: 127–32 [in Chinese].

Dangschat, J., and J. Blasius. (1987) "Social and Spatial Disparities in Warsaw in 1978: An Application of Correspondence Analysis to a 'Socialist' City." *Urban Studies* 24: 173–91.

Danielson, M. (1976) *The Politics of Exclusion*. New York: Columbia University Press.

Enyedi, G. (1992) "Urbanisation in East Central Europe: Social Processes and Societal Responses in the State Socialist Systems." *Urban Studies* 29: 869–80.

Farley, R. (1970) "The Changing Distribution of Negroes Within Metropolitan Areas: The Emergence of Black Suburbs." *American Journal of Sociology* 75: 512–29.

Feng, J., and Y.X. Zhou. (2002) " The Spatial Redistribution and Suburbanization of the Urban Population in Hangzhou City." *City Planning Review* 26, 1: 58–65 [in Chinese].

Feng, J., and Y.X. Zhou. (2003a) " The Latest Development in Demographic Spatial Distribution in Beijing in the 1990s." *City Planning Review* 27, 5: 55–63 [in Chinese].
Feng, J., and Y.X. Zhou. (2003b) " The Social Spatial Structure of Beijing Metropolitan Area and Its Evolution: 1982–2000." *Geographical Research* 22, 4: 465–83 [in Chinese].
Feng, J. et al. (2004) "The Development of Suburbanization of Beijing and Its Countermeasures in the 1990s." *City Planning Review* 28, 3: 13–29 [in Chinese].
Foley, D. (1973) "Institutional and Contextual Factors Affecting the Housing Choices of Minority Residents." In A. Hawley and V. Rock (eds.), *Segregation in Residential Areas*. Washington, DC: National Academy of Sciences.
Frey, W., and A. Speare. (1988) *Regional and Metropolitan Growth and Decline in the United States*. New York: Russell Sage.
Gao, X.D., and Q.Z. Jiang. (2002) "The Redistribution of the Population and Suburbanization in Shanghai Municipality." *City Planning Review* 26, 1: 66–89 [in Chinese].
Garreau, J. (1991) *Edge City: Life on the New Frontier*. New York: Doubleday.
Gaubatz, P. (1999) "China's Urban Transformation: Patterns and Processes of Morphological Change in Beijing, Shanghai, and Guangzhou." *Urban Studies* 36: 1495–521.
Giusti, J. (1968) "Rasgos Organizativos en el Poblador Marginal Urbano Latinoamericano." *Revista Mexicana de Sociologia* 30: 53–78.
Guest, A.M. (1980) "The Suburbanization of Ethnic Groups." *Sociology and Social Research* 64: 497–513.
Keivani, R., A. Parsa, and S. McGreal. (2001) "Globalisation, Institutional Structures and Real Estate Markets in Central European Cities." *Urban Studies* 38: 2457–76.
Logan, J.R., and R. Golden. (1986) "Suburbs and Satellites: Two Decades of Change." *American Sociological Review* 51: 430–7.
Logan, J.R., and H.L. Molotch. (1987) *Urban Fortunes: The Political Economy of Growth*. Berkeley: University of California Press.
Logan, J.R., and M. Schneider. (1984) "Racial Segregation and Racial Change in American Suburbs: 1970–1980." *American Journal of Sociology* 89: 874–88.
Massey, D., and N. Denton. (1987) "Trends in the Residential Segregation of Blacks, Hispanics, and Asians: 1970–1980." *American Sociological Review* 52: 802–25.
Muller, P.O. (1981) *Contemporary Suburban America*. Englewood Cliffs, NJ: Prentice Hall.
Musil, J. (1993) "Changing Urban Systems in Post-communist Societies in Central Europe: Analysis and Prediction." *Urban Studies* 30: 899–905.
Ning, Y.M., and Y.C. Deng. (1996) "A Study of Shanghai's Suburbanization". In S.M. Li et al., (eds.), *Aspects of China's Regional and Economic Development*. Taipei and Hong Kong: Population Research Center of National Taiwan University and Hong Kong Baptist University. (In Chinese): pp. 129–53.
Preteceille, E. (2003) *La division sociale de l'espace francilien. Typologie socioprofessionnelle 1999 et transformations de l'espace résidentiel 1990–99*. Paris: Observatoire Sociologique du changement.
Reardon, K.M. (1997) "State and Local Revitalization Efforts in East St. Louis, Illinois." *Annals of the American Academy of Political and Social Science* 551: 235–47.
Ruoppila, S. (2004) "Processes of Residential Differentiation in Socialist Cities: Literature Review on the Cases of Budapest, Prague, Tallinn and Warsaw." *European Journal of Spatial Development* 9: 1–24. Online at: http://www.nordregio.se/EJSD

Sjoberg, O. (1992) "Under-Urbanisation and the Zero Urban Growth Hypothesis: Diverted Migration in Albania," *Geograska Annaler* 74B: 3–19.
Tammaru, T. (2001) "Suburban Growth and Suburbanisation under Central Planning: The Case of Soviet Estonia." *Urban Studies* 38: 1341–57.
Taschner, S.P., and L.M.M. Bogus. (2001) "*Sao Paulo*, uma metropole desigual EURE." *Revista latinoamericana de estudios urbano regionales* 27, 80, May: 87–120.
Wacquant, L. (1993) "Urban Outcasts: Stigma and Division in the Black American Ghetto and the French Urban Periphery." *International Journal of Urban and Regional Research* 17, 3: 366–83.
Wang, D., and Y. Zhou. (2002) "Study on Shanghai Consumers' Choices for Great Supermarket." *Urban Planning Forum* 4: 46–50 [in Chinese].
Wang, F.H., and Y.X. Zhou. (1999) "Modeling Urban Population Densities in Beijing 1982–1990: Suburbanization and Causes." *Urban Studies* 36, 2: 263–79.
Wu, F. (1995) "The Changing Urban Process in the Face of China's Transition to a Socialist Economy." *Environment and Planning C: Government and Policy* 13: 159–77.
Wu, F., and A.G.-O. Yeh. (1997) "Changing Spatial Distribution and Determinants of Land Development in Chinese Cities in the Transition from a Centrally Planned Economy to a Socialist Market Economy: A Case Study of Guangzhou." *Urban Studies* 34: 1851–80.
Xie, S.H., and Y.M. Ning (2002) "Urbanization and Suburbanization: The Dual Engines to Spatial Change of a Chinese Metropolis during the Transitional Era: A Case Study of Guangzhou." *City Planning Review* 27, 11: 24–38 [in Chinese].
Yeh, A.G.O., X.Q. Xu, and H.Y. Hu. (1995). "The Social Space of Guangzhou City, China." *Urban Geography* 16, 7: 595–621.
Zhang, Y. (1998) "A Study of the Suburbanization of Population in Suzhou, Wuxi and Changzhou." *Economic Geography* 18, 2: 35–40 [in Chinese].
Zhou, C.S., and X.Q. Xu. (1996) "A Study of the Spatial Characteristics of Population Change in Guangzhou City." *Economic Geography* 16, 2: 25–30 [in Chinese].
Zhou, M. (1997) "A Preliminary Analysis of the Question of Hangzhou's Suburbanization." *Economic Geography* 17, 2: 85–8 [in Chinese].
Zhou, Y.X. (1992) "On the Guiding Ideology of Urban Planning in Beijing." *Beijing City Planning and Construction Review* 4: 14–16 [in Chinese].
Zhou, Y.X. (1995) "Suburbanization of Beijing and Settlements Problems in Metropolitan Core Area." In National Land Agency of Japan and United Nations Center for Regional Development (eds.), *Settlements Problems in Metropolitan Core Area : Knowing the Facts and Finding A Solution*, Proceeding of the International Seminar on Human Settlements Policy '95: pp. 299–322.
Zhou, Y.X., and L.J.C. Ma. (2000) "Economic Restructuring and Suburbanization in China." *Urban Geography* 21: 205–36.
Zhou, Y.X., and Y.C. Meng. (1997) "Suburbanization in China – A Case Study of Shengyang." *China City Planning Review* 13, 2: 50–7.
Zhou, Y.X., and Y.C. Meng. (1998) "The Tendency of Suburbanization of Big Cities in China." *Urban Planning Forum* 3: 22–7 [in Chinese].
Zhou, Y.X. et al. (2000) "Report on the Survey of a Thousand Relocation Households in Beijing." *Planners* 16, 3: 86–95 [in Chinese].

7

Mirrored Reflections: Place Identity Formation in Taipei and Shanghai

Jennifer Rudolph and Hanchao Lu

Introduction

Shanghai and Taipei, two of Greater China's most vibrant cities, increasingly vie with one another for recognition as international hubs and regional economic powers. Physically divided by 428 miles, political splits marked by the Taiwan Strait have kept the two cities separated economically and politically for over a century. Within the comparative realm, the two cities have remain divided as well, with very little attention paid to how their shared nineteenth- and twentieth-century experiences illuminate our understanding of the paths the cities are taking into the twenty-first century. By focusing on colonial domination and the guiding hand of nationalist states as underlying historical commonalities, this chapter examines the place identities that have emerged in each city and assesses the roles played by imperialism and the State in defining the cities' characters and identities. Finally, it explores the current migration patterns between the two cities to address the historical and contemporary resonance of their identities.

Comparing Taipei and Shanghai adds to our understanding of urban Chinese environments in two ways: it provides a historical examination of two Chinese cities with shared legacies over time; and it provides a counterpoint to the assessments in other chapters that turn to non-Chinese cities as points of comparison. Taipei is, of course, firmly located within the Chinese diaspora, but the multi-phased and multi-state creation of Taipei's Chinese identity makes it somewhat of an anomaly within Greater China. Of great relevance for this volume is the fact that the political divide of the Taiwan Strait has meant that Taipei did not experience socialism. Thus, although the city is Chinese, it does not have a socialist history, providing a contrast to the socialist development of Chinese cities on the mainland. Nonetheless, the complex of identities that developed and were constructed in each of the cities mirrors the other because of the similarities of their colonial legacies, paths to economic development in a global world, and large migrant populations.

Thus, together Taipei and Shanghai present an opportunity to explore historical influences and patterns in two particular and distinct Chinese urban environments. From the perspective of rural-urban relations in Chinese history, the two cities (Shanghai in particular) represented the major breakthrough of what some scholars have termed as the rural-urban continuum in late imperial China, in which cities and the countryside integrated with each other culturally and in socioeconomic terms (Skinner 1977). Such a breakthrough in rural-urban relations was first and foremost brought forth by the extensive and, in many ways, extraordinary colonial experiences of both cities in modern times. By the middle of the twentieth century, Shanghai and Taipei had become the most prominent urban centers in their respective settings – the mainland and the island of Taiwan – surrounded by vast countryside areas that lagged behind. In addition, both cities bore the brunt of severe state developmental policies imposed by repressive regimes in the post-war era. By the same token, both cities were at the forefront of reform when political restrictions gradually faded away. None of these elements was absolutely unusual in modern China, but their combination makes Taipei and Shanghai a unique pair for comparing the formation of place identity in modern China. Such a comparison may also help explain, from a sociological perspective, why an estimated expatriate community of a quarter million Taiwanese – the majority of whom have come from the Taipei area – have found homes in Shanghai at the turn of the twenty-first century.[1]

More specifically, Shanghai and Taipei share variations of colonial and post-colonial legacies and distinctly Chinese migration patterns that have effected the formation of multi-layered identities centered on the urban place and dependent on the adaptation and layering of individual identities into situational and complex ones. Shanghai's colonial experience came in the form of the semi-colonial designation of Chinese treaty port. Taipei, too, had the treaty port experience of semi-colonialism. However, Taipei was also subjected to the earlier colonialism of an expanding Qing Empire and then served as Tokyo's local seat of power during Japan's possession of Taiwan from 1895 to 1945. Lastly, once the Nationalist Party (Kuomintang or KMT) found haven in Taiwan following the Chinese Civil War, Taiwan arguably underwent a third form of Nationalist colonial experience.

Although the colonialisms of Shanghai and Taipei differ distinctly from formal British colonialism in places like India – the imperialism inspired by the mercantilist policies of an expanding European power that led to direct colonization – the modern histories of the cities have been shaped by their colonial experiences and their efforts to escape them. Significantly, their decolonization processes took the form of anti-imperialism and movements for social justice, rather than the formal transference of legal control over territories. And because an Asian power, Japan, was the main protagonist of colonialism in Asia, decolonization in China did not involve a cultural critique of the West or an "ideological de-colonization" (Duara 2004), allowing for a more positive relationship in cities like Shanghai and Taipei with their colonial pasts.

In cities in the post-colonial order actively participating in the globalizing world, local cultures have interacted with non-indigenous influences with varying results. Schiller (1976) has proposed that regional culture can become homogenized as societies absorb the pervasive dominant Western or American cultural forces at the forefront of globalization. Others hold that the particular patterns of interface between the local and the global in a particular setting hold sway to form a local interpretation of the global (Massey 1993; Watson 1997). According to this view, globalization encourages cross-fertilization and results in variations of localization of culture and new forms of local identities, the result of a mixing of cultures or "hybridization" (Bhabba 1994). Shanghai and Taipei have experienced the localization of culture while the cities have participated in the process of globalization. Their place identities reflect the legacies of the interlocking forces of complex migration patterns and imperialism, as well as the impact of globalization on the local and on consciousness. Globalization also involves a consciousness of increasing social interdependence and the decreasing meaning of distance and nation (Robertson 1992). Although globalization accelerated in the late twentieth-century world, the early twentieth century set the stage in Shanghai and Taipei for later developments. As faster means of travel and the knowledge of a broader world developed, so did commerce expand and people migrate. With the establishment of East-West trade networks and the infrastructure to facilitate that trade, industrialized countries provided the means for spreading not only their goods around the world, but also their people and their culture.

Background

China's most developed urban centers in the nineteenth century were coastal cities or cities on major waterways. With the onset of Western imperialism, treaty ports with their, at times, heavy Western and Japanese presence, emerged as a distinct Chinese urban type (Esherick 1999). The commercial and financial development of the treaty ports contributed to a financial and commercial primacy that often overshadowed the traditional administrative centers favored within the Qing administrative hierarchy. The dozens of treaty ports that dotted the Chinese coast and inland areas at the start of the twentieth century contributed not only to the cause of Western imperialism in China, but also to the formation of distinct Chinese urban identities that emerged as a result of the interactions among Westerners (including Japanese as fellow imperialists) and Chinese in those ports. By providing urban spaces in which Chinese officials, merchants, and Western residents all interacted, the treaty ports presented a transitional stage for the forging of new urban and global identities. Shanghai, treaty port since 1843, and Taipei, first as a treaty port (1860) and then as colonial seat, absorbed and adapted the influences of imperialism and industrial development and led the way in China's urbanization process into the 1890s (Rozman 1990). Treaty ports came to be viewed simultaneously as outposts of imperialism, with all the associated negative connotations, as well as beacons of modernity and cosmopolitan internationalist verve.

The development of foreign concessions and international settlements in Shanghai compounded foreign influence. Although the Chinese government retained jurisdiction over the city as a whole, foreign treaty powers maintained control over the concessions through consular authority and the Euro-American dominated governing body known as the Municipal Council. The state of flux resulting from different polities competing to assert jurisdiction over various areas and aspects of treaty port life created possibilities for new communities and new identities formed by the migration that accompanied imperialism and new economic opportunities.

Although Taipei's treaty port experience lacked the entrenchment of Shanghai's foreign concessions, Taiwan experienced a longer period of imperialism. First fully incorporated into China during Qing expansionism, Taiwan underwent a process of integration that some scholars view as comparable to European imperialism (Harrell 1995; Millward 1996; Perdue 1998), countering the traditional interpretation of Sinification as slow assimilation rather than active colonialism. To further integration, the Qing government fostered a Han Chinese and Qing identity on the island by transplanting institutions and customs from the Mainland. (Li Guoqi 1975) Because of the diverse population groups on the island, forging a political Chinese identity that tied the population to the state proved difficult, but the Qing regime did succeed in imprinting such an identity on the main ethnically Chinese population groups (Hakka and Hokla or Fujianese). These groups continued to experience indigenization and formed local ties based on pragmatism and territorial proximity that outweighed their ties to their mainland hometowns, but they did so within the framework of a Han Chinese identity (Chen 1984). Mao-kuei Chang (2003, p. 27) claims that based on this process of indigenization and localization, there was no "collective imagination of 'being Taiwanese'" before 1895.[2] Thus, Qing colonization efforts on the island significantly shaped the Chinese identity on Taiwan that prevented the formation of a local identity distinct from Mainland Chinese identity.

Both cities experienced forms of colonial transplantation of institutions and practices. The processes of adaptation and transformation that occurred in Shanghai and Taipei as a result of new social, economic, political, and environmental conditions helped determine each city's distinctive urban forms and resulted in a "colonial third culture" (King 1990). The legacy of the colonial culture persists today. In fact, themes, conflicts, and attitudes underlying the colonial culture have re-emerged.

Impacts of Colonialism

Colonialism and urbanity

In the West, the sense of urban superiority resulting from rural-urban differentiation has a long history: ancient Rome served as the enduring symbol of Western civilization, medieval city states divided Europe into two distinctive worlds in which scattered and isolated walled cities were surrounded by vast "barbarian" rural areas; London and Paris could be seen as representatives of British and French culture, respectively, and so on. In comparison, culture flowed quite freely between the city and the countryside in traditional China, and no single city stood out as the epitome of Chinese civilization. Even Sui-Tang Chang'an could hardly be taken as an enduring symbol of this immense empire (Xiong 2000). Moreover, in a nation that was consciously built on agriculture, cities in Chinese history did not have obvious advantages over the countryside, as demonstrated in Frederick Mote's (1977) discussion of the urban-rural continuum in Nanjing during the Ming transition. Quite the opposite: virtually all Chinese had rural roots, as Max Weber (1958, pp. 81–2) pointed out: "The Chinese urban dweller legally belonged to his family and native village in which the temple of his ancestors stood and to which he conscientiously maintained affiliation."

This situation dramatically changed in the late nineteenth century. Treaty ports and the concentration of modern industries and commerce in urban areas brought job opportunities and material comfort as well as progressive ideas and advanced education to the city. By the early twentieth century, the old Western cliché of urban superiority became a new trend in Chinese society. The Shanghai identity was first of all built on this newly formed urban superiority. Being an "urbanite" (*shimin*) now had many advantages over being a "villager" (*xiangmin*).

But the Shanghai identity differs significantly from the urban identity of other Chinese cities. It wove a sense of being associated with the West (therefore superior or sophisticated) into the more conventional sense of urban superiority that existed elsewhere in the country. This sense of superiority was not necessarily a self-glorification of the Shanghainese and in fact often it was not even consciously held by them. It was more a stereotype of the Shanghainese in the minds of the people outside the city. Shanghai no doubt was the most Westernized city in modern China. With China's self-esteem and confidence badly damaged in modern times, to say a place was Westernized implied that it was somehow superior to an indigenous locality.

As in the case of Shanghai, treaty port status and modernization efforts contributed directly to Taipei's development. A relatively new city, Taipei grew in the nineteenth century as a result of the global economy and the Qing government's administrative manipulations to incorporate Taiwan's increasingly profitable commodities trade more fully into the government's modernization efforts. Starting as a commercial center, Taipei gained administrative importance when the Qing conducted a spatial reordering and upgrading of treaty ports in the late nineteenth century to bring them all to at least the circuit seat level within the administrative hierarchy. (Leung 1990; Rudolph 2007) Through this reorganization, the Qing government helped emphasize the importance of Taipei and other treaty port cities and contributed to an emerging urban identity that emphasized trade and interaction with the West in an effort to balance commercial and administrative powers within the ports.

Taipei's residents or *Taibei ren* like their Shanghai counterparts, locate part of their identity in a sense of urban superiority. Historically, the distinction between city and hinterland for Taipei urbanites was in stark contrast to situations in most of the Mainland. Aboriginal tribes populated the hills around the city, and even in the late 1800s, Qing officials were reluctant to take responsibility for actions outside of the urban areas, calling the aborigines "lawless savages," creating a sharp distinction between hinterland and city (Rudolph 2007, ch. 5). Urban Taipei quite literally represented the civilized during the late Qing. As the aborigine population came increasingly under the jurisdiction of the governing regime, the stark contrast between rural and urban in the Taipei basin receded, but did not disappear. Urban superiority, for instance, extended into the commercial and administrative realms, first through trade and then through commerce and administration as Taipei became the island's center for those activities.

Treaty port status encouraged migration to Shanghai and Taipei. Shanghai's increasingly important role in Western commerce and Qing China's modernization efforts provided economic opportunities for interested Chinese. In addition, the city's International Settlement and French Concession and Western presence contributed to the perception of the city as a safe haven from internal chaos, contributing to internal migration during the Taiping Rebellion. Taipei's city boundaries continued to be readjusted over the course of the late nineteenth and early twentieth century to reflect the city's increasingly important administrative and commercial roles. The Japanese and then the Kuomintang chose to administer the island from Taipei, and not Tainan, reflecting the city's ascendance. Migration to the city followed. Today, the historic central business area of contemporary Taipei contains the areas chosen by both the Qing and Japanese authorities as the seat of their administrations and reflects the layered identity of the city (Tsay 2001).

Imperial and colonial legacies

Throughout the first half of the twentieth century, Taipei residents confronted the conflict between a cultivated Chinese identity and a forced Japanese association during Japanese colonial rule. For many living under the Japanese, mandatory Japanese schooling and customs reinforced the Chinese identity that the Qing government had tried to impose on the island's residents by forcing them to confront an "Other." At the same time, the structure and span of the colonial period distanced Taiwanese, especially those in urban Taipei benefiting from the Japanese infrastructural and educational improvements, from their Chinese identity. The Japanese policies of assimilation (*doka*) and imperialization (*kominka*) required residents of Taipei (and Taiwan) to "become Japanese" (*nihon minzoku*) and threatened local identities, forcing them to become compartmentalized (Ching 2001).

The physical structure and layout of the city reflected the two identities. Japanese residents chose to live in the modern western city built by the Japanese administration in the inner city of Taipei. Dihua Street in Dadaocheng, the commercial center of old Taipei and an active market since the 1850s, constituted the center of local efforts to maintain a separate Taiwanese culture and space and illustrates the evolution of a Taipei identity that reflects the physical separation of identities. Taipei residents who resisted the imposition of Japanese lifestyle, for instance, gathered there to discuss the reconstruction of a Taiwanese identity and spawned new types of urban spaces with their coffee shops and restaurants. Nonetheless Dihua Street also manifests the absorption of Japanese influences, with examples of new modernist architectural forms of Japanese as well as Taiwanese construction, making it an example of Taipei's layered character and the emergence of a third culture (Yen 2003).

Dihua Street continues to play a central role in traditional market activity, especially at the Chinese New Year, and holds a key position in efforts to preserve historic Taipei. As Yen so adeptly demonstrates, the efforts in the 1980s and 1990s to preserve Dihua capture the multi-level nature of place identity in Taipei. It is the only area in Taipei that reflects the city's native Taiwanese, Qing, Japanese, and Chinese influences in a "live" settlement. Dihua Street's role as traditional commercial center and as one of the remaining historic areas has elevated its importance to "the most important urban symbol for the middle strata" because, to some, "it was the only place in Taipei that could make Taipei different from other cities" (Yen 2003, p. 164).

For many Taiwan residents, the layered forces create an identity crisis. Lee Teng-hui (1923–), Taiwan's first native Taiwanese president, captured the conflict in a 1994 interview printed in the *Independence Evening Post* in Taiwan when he publicly identified the dilemma of Taiwanese raised during the Japanese colonial period by stating, "I thought that I was Japanese until I was twenty-one years old" (Chang 2003, p. 39). To this day, a strong identification with Japan has left some older Taiwanese with a sense of nostalgia for the formal colonial days under the Japanese, viewing their rule as a period of efficiency and infrastructural development that strongly contrasts with the governance of the Qing Dynasty and the Kuomintang (Cumings 2004), contributing to the complexity of national and local identities for Taiwanese.

Nationalist Imprints: CCP and KMT Rule

After the active Civil War ended in China and the Chinese Communist Party (CCP) came to power on the Mainland and the KMT retreated to Taiwan, Shanghai and Taipei became subject to their respective government's nationalist agendas. In the case of Shanghai, national policies concerning migration and urban areas greatly benefited and shaped the outlook of city residents, making them keenly aware of Shanghai's privileged position as a city. Taipei residents experienced a nationalist molding of identity to fit the KMT agenda of once again ruling the Mainland, creating an atmosphere that focused on national identity within the local.

Communist Shanghai

A well-known slogan after the Communist revolution was "Wipe out the three great distinctions," namely, the distinction between the city and the countryside, the distinction between industry and agriculture, and the distinction between manual and mental labor. The reality was, however, that the rural-urban distinction widened in the half century after the Communist revolution. The most significant development in that regard was the household registration system (*hukou*) established in the 1950s. In Shanghai, the system was firmly enforced in 1958 in order to eliminate migration from other parts of China. For decades having the status of a permanent resident of Shanghai was a legal basis for benefits in employment, medical insurance, education, housing allocation, ration coupons, and other supplies. Except for an extremely small number of college graduates and relocated cadres, Shanghai was virtually closed to newcomers. Although the system began to loosen in the 1990s, to this day it has not been officially abolished, and migrant workers are still legally handicapped in the city with regard to benefits and privileges provided by *hukou* (most importantly, provisions of housing and education).

Meanwhile, generations of Shanghai-born residents greatly enhanced the notion of being Shanghainese in the years after the revolution. Unlike the approximately 80 percent of the residents in pre-1949 Shanghai, who were immigrants from elsewhere in China, the younger generations were virtually all Shanghai-born. These people naturally identified themselves with the city where they were born and grew up. Unlike the older generations, who had lived in a highly mobile city, the younger generations lived in a "frozen" city to which few people could move. They found they had been born with the privilege of being Shanghainese, a status millions of people outside the city desperately coveted. The Shanghai dialect played an important role in forging their sense of identity. Immigrants usually spoke the Shanghai dialect with an accent (and many indeed spoke their native tongue at home); younger generations, however, spoke perfect Shanghainese and lost their parents' dialect if they had ever learned to speak it at all. The sense of the native place of their parents (or, more precisely, their fathers) was still there, but it became secondary and relegated to the *jiguan* (ancestor's native place) category used mainly for filling out bureaucratic paperwork.

One should also note that a sense of Western superiority did not totally die out in Mao's China. The West – the term at that time was used almost solely in the context of "Western imperialism" and "Western capitalism" – was officially criticized and condemned, but amidst the bombastic propaganda in Mao's years, things Western were not necessarily devaluated in the mind of the ordinary people. To some extent, just because the West was now taboo, it became attractive. For the younger generation who had never experienced the age of pre-1949 "semi-colonialism," things Western were mysterious and, therefore, charming and appealing. For those who experienced the "old society," things Western were often associated with nostalgia. This was particularly true for those who disliked the Communist regime. In their minds the West had become a kind of spiritual sustenance in an arid social and political environment. Even in the most radical years of the Cultural Revolution, the expression *Yang pai* (Western style) never became a derogatory term.

Nationalist Taipei

As with Shanghai, there remains a distinction between Taipei and its region that contributes to a north-south divide within the island and a sense of privilege for Taipei residents. In terms of resources, Taipei has a disproportionate advantage over the rest of the island, largely due to its position as governmental seat and commercial center. Taipei comprises only 0.8 percent of Taiwan's land, yet holds roughly 12 percent of its population. Moreover, Taipei is home to one-sixth of the island's secondary schools and one-third of its universities. In terms of business advantage, over 90 percent of multinational companies in Taiwan have their headquarters in Taipei, and 70 percent of the largest domestic businesses make Taipei their home. Although an increasing number of factories are locating in central and southern Taiwan, local residents do not perceive them as benefiting their region, since the factories are geared towards producing for elsewhere. In fact, local Taiwanese capture the disparity between Taipei and the rest of the island with the expression, "The hens produce droppings but lay no eggs."

Civil War-related migration patterns compound the divide between Taipei and the rest of the island. Unlike Shanghai, where the official population became less oriented toward migrants after World War II, Taipei's population shifted heavily toward a migrant population in the late 1940s. The Nationalist or KMT migration to Taiwan that accompanied their 1949 loss in China's Civil War centered on northern Taiwan. About one million military personnel and one million civilians fled to the island, inundating it, with Taipei receiving the highest percentage of migrants. That migration, followed by a high birth rate, swelled the island's population from 5.8 million in 1940 to more than 10 million by 1960. Likewise, Taipei's population grew from less than 326,000 in 1947 to more than one million in 1966.[3] More Mainlanders (*waisheng ren*) chose to reside in Taipei proportionally than any other part of Taiwan, due in part to the urban origins of many of the migrants. During the Japanese period, Mainlander presence in Taipei did not exceed 7 percent, but in 1962, it peaked at 38 percent, with certain areas of Taipei becoming known as Mainland domains. Today, Mainlanders and their descendants constitute 30 percent of Taipei's population, compared to the 13 percent of the rest of the island. Because of the political nature of the migration and because of the goals of the KMT to regain power in China, a strongly nationalist orientation permeated KMT policies on the island in terms of policy and in relation to the native population; this orientation became distilled in Taipei, with tension emerging between Mainland immigrants and native Taiwanese Taipei residents.

In an effort to impose a Nationalist identity that responded both to KMT political rhetoric toward Mainland China and Chinese re-possession of the island from the Japanese, the Nationalist regime emphasized Chinese national identity and tried to eliminate local identity. The government strove to preserve Chinese culture and traditions as a matter of self-identification. As a result, for the next four decades Taipei's public spaces were dominated by an enforced Chinese identity, not unlike the enforced Japanese identity in the first half of the century. By government order, Mandarin Chinese became the national language (*guoyu*) and was the exclusive language for education, even though the majority of the population spoke native Taiwanese (*minnan hua*). The media and government also solely used Mandarin. In fact, the government banned the use of the native Taiwanese language in public forums in an effort to force the adoption of Mandarin. Most native Taiwanese (*bensheng ren*) did not speak Mandarin, and most Mainlanders did not speak Taiwanese, exacerbating communication difficulties and reinforcing cultural divides.

The imposition of martial law throughout the island from 1947 until 1987 demonstrated to the native population the heavy-handed imposition of a Chinese identity. The division and tension between Mainlanders and Taiwanese manifested itself not only through protests such as the 1947 February 28 Incident which led to martial law, but also through social practices, such as discouragement of marriage between Mainlanders and native Taiwanese. Consequently, separate political identities ran through the city: marketplaces and unofficial public spaces were venues for a suppressed Taiwan Chinese identity in Taipei and government and official public spaces were showcases for the Mainland Chinese identity of Taipei residents. The nationalist orientation of Mainland Chinese KMT followers and their identification with Chinese culture combined with a sense of cultural superiority associated with China. Conversely, they viewed Taiwanese *Taibei ren* as residents of the Japanese seat of power and therefore as "Japanized" and worthy of contempt.[4] Efforts were made to eliminate remnants of Japanese culture and replace them with Chinese culture. The divide between Taipei's native Taiwanese residents and its Mainland Chinese migrants persisted at least until the late 1980s when political liberalization began.

In contrast to Shanghai-born children of migrants in the post-1949 period, mainland refugees who settled in Taipei and their Taipei-born children preferred to retain their native place and national identities and did not see themselves as becoming native *Taibei ren*. Popular literature of the 1950s to 1970s authored by Mainland residents of Taipei portrays the preservation of native identity at the expense of an assimilated identity centered in Taipei. In the short story collection *Taibei Ren*, the Mainlander turned Taipei migrant Bai Xianyong (Pai Hsien-yung) perhaps most famously captures the wounded pride and the desire of Nationalist Mainland migrants' to maintain a superior identity to Taiwanese Taipei residents through their alternative Mainland identities. In the story "The Last Night of Daban Jin," Bai portrays Jin Zhaoli, a Shanghai native, venting frustration against her boss by using Shanghainese profanity to compare obliquely the status of the two places by implying that her Taiwanese boss was not even worthy of washing the toilet bowls at the dancehall in Shanghai where she had been a popular hostess (Pai 2000, pp. 117–19).[5] The portrayal of characters like Ms Jin demonstrated the perceived spiritual superiority of Mainlanders over native Taipei residents. The persistent and superior identification with native place that shadowed Mainland migrants' presence in Taipei contributed to the continued acute awareness among Taipei residents of the divide between native *Taibei ren* and Mainland *Taibei ren*.

Liberalization and the resurgence of the old patterns

Resurging sense of urbanity. Old patterns of Shanghai identity are resurgent in the post-Mao reform era, especially since the late 1990s. Shanghai is a rapidly growing city. The population holding official permanent resident status remained about 13 million in the 1990s, but by the late 1990s there were more than 3.5 million long-term residents living in the city without official *hukou*. This means one out of every five Shanghai residents was a recent migrant (Kang 2001, pp. 33, 383, Table 1).[6] The newcomers included small-time entrepreneurs and adventurers of all sorts, but the majority of them were migrant workers who, as in the pre-1949 era, almost all came to Shanghai directly from villages. At the beginning of their urban adventure, they were part of the so-called floating population, which numbered about 80 million nationwide in July 1995. Close to 60 percent had moved from rural to urban areas (Solinger 1999, p. 21). But by the late 1990s they had become settled in the city and, although still without a *hukou*, had become permanent residents of Shanghai. Like their predecessors in the Republican era, these people are in the process of becoming urban and they gradually identify themselves as Shanghainese. Some adapted to the city so quickly that they acted like "old time Shanghainese" (*lao Shanghai*), while others modestly called themselves "quasi Shanghainese" (*zhun Shanghairen*) (Chen 2003). But for those peasants of yesterday, the main component of this identity is that now they are no longer "country people" but "city people."

The lifting of martial law in Taiwan in 1987 and the subsequent democratization movement has eased the tension between Mainlanders and Taiwanese on the island and in Taipei, but interestingly, it has brought to the fore another, related cultural conflict. Are the residents of Taipei culturally Chinese or Taiwanese? In today's democratized Taipei, both sets of identity conflict persist, but the recent struggle for a national Taiwanese identity for Taipei supersedes and further complicates the previous battles for the city's identity. As a result of this confusion, compounded by the nationalist policies and the North-South divide, Taipei migrants' identification with their new hometown does not follow the same pattern as in Shanghai. Instead of slowly identifying with Taipei, the migrant mentality holds. Tseng Shu-cheng, head of the Urban Reconstruction Center and professor of architecture at Tamkang University sums up the dilemma of Taipei migrant residents: "While many cities want to reconstruct or rediscover an older identity, Taipei never had one and must create a new identity." (Li 2000, p. 6) Although Tseng's statement contains a level of dismissal of Taipei's own history, it reflects city residents' concern with being part of the globalized world in their own right. The historical identity of Shanghai as the "Paris of the East" and "Pearl of the Orient" and its current identity of international gateway to Asia have great appeal to Taipei residents and contribute to a fascination among Taipei residents with Shanghai. Likewise, Shanghai's strong historical identity contributes to its own residents' return to old patterns.

Westernization and globalization. Westernization is once again central to the notion of the Shanghai identity, but unlike the sway of the West in the old days that brought national humiliation, current influences dovetail with the strong desire to modernize and hence they are welcomed. The Chinese quest in the post-Mao era to reform, to open up, and to keep abreast of advanced technologies to participate in vibrant globalization, all concern the West. But in Shanghai, such a connection seems particularly strong.

In Beijing, the Communist leaders since 1995 have invited academicians from China's most prestigious academic institutes to give lectures on science and technology to top leaders in Zhongnanhai. By comparison, this kind of "enlightenment" in Shanghai started earlier and has been at the grassroots level. In 1992 academicians from the same institutions were invited to lecture in ordinary neighborhoods to the common people. From 1994 to 2000, Shanghai had two million residents who earned certificates in computer and foreign languages (predominantly, English) (Kang 2001, pp. 294–5, 299). Activities like these are known in the city as "connecting the rails" (*jiegui*), that is, connecting the Chinese "railroad" with that of the outside world.

The "outside world" is understood essentially as the Western world. The highlights of Shanghai's new urban landscape – what have been called the city's "bright spots" – have an unmistakable Western flavor. The Xintiandi complex, the coffee shop area of Huaihai Road, the "leisure street" on Hengshan Road, and the new urban jungle in Pudong, are all modeled on the West. Shanghai's Toy City on Henan Road is stuffed not with Chinese products but foreign brands such as Lego and Toys "R" Us. A journalist, has commented:

> In comparison, the toys offered by our native merchandisers are mediocre. What it reflects is the reality of the First World and the Third World and a choice you cannot resist, that is, our children are immersed in foreign culture and have accepted a non-native lifestyle. What impact will this type of "immersion" and "acceptance" have on our national tradition and cultural heritage? What impact will a fully Westernized new generation have on the regeneration of our nation? We do not know. We can only wish that the notion of global village will not cause us to lose the diversity and colorfulness of our indigenous culture, and the establishment of Toy City will not lead our children into valuing only one culture and one lifestyle (Wang 1996, p. 185).

These concerns about excessive Western influence in a toy store reflect, in a small but exemplifying way, the overwhelming presence of Western commercial culture in the city. Shanghai is of course not alone in this regard, but it certainly is a most receptive city to Western-led cultural globalization.

Ironically, while the Shanghai identity contains this component of renewed Westernization, it may also be threatened by that very component, as this journalist has suggested. In the long run, such a dilemma may not just be limited to Shanghai but may constitute a challenge to all Chinese cities on the road to modernization.

*"A presumptuous guest usurps the host's role" (*xuan bin duo zhu*)*

One distinguished tradition of Shanghai was that the city was very receptive to outsiders, to the extent that native people were often overshadowed by newcomers. The prosperity of Shanghai in the Ming-Qing period as a "cotton town" renowned for the trade of nankeens was mainly created by the so-called guest merchants who came from outside Shanghai to do business in the city; among them the merchants from Anhui were the most prominent. At least 70 percent of the business community in Shanghai consisted of these "guest merchants." A local Shanghainese commented in early nineteenth century: "The profits of the Huangpu River are mainly taken by the merchants. However, only 20 to 30 percent of them are native tradesmen" (Zhang 1936). The situation became more obvious – even striking – after Shanghai was opened as a treaty port. The rapid growth of Shanghai thereafter was orchestrated almost entirely by outsiders. The domination of the Westerners in the city is an often-told story, but the roles played by Chinese migrants to the city were equally important. The population of Shanghai in the early twentieth century was consistently 80 percent non-native (Zou 1980, ch. 4). According to a collection of biographies of the so-called Shanghai celebrities published in 1930, only 10 percent of the celebrities (that is, the most prominent persons who played a leading role in various fields in Shanghai) were natives (Haishang mingren zhuan bianjibu 1930). The situation perhaps can be described by the Chinese metaphor, *xuan bin duo zhu* (presumptuous guests usurp the host's role).

Presumptuous guests continue to usurp the host's role in Shanghai in the reform era. After 40 years of government restriction on population mobility to the city, Shanghai once again has become a favored spot for China's migrants. Although the *hukou* system is still in effect, the market economy has made it largely irrelevant for the millions who find jobs in the city. From the mid-1980s on, every four or five years the number of the people who lived in Shanghai without *hukou* status has doubled.[7] Like the social milieu in the 1920s–40s, "going to Shanghai" is once again a trend. Newcomers have significantly changed the city's cultural landscape, much like the migrants to Shanghai did early in the twentieth century.[8] For instance, to accommodate the millions of people whose native tongue is not Shanghainese or the Wu dialect, Mandarin (or *Putonghua*, the "common language") to an unprecedented degree is quickly becoming a major language among the public. This trend has not been promoted by an official order or government programs but driven by the practical need for communication in a city that is now full of people who come from different backgrounds. But the city had been adaptive to language before. In the early twentieth century, pidgin English – commonly used by shrewd business compradores and illiterate rickshaw pullers alike – was characteristic to Shanghai. The Shanghai dialect itself was a product of the integration of a variety of dialects and is arguably the youngest dialect in China. (Lu 1999, pp. 54–5).

A prominent group among the newcomers is people from Taiwan, who in a significant way help promote Mandarin since as a result of Taiwan's identity politics they are all Mandarin speakers, and the Taiwanese as a group are a major player in the city's economy. The fascination for Shanghai among Taiwanese, a phenomenon known as "Shanghai fever," has spread since the mid-1990s. The attraction of Shanghai is, of course, mainly due to the business opportunities that the city provides. Taiwan was among Shanghai's top five trading partners in the late 1990s. By the end of October 2001, Taiwanese investment in Shanghai counted 4,027 projects with a capital of US$6.58 billion in terms of contract value, accounting for about 15 percent of total foreign investment in the city.[9] But the allure of Shanghai goes beyond business opportunities. For many Taiwanese, the fascination with the city has become the "Shanghai dream." As a journalist describes, from fashion to business to entertainment, Shanghai dominates every topic on the island: "According to the words across the Taiwan Straits, everything is better in Shanghai; everything is bigger; everything is grander; and everyone wants to be here." (*Shanghai Star*, December 20, 2001). At the turn of the twenty-first century, it was estimated that about 300,000 Taiwanese were working and living permanently in metropolitan Shanghai. In addition to these new residents, approximately 10,000 business persons commute between Taiwan and Shanghai per month. (*Xinmin zhoukan* July 15, 2001, p. 9).

These newcomers have a great impact on Shanghai. The term *Taishang* ("Taiwan merchants") has been associated with new businesses and fashions in the city. Taiwanese chain stores such as Yonghe Doujiang (a fast food store) and Taipingyang Baihuo (a department store), among others, have to some extent changed the daily lives of ordinary residents and influenced lifestyles. Even vegetables from farms run by Taiwanese in suburban Shanghai are regarded as of higher quality (Pan 1995, p. 201). Media jargon describes these enterprises as having "formed a line of bright scenic spots" (*zucheng yidao liangli de fengjing xian*) in the city.

While the Taiwanese impact on the city is obvious, in a less tangible way the city also changed its "guest merchants." Many Taiwanese have integrated into Shanghai society and started to see themselves not just as sojourners or investors but an indispensable part of the city. They call themselves "Shang-Tai-nese" (*Shang Tai ren*). The coined term is a pun. It means "Taiwanese in Shanghai" as well as the "man on the stage" (i.e., "a person in power"). This name quite powerfully indicates the dual identity that has emerged among the Taiwanese in the city. It also reflects a sense of pride among the Taiwanese on the role that they have played in the development of the city. Nearly a century ago, business people from Ningbo, Guangdong, and elsewhere came to Shanghai and significantly changed the city; now "presumptuous guests" are once again usurping the host's role. They are generally welcomed, for Taiwanese are regarded as pioneers who have advanced themselves to modernity. The integration of Taiwanese into the city can be seen as the revival of a tradition of openness and acceptance that Shanghai was renowned for in the early twentieth century.

Not only is Shanghai Fever manifest in Shanghai, but also in Taipei. Shanghai-themed restaurants and clubs proliferate, catering to both an older nostalgic clientele and the hip crowd seeking the atmosphere and style of Shanghai. In addition, myriad TV shows based in Shanghai showcase the city's history and culture. The popularity of the Taiwan mini series "April Rhapsody" about Shanghai's famous early twentieth-century poet Xu Zhimo and Hong Kong's Wong Kar-wai's feature film "In the Mood for Love" set in Shanghai are only two examples of the popularity of all things Shanghai. Shanghai's allure is located in its 1930s cosmopolitan romantic retro appeal and in its current international sparkle.[10] For many residents of Taipei who see Taiwan as caught between China's nationalism and Taiwan's nationalism, the city provides a gateway for advancement in an increasingly internationalized environment. Considering that more than 30 percent of Taipei's population (or more than 900,000) travel overseas each year (Li 2000),[11] the fascination with Shanghai over other cities reveals a connection to the globalized Chinese diaspora.

Conclusion

An interview dialogue in *Shijie Zhoukan* (2005) with two prominent authors, Wang Anyi (a *Shanghai ren*) and Long Yingtai (a *Taibei ren*) provides a meaningful glimpse into representative perceptions each has of her own and her counterpart's city and can act as a way of placing historical observations into contemporary context. Wang Anyi sees Shanghai as developing along American lines, much like Los Angeles, equating infrastructural development and Westernization with modernization. She emphasizes Shanghai's globalization as part of a quest for the latest material culture. Long, on the other hand, while recognizing Shanghai's upper hand in terms of international experience and in its status as an economic powerhouse, believes that Taipei residents' less tangible assets, such as active participation in citizenship and democracy and the accessibility of international travel, belie a level of modernity that Shanghai residents have yet to see. Shanghai's current status reflects its colonial emphasis on trade and international interaction. Taipei's international strength resulted from its democratization, which can be attributed to the conflicting identities within the city that the KMT had to finally take into account when the geopolitical arena created by the Chinese Civil War and the Cold War shifted in the 1970s.

Although the authors mark modernity very differently, they interestingly both take coffee as a yardstick of maturity in the process of modernization. Coffee drinking is a common, affordable habit among the people of Taipei, while in Shanghai, the city that once-upon-a time was regarded as the most Westernized city in China, coffee drinking has been quickly resurging, but nevertheless is still regarded as a "high class" activity and a token of elite culture. The comparisons of coffee drinking made by the two writers are not coincidental, but reveal the popular image of modernization as primarily "to be like the West." Both consider coffee to be a symbol of western lifestyle; whether or not it is an ordinary part of urban life is an indicator of the maturity of modernity. Such a simplistic format – that modernization and westernization come in pairs, if not equated – has been questioned in academia. But Wang and Long's comments indicate that such a formulation holds much truth in the popular mind.[12] Coffee thus becomes a manifestation of the diminishing meaning of nation and distance that is part of globalization.

In recent Chinese history, Taipei and Shanghai both started from relative obscurity and developed national and international prominence in relatively short periods of time. Shanghai had in the 50 years following the Opium War developed from a county seat into China's most prominent and modernized city. Taipei had in a few decades after World War II developed from a peripheral town into a major political center and economic hub in Pacific Asia. They are arguably the two cities that have most dramatically climbed the Chinese urban hierarchy in recent history.[13]

Modern Shanghai and Taipei might be considered unexpected children of China's turbulent recent history. It is not an exaggeration to say that without a century of Western intrusion and domination Shanghai would not be an international hub and that without the various forms of colonialism and the communist victory, Taipei would have developed very differently. Imperialism, colonialism, and communism played definitive roles in each city's development. The theoretical frameworks outlined in the introduction to this volume help frame the discussion of the cities' experiences. The developmental state, modernization theory, and socialism and post-socialism have certainly impacted one or both cities at different times; however, no one of these theories seems to fully explain the complexities of the experiences of either Shanghai or Taipei. For this, an understanding of the shared forces and historical particulars that have contributed to the parallel constructions of their place identities and urban environments is arguably of equal or greater importance.

The shared modern fate of the two cities has rendered them some characteristic similarities. Colonial experience, immigration, commercialization, openness, and acceptance are among the most obvious and comparable aspects of the cities. Some of these phenomena are general features of urban society, while others are more characteristically associated with these cities; together they have shaped the identities of each of the cities in a way that resonates with the other. Residents of Taipei and Shanghai are to some extent alienated from their broader societies. *Taibei ren* are seen by "indigenous" Taiwanese in the south as consisting of intrusive mainlanders and their liberal descendants. *Shanghai ren* have been stereotyped for nearly a century by the Chinese in hinterlands as being arrogant and "Westernized." Such alienation of the two cities from their "own people" and recent interactions with each other reflect the avant-garde nature of the two cities in terms of globalization.

NOTES

1 The exact number of Taiwanese in Shanghai is unknown even to officials in charge of Taiwan affairs. The Shanghai municipal office for Taiwan affairs stated in 2002 that 220,000 Taiwanese had been staying in Shanghai for more than three months, and 470,000 entered or left the mainland through Shanghai in 2001. See *People's Daily*, July 2, 2002, for a discussion of the topic.
2 For an argument that Taiwan's colonial status persisted even after 1945 when the island was returned to Kuomintang rule, see Gold (1986); Cumings (1999).
3 Selya puts Taipei's population at the end of World War II at 600,000. By 1974, Taipei's population was more than two million. Currently, it is 2.6 million.
4 We are using the spelling "Taipei" in this chapter when referring to the city, because that is how it is commonly known in both the West and in its Romanized form on Taiwan. For most other Chinese words and names, including referring to the residents of Taipei in Chinese, we are using the pinyin Romanization system used on Mainland, which would render the city's name as Taibei and the residents as Taibei ren. Exeptions are made for the Nationalist Party, for which we use Kuomintang (pinyin Guomindang) and for the Lee Teng-hui, the former President of Taiwan (pinyin Li Denghui), as they are better known in the West by their alternative spellings.
5 For a discussion of representations in *Taipei Ren,* see Lau (1975).
6 See also an analysis on Shanghai's floating population in Ying Jizu (ed.), *Tigao chengshi de shenghuo ziliang*, 226. "Long term resident" (*changzhu renkou*) is defined as someone having continuously lived in the city for at least six months.
7 The following are numbers for long-term residents without *hukou* status in Shanghai between 1984 and 1993: 1984: 750,000; 1988: 1,250,000; 1993: 2,810,000 (see Yin and Lu 2001, p. 24).
8 For sociological survey and interviews of the new residents of Shanghai from all walks of life, see Chen (2003).
9 Statistics provided by the Commercial Section of the Consulate General of Switzerland in Shanghai.
10 For a discussion of Shanghai's appeal in this regard, see American Chamber of Commerce, Taipei (2005).

11 The article also points out that Taipei is unique for Taiwan in this regard; although 28 percent of Xinzhu (Hsinchu) residents, a Taipei suburb, travel overseas each year, only 6 percent of Tainan's and 5.4 percent of Kaohsiung's do.

12 For the resurgence of the "coffee culture" in present-day Shanghai, see also Lu (2002).

13 An exception is perhaps Hong Kong, if one considers Hong Kong as an essential Chinese city. The rise of Shenzhen in the past two decades is not included in this analysis. We generally consider Taipei to be in the greater China context, regardless of the political controversies over the status of Taiwan.

REFERENCES

American Chamber of Commerce, Taipei. (2005) "Looking for Shangri-La." *Taiwan Business Topics* 31, 10.

Anderson, B. (1991) *Imagined Communities: Reflections on the Origins of the Spread of Nationalism* (revised edn). New York: Verso.

Bhahba, H.K. (1994) *The Location of Culture*. Routledge, New York.

Chang, M.K. (2003) "On the Origins and Transformation of Taiwanese National Identity." In P.R. Katz and M.A. Rubinstein (eds.), *Religion and the Formation of Taiwanese Identities*. New York: Palgrave: pp. 23–58.

Chen, Q. (1984) "Tuzhuhua yu neidihua; lun qingdai Taiwan hanren shehui de fazhan." *Zhongguo hai yang fazhan shilun wenji*: 335–66.

Chen Y. (2003) (comp.) *Yimin Shanghai*. [Migrating to Shanghai]. Shanghai.

Ching, L.T.S. (2001) *Becoming Japanese: Colonial Taiwan and the Politics of Identity Formation*. Berkeley: University of California Press.

Chow, P.C.Y. (ed.) (2002) *Taiwan in the Global Economy: From an Agrarian Economy to an Exporter of High-Tech Products*. Westport, CT: Praeger.

Clark, C. (1989) *Taiwan's Development: Implications for Contending Political Economy Paradigms*. New York: Greenwood Press.

Cumings, B. (2004) "Colonial Formations and Deformation: Korea, Taiwan, and Vietnam." In: P. Duara (ed.), *Decolonization: Perspectives from Now and Then*. London and New York: Routledge: pp. 278–98.

Duara, P. (ed.) (2004) *Decolonization: Perspectives from Now and Then*. London and New York: Routledge.

Esherick, J.W. (ed.) (1999) *Remaking the Chinese City: Modernity and National Identity, 1900–1950*. Honolulu: University of Hawaii Press.

Gold, T.B. (1986) *State and Society in the Taiwan Miracle*. Armonk, NY: M.E. Sharpe.

Haishang mingren zhuan bianjibu. (comp.) (1930) *Haishang mingren zhuan* [Biographies of the Shanghai celebrities]. Shanghai.

Harrell, S. (1995) "Introduction: Civilizing Projects and the Reaction to Them." In *Cultural Encounters on China's Ethnic Frontiers*. Seattle: University of Washington Press: pp. 3–36.

Kang. Y. (2001) *Jiedu Shanghai: 1990–2000* [Understanding Shanghai: 1990–2000]. Shanghai: Shanghai renmin chubanshe.

King, A.D. (1990) *Urbanism, Colonialism, and the World-Economy*. New York: Routledge.

Lau, J. (1975) "'Crowded Hours' Revisited: The Evocation of the Past in Taipei jen." *Journal of Asian Studies* 35, 1, Nov: 31–47.
Leung, Yuen-sang. (1990) *The Shanghai Taotai: Linkage Man in a Changing Society, 1843–90*, Honolulu: University of Hawaii Press.
Li, G. (1975) "Qingji Taiwan de zhengzhi xiandaihua-kaishan fufan yu jiansheng (1875–1894)." *Zhonghua wenhua fuxing yuekan* 8, 12: 4–16.
Li, G. (2000) "Jiao 'diyi ming' tai chenzhong? – que li ji qun taipei cheng." Online at: http://www.sinorama.com.tw. (Nov.): 6.
Lu, H. (1999) *Beyond the Neon Lights: Everyday Shanghai in the Early Twentieth Century*. Berkeley: University of California Press.
Lu, H. (2002) "Nostalgia for the Future: The Resurgence of an Alienated Culture in China." *Pacific Affairs* 75, 2: 169–86.
Marsh, R.M. (1960) *The Great Transformation: Social Change in Taipei, Taiwan Since the 1960s*. Armonk, NY: M.E. Sharpe.
Massey, D. (1993) "Power-Geometry and a Progressive Sense of Place." In J. Bird (ed.), *Mapping the Futures: Local Cultures, Global Change*. New York: Routledge.
Morse, H.B. (1910–18) *The International Relations of the Chinese Empire* 3 vols. London: Longmans, Green, and Co.
Mote, F.W. (1977) "The Transformation of Nanking, 1350–1400." In G.W. Skinner (ed.), *The City in Late Imperial China*, Stanford: Stanford University Press: pp.101–53.
Millward, J.A. (1996) "New Perspectives on the Qing Frontier." In G. Hershatter, E. Honig, L. Lipman, and R. Stross, (eds.), *Remapping China: Fissures in Historical Terrain*. Stanford: Stanford University Press: pp. 113–29.
Pai, H.-Y. (Bai Xianyong). (2000) *Taipei Ren* (trans. H.-Y. Pai, and P. Yasin, ed. G. Kao). Hong Kong: The Chinese University Press.
Pan, X. (comp.) (1995) *Dushi rixian* [Urban Hotlines]. Shanghai.
Perdue, P.C. (1998) "Comparing Empires: Manchu Colonialism." *International History Review* 20, 2, June: 255–62.
Robertson, R. (1992) *Globalization: Social Theory and Global Culture*. London: Sage.
Rozman, G. (1990) "East Asian Urbanization in the 19th Century, Comparisons with Europe." In *Urbanization in History: A Process of Dynamic Interactions*. Oxford: Clarendon Press.
Rudolph, J. (2007) *Negotiating Power in Late Imperial China: The Zongli Yamen and the Politics of Reform*. Ithaca: Cornell East Asia Series.
Schiller, H.I. (1976) *Communication and Cultural Domination*. White Plains, NY: International Arts and Sciences Press.
Selya, R.M. (1995) *Taipei*. Chichester: John Wiley & Sons.
Shijie zhoukan (Supplement to *World Journal*), January 16, 2005: pp. 12–15.
Skinner, G.W. (1977) "Introduction: Urban and Rural in Chinese Society." In G.W. Skinner (ed.), *The City in Late Imperial China*. Stanford: Stanford University Press.
Solinger, D.J. (1999) *Contesting Citizenship in Urban China: Peasant Migrants, the State, and the Logic of the Market*. Berkeley: University of California Press.
Teng, E. (2004) *Taiwan's Imagined Geography: Chinese Travel Writing and Pictures, 1683–1895*. Cambridge, MA: Harvard Asia Center.
Tsay, C.-L. (2001) "Urban Population in Taiwan and the Growth of the Taipei Metropolitan Area." In F.-C. Lo and P. Marcotullio (eds.), *Globalization and the Sustainability of Cities in the Asia Pacific Region*. Tokyo: United Nations University Press.

Wang, W. (1996) *Yuwang de chengshi* [The City of Desire]. Shanghai: Wenhui chubanshe.

Watson, J.L. (1997) *Golden Arches East: MacDonald's in East Asia.* Stanford: Stanford University Press.

Weber, M. (1958) *The City* (trans. D. Martindale and G. Neuwirth). Glencoe, IL: Free Press.

Xiong, V. (2000) *Sui-Tang Chang'An: A Study of the Urban History of Medieval China.* Ann Arbor: University of Michigan, Center for Chinese Studies.

Yin, J., and Lu, H. (eds.) (2001) *A Report of Social Development in Shanghai,* 2001. Shanghai.

Yen, L.-Y. (2003) "Heterotopias of Memory: The Cultural Politics of Historic Preservation in Taipei." PhD Dissertation in Urban Planning, UCLA.

Zhang, C. (1936) *Hucheng suishi quge* [The Seasonal Folk Songs of Shanghai]. Shanghai.

Zou, Y. (1980) *Jiu Shanghai renkou bianqian de yanjiu* [Research on Population Changes in Old Shanghai]. Shanghai.

8

Is Gating Always Exclusionary? A Comparative Analysis of Gated Communities in American and Chinese Cities

Youqin Huang and Setha M. Low

Introduction

As part of the overall transition toward a market economy, housing reform was launched nationwide in 1988 to introduce market mechanisms into a welfare-oriented housing system. Since then, there have been significant transformations in the provision and consumption of housing. One of the most important and visible changes in Chinese cities is the massive construction of private housing mainly in suburbs, providing shelters to the emerging middle class and the new rich. As a result, there has been increasing housing inequality and residential segregation, and wealthy communities and dilapidated migrant enclaves are emerging side by side in Chinese cities, in sharp contrast to the socialist residential pattern that was characterized by uniform public housing and homogeneous neighborhoods (Hu and Kaplan 2001; Huang 2005). This phenomenon is closely related to spatial inequality documented by White et al. and rapid suburbanization by Zhou and Logan in this volume. These new private homes are mostly built in the form of walled and gated complexes. While those for the middle class usually have basic guarding systems such as security personnel standing at the gate to monitor the entrance, the high-end housing complexes for the elite often have sophisticated security systems such as card-activated entrance, infrared alarm systems, intercom, surveillance camera, and security personnel on patrol, which are comparable to those in wealthy gated communities in the West. Many of these housing estates are funded with foreign investment, have Western style architecture and even names like "Orange County" (in Beijing) and "the Fontainebleau Villas" (in Shanghai) to project an image of Western lifestyle (e.g. Giroir 2005b). The numbers of these gated developments is also growing rapidly, for example in Guangdong Province there are 54,000 gated communities covering over 70 percent of the residential areas (Miao 2003). Rapid globalization, market liberalization, privatization in service provisions, and increasing social inequality and security concerns are often considered driving forces for these walled communities (Giroir 2005a, b; Miao 2003; Wu 2005). It seems that the "global spread" of "gated communities" has reached China (Webster, Galsze, and Frantz 2002).

Yet, gated and walled communities have always existed in Chinese cities (Knapp 2000). The traditional Chinese house – courtyard house (*siheyuan*) – was built in an enclosed form, and elaborate courtyard house complexes such as the Qiao Family Manor in Pingyao and the Forbidden City in Beijing are walled communities. Most housing built in the socialist era was in the form of "work-unit compounds"(*danwei dayuan*), which were often walled, gated and guarded. Thus gating has its Chinese roots. While globalization and market transition help to explain the latest wave of gating especially in high-end private developments, they cannot fully explain the widespread gating in Chinese cities, especially those in socialist work-unit compounds and traditional housing complexes. We argue that American-based theories are not adequate in explaining gating in China. While the physical form of the gating may look similar across nations, its sociopolitical construction can vary significantly, and gating has to be understood in its specific social, cultural, and economic context. Even within a nation, different types of gating in different historical periods can have rather different dynamics.

With a historical and cultural perspective, this chapter aims to understand gating better by studying different types of neighborhood enclosure in Chinese cities in comparison to that in the US. We argue that while each type of enclosed neighborhood has its own social, cultural, and historical backgrounds, they all embody the collectivism-oriented culture deeply embedded in Chinese society, and gates and walls help to define collectives and foster social cohesion and solidarity. In traditional Chinese cities, the *jiefang* system in urban planning and enclosed courtyard houses divided the city into walled compounds. Walls were built not only to delineate private property, but also to define collectives based on family/clan lineage and occupation, and a space where Confucian values and orders prevailed. In the socialist era, public housing was provided in the form of work-unit compounds with surrounding walls and guarded gates. In addition to economic and political reasons, gating helped to define socialist collectives based on common work-unit affiliation, and created a realm where work-units reigned and socialist collectivist living was promoted. Since the late 1980s, however, Western-style gated communities have mushroomed. The American-based theories such as the discourse of fear, privatization in governance and social services, and desire for exclusivity and prestige help to explain the emergence of these high-end gated communities. Yet we argue that collectivist tradition and culture continues to contribute to the prevalence of gating in the reform era, and multiple agents including the government, developers, and residents all prefer enclosed neighborhoods for their own interests. Instead of family or work unit connections, socioeconomic status, life style, and property-related interests are now the main factors for people to live collectively behind walls. Thus, despite profound socio-political changes in Chinese society, gating and neighborhood enclosure remain prevalent in the urban residential landscape.

This emphasis on collectivist tradition in China points to a very different social construction of gating from those in the West, which is centered on individualism, privatization, social segregation, and exclusion (e.g. Blakely and Snyder 1997; Brodie 2000; Caldeira 1996, 2000; Davis 1990). It illustrates how the built form of gated housing may look quite similar across nations, yet evolve from distinct historical, cultural, and architectural traditions. Understanding these distinct contexts generates new interpretations, both symbolic and political economic, of the growth and spread of gated communities. Yet, emphasizing the role of culture is not intended to replace or downplay Western theories in explaining gating. In fact, these approaches are not mutually exclusive, as the latter help to understand the rise of the latest wave of gating in Chinese cities. But we hope to contribute to the emerging literature on gating by adding the cultural and historical dimensions in explaining gating in Chinese cities.

After defining terminologies, we will review literature on gated communities in both China and the United States. Then we will examine gating in different historical eras in China, and findings and discussions will be provided in the final section.

Definitions and Literature Review

In this chapter a "gated community" is a walled or fenced housing development with secured and/or guarded entrances, to which public access is restricted. Inside the development there is often a neighborhood watch organization or professional security personnel, and there are often legal agreements (tenancy or leasehold) that tie residents to a common code of conduct (Blakely and Snyder 1997; Blandy et al. 2003). Gated communities vary in size from a few homes in very wealthy areas to as many as 21,000 homes in Leisure World in Orange County, California. Many include golf courses, tennis courts, fitness centers, swimming pools, lakes, or unspoiled landscape as part of their appeal, while commercial or public facilities are rare. This definition is based mainly on gated residential developments in the US; yet, similar residential developments have been proliferating in every continent in recent decades (Webster et al. 2002). According to Blakely and Snyder (1997), there are three types of gated communities in the US – lifestyle, elite, and security zone – each categorized by income level, amenities, aesthetic control, and location in the region.

While the use of gates and walls in residential development dates back to the ancient time, "gated communities" with their unique legal framework is a relatively new phenomenon. With their symbolic physical design and complex dynamics and urban impact, gated communities have generated many debates. While some scholars argue that gated communities are a response to the failure of urban governance and embody an innovative method of service provision (e.g. McKenzie 1994; Webster 2001), others are more concerned about its urban impact as it reinforces residential segregation, leads to the loss of public space, and introduces "cities of walls" and "urban fortress" (e.g. Caldeira 2000; Davis 1990; Foldvary 1994). The fear of crime (real and perceived) is often considered the main reason for people to live in gated communities; yet, gated communities may contribute to improved community safety through their physical design and like-minded neighbors (Blandy et al. 2003; Low 2003). They may also serve an important function in encouraging middle and upper income groups back into the inner cities. Thus gated communities have many economic, social, and political implications, which deserve scrutiny.

To understand gating better, we choose the US and China as our comparative cases, as gated communities are most popular in the US and most existing studies on gated communities are based on American experience. China has had a long history of using gates and walls in residential development, and is now experiencing a new wave of gating as its socialist housing system privatizes. Thus studying gating in the Chinese context in comparison to the American context should provide new insights on gating.

Scholars have been using the same term "gated communities" to refer to gated private housing developments in Chinese cities (e.g. Giroir 2005a, b; Miao 2003; Wu 2005; Wu and Webber 2004). However, the socioeconomic connotation of gated communities in the US is not applicable to many gated housing complexes in Chinese cities including private housing complexes for low-medium income households (affordable housing), work-unit compounds, and traditional housing complexes. In the context of China, the term "enclosed neighborhoods" is used for residential developments with gates and surrounding walls or fences. The entrances may be guarded by security personnel and/or card activated gates, or by the watchful eyes of senior members in the neighborhood. While some enclosed neighborhoods include amenities such as swimming pools and legal agreements among residents, others may have only a small patch of green land and no legal agreement. In other words, an enclosed neighborhood in China is a broader concept than a gated community in the US, and it emphasizes the physical form of enclosure, not the legal and social aspect, of the housing development. Gated communities in China refer to high-end private housing estates built in recent decades that share many similarities with gated communities in the US. Gated communities in China are usually much larger in population and land area, and have a more standardized layout than their American counterparts (Miao 2003).

There are different types of enclosed neighborhoods in Chinese cities: elaborate courtyard house complexes for extended families or clans in inner cities that were built in the pre-1949 period, work-unit compounds built in the socialist era for state employees in near suburbs, alleyway houses (*linong fang*) built in colonial Shanghai, and newly built private housing estates for the middle class and the elite mostly in suburbs. These enclosed neighborhoods are built in different socioeconomic contexts and have different dynamics. Despite the long history and the prevalence of neighborhood enclosure, there has been limited research on gating, and most of existing studies focus on private gated communities built in the past two decades. Scrutinizing the internal space of some of the most upscale gated communities such as the Purple Jade Villas in Beijing and the Fontainebleau Villas in Shanghai, Giroir (2005a, b) demonstrates how Western architectural and Chinese cultural and landscape elements are combined in these estates to create "golden ghettos" for the elite. Residents in these gated communities live a "club" like lifestyle that has very different social spaces from the rest of the urban society. Wu and Webber (2004) focus on gated communities built exclusively for foreign expatriates in Beijing, and argue they are a result of economic globalization and the unique housing system in China. As foreigners cannot access housing within the socialist housing system, private housing was built in gated communities to meet their housing needs. Comparing gated commodity housing enclaves with enclosed work-unit compounds, Wu (2005) argues that while gating is not new in China it has been "rediscovered" as an instrument for the partitioning of derelict socialist landscapes and a post-socialist imagined "good life." The shift toward market-oriented service provision is the distinguishing characteristic that Chinese gating shares with Western housing developments (Wu 2005), and increasing social inequality and security concerns during market transition contribute to the widespread gating (Miao 2003). To understanding gating in China better, we need to study different types of enclosed neighborhoods and understand their sociopolitical construction. After reviewing the literature on gated communities in the US, we will examine neighborhood enclosure in different historical eras in China.

Gated Communities in the US

Contemporary gated communities in the US first originated for year round living on family estates and in wealthy communities such as Llewellyn Park in Eagle Ridge, New Jersey built during the 1850s, and as resorts exemplified by New York's Tuxedo Park developed as a hunting and fishing retreat in 1886 (Beito 2002; Hayden 2003). Planned retirement communities such as Leisure World which emerged in the 1960s and 1970s, however, were the first places where middle class Americans walled themselves off. Gates then spread to resort and country club developments, and finally to suburban developments. In the 1980s, real estate speculation accelerated the building of gated communities around golf courses designed for exclusivity, prestige, and leisure. In 2001, the US Census Household Survey found that 7 million households (6 percent of all households), approximately 16 million people, were living in gated communities.

There is an extensive debate as to why gated communities have become so popular in the US. Arguments range from supply-side claims that financial benefits to developers, builders, and municipalities drive gating's success, to demand-side proposals that home buyers preferences and a discourse of fear are the principal motivating factors. From a broad political economic perspective, the gated community is a response to transformations in the political economy of late twentieth-century urban America (Davis 1990; Devine 1996; Low 1997). Racism is another major contributor to patterns of urban and suburban separation and exclusion in the US. Cities continue to experience high levels of residential segregation based on discriminatory real estate practices and mortgage structures designed to insulate Whites from Blacks (South and Crowder 1997).

Residents of middle-class neighborhoods also cordon themselves off as a class by building fences, cutting off relationships with neighbors, and moving out in response to problems and conflicts. At the same time governments have expanded their regulatory role through zoning laws, local police patrols, restrictive ordinances for dogs, quiet laws, and laws against domestic and interpersonal violence that narrow the range of accepted behavioral norms. Indirect economic strategies that limit the minimum lot or house size, policing policies that target non-conforming uses of the environment, and social ordinances that enforce middle-class rules of civility further segregate housing and neighborhood life (Merry 1993, p. 87).

The creation of "common interest developments" (CIDs) provided a legal framework for the consolidation of suburban residential segregation. CIDs describe "a community in which the residents own or control common areas or shared amenities," and that "carries with it reciprocal rights and obligations enforced by a private governing body." (Judd 1995, p. 155). Specialized "covenants, contracts, and deed restrictions" that extend forms of collective private land tenure and the notion of private government were adapted to create the modern institution of the homeowner association in 1928 (Judd 1995; McKenzie 1994).

Gated communities are considered "new spatially defined markets in which innovative neighborhood products are supplied by a new style of service producer." (Webster 2001). For people living inside gated communities it is an efficient economic solution because of the legal requirement to pay fees, homogeneity of community needs and desires, and because residents can choose their package of communal goods according to their personal preferences. Rather than being "taxpayers" who pay for public goods and services that are not always available, they become "club" members who pay fees for private services shared only by members of community. Supply-side economic factors figure prominently in understanding the widespread expansion of these communities. Developers want to maximize profits by building more houses on less land and incentive zoning packages for common interest development housing allow clustering units to achieve this higher density (McKenzie 1998).

Crime and the fear of crime also have been connected to the building of gated communities (Low 2003). Oscar Newman, who uses the term "defensible space" for this idea, argues that high-rise buildings are dangerous because the people who live in them cannot defend – see, own, or identify – their territory. He proposes that gating city streets can promote greater safety and higher home values as long as the percentage of minority residents is kept within strict limits. Unfortunately, there is no evidence that homes in gated communities maintain their value better than those in non-gated ones in comparable suburban areas (Blakely and Snyder 1997). Nor is there evidence that gated communities are safer, although in Southern California they are associated with perceptions of greater safety among the upper middle class (Wilson-Doenges 2000). What studies identify, though, is a desire for security and safety even in the face of low crime rates in suburban areas in the US (Low 2003). There is an emphasis on a discourse of fear that legitimates and rationalizes the desire to live behind gates, and the use of privatization to maintain greater social control and a redistribution of public goods. In addition, some gating is pursued purely for its symbolic values of prestige and exclusivity (Romig 2005).

Collectivism and Neighborhood Enclosure in Urban China

While scholars argue that gated communities in China and other developing countries are a result of globalization and increasing social inequality and segregation (Giroir 2005a, b; Miao 2003; Wu and Webber 2004), enclosed neighborhoods have been a widespread phenomenon in Chinese cities since the beginning of urban history. Walls and gates have always been integral components of urban construction in China, such that the same word, *cheng*, is used for both walls and cities (Knapp 2000). While walls serve multiple purposes, such as defense, protection from harsh weather, markers for private property, and tools demonstrating imperial authority and Chinese cosmology, they are also "symbolic markers separating inside and outside, family and stranger" (Knapp 2000, p. 1).

In Confucian China, collectives have always been more important than individuals, and collective identification and group recognition are highly pursued by Chinese. The physical walls surrounding residential compounds help to define these "collectives" for people living inside. In addition, the space inside the walled compound is organized in such a way to promote the production of "collective-oriented subjectivities" (Bray 2005, p. 13). Enclosed neighborhoods allow people to identify themselves with a group other than their immediate nuclear families, whose protection and help can be essential especially during difficult times. The code of conduct for members within the collective and the social governance in the enclosed space are often different from those outside. Thus a collective identity is developed for members within, which distinguishes them from the rest of the society and reinforces cohesion and solidarity within the collective. Despite various socioeconomic transformations in Chinese history, collectivism remains central, and walls have maintained their symbolic values in defining collectives although what constitutes a collective has changed over time.

The physical form of cities, the philosophy of urban planning, and the basic principles of residential development have remained essentially unchanged throughout Chinese history (Knapp 2000; Sit 1995; Skinner 1977; Zhang 2002). In contrast to Western cities that may serve as the museums of changing architectural styles, "the cultural landscapes of China are remarkably ahistorical" (Knapp 1990, p. 1). In the following sections, we examine neighborhood enclosure in three major historical periods: the pre-socialist, the socialist, and the reform eras.

The pre-socialist era (pre-1949)

Despite the long history of urban development, the *jiefang* system composed of courtyard houses had been widely implemented in Chinese cities since the Shang (1700–1027 BC) and Zhou dynasty (1027–771 BC) till the Ming (AD 1368–1644) and Qing (AD 1644–1911) dynasty (Zhang 2002). With large avenues and narrow lanes, the *jiefang* system divided the residential area into smaller wards, called *fang*, whose size varied from 400 × 400 m to 800 × 800 m. Each *fang* was wrapped with tamped-earth walls and guarded with gates and sentries (Knapp 2000; Zhang 2002). People living in the same or nearby *fang* often had similar occupations, forming enclosed neighborhoods with occupational homogeneity but with significant personal-wealth heterogeneity (Belsky 2000; Skinner 1977). While the physical form of the *jiefang* system discouraged interactions between "insiders" and "outsiders," it helped to foster a strong sense of internal community and establish an identity often based on occupation through the popular guild system.

Within each *fang*, courtyard houses further divided the space into enclosed units often based on family/clan lineage. The courtyard house was a traditional, rectangular housing complex, with multiple bungalows and walls surrounding a courtyard in the center (Figure 8.1). Wealthier families often had large courtyard houses that might contain several smaller ones such as the well-known Qiao Family Manor in Pingyao, Shanxi province (Knapp 2000) and the Forbidden City in Beijing. The walled enclosure was one of the four basic design principles of Chinese architecture (Boyd 1962; Liu 1990). Even in coastal cities that were under foreign influence since the middle of the nineteenth century, new housing development continued to reflect the traditional courtyard house design. For example, in Shanghai, rapid industrialization during the 1840s to 1940s had led to massive construction of alleyway house, which accounted for 72 percent of the city's dwellings by the end of the 1940s

Figure 8.1 A recently renovated courtyard house in downtown Beijing

(Lu 1999). These multi-story houses were built in rows, with surrounding walls and often a stone-framed entrance to form an enclosed compound (Lu 1999).

In addition to delineating private property and providing protection, the surrounding walls of courtyard houses helped to define the extended families living inside as a distinctive collective. For example, the walls of the Qian Family Mansion clearly defined the Qiao family as a prosperous collective in Pingyao, while those for the Forbidden City were not only to demonstrate the emperor's authority and legitimacy to rule, but also to define a collective superior to the rest of the population. Furthermore, the surrounding walls defined a spatial realm of Confucian family where Confucian values were embodied in spatial and social practices (Bray 2005).

In summary, traditional Chinese cities had been characterized by enclosed courtyard houses within walled *fang*. While the physical walls and gates divided and fragmented urban space, they helped to define the spatial realm of a Confucian family, form occupation and family/clan-based collectives, establish a collective identity, and foster extremely close ties among members. There were high expectations for members inside to help each other and fulfill guild and family/clan obligations. Thus, in pre-socialist China, neighborhood enclosure was a result of guild and family/clan-based collectivism deeply embedded in Chinese culture. Despite spatial division, there was enough economic heterogeneity within enclosed neighborhoods, and the socioeconomic disparity between enclosed neighborhoods was not obvious.

Socialist urban China (1949–87)

Since the Chinese Communist Party (CCP) took power in 1949, the urban transformation has been profound and multifaceted; yet, gating and enclosure, albeit in a different form, have remained in residential developments. Faced with severe housing shortage, poor maintenance, and unequal housing distribution, the socialist government launched the Socialist Transformation of private housing, and invested heavily in the construction of public housing (Wang and Murie 1999; Zhang 1998). The Socialist Transformation was to convert most existing private housing into public, and then lease it to households with nominal rents (Zhang 1998). Courtyard houses that previously were occupied by extended families, now were often shared by unrelated families. Because of poor housing conditions, households often had to share private space such as kitchens in addition to public space such as the courtyard. With intensive daily interactions and proximity, households in the same and nearby complex knew each other very well. While the lack of privacy was often a source of complaint, neighbors often developed a strong sense of connection with each other and a pseudo-family/clan relationship (e.g., seniors taking care of neighbor's children, the young helping the senior, borrowing/lending money, and sharing food), despite the lack of actual family/clan ties and occasional interest conflicts. The sense of community is usually the strongest among households in courtyard houses compared to the rest of the city.

At the same time, new apartment buildings were built mostly in suburbs to meet growing housing needs. With the socialist philosophy of combining work and residential space, public housing complexes were often built next to large employment centers such as factories and government agencies, forming so-called work-unit compounds (Ma 1981; Sit 1995). These compounds were mostly enclosed territories with surrounding walls, and provided not only housing but also public services such as clinics, schools, grocery stores, canteens, and public bathhouse to employees from the same work unit (Bian et al. 1997) (Figure 8.2). They were also guarded, some with uniformed security personnel, and others with vigilant senior residents watching at the complex entrances or nearby convenience stores. Even in workers' villages that provided shelters to employees from different work units and where restricted entrance was hard to enforce, a symbolic gate was often built (Wu 2005) and internal roads were deliberately enclosed to prevent through traffic.

As urban land was owned by the state and allocated to work-units, walls were often built by work-units to mark the boundary of the territory and

Figure 8.2 A work-unit compound in Shenzhen

claim their "ownership" of the property before housing and facilities were constructed. Yet, walls were maintained after the construction period for both practical and symbolic purposes; for example, walls were used to reduce external traffic, and restrict outsiders' access to facilities, more stringently in some work-unit compounds (e.g., military and party) than others. Walls were also used to economize urbanization given the lack of resources and emphasis on production in the socialist era (Wu 2005). More importantly, walls had symbolic values in creating a unique space, promoting a collective lifestyle, and defining a collective identity. Just like the surrounding walls of courtyard houses that defined the realm of the Confucian family within which the family patriarch ruled, the surrounding walls of work-unit compounds marked the realm of work-units within which work-units reigned (Bray 2005). Virtually every aspect of private life (e.g., employment, marriage, divorce, children's education and employment) needed approval from one's work-unit, and even neighborhood disputes and household incidents were brought to and resolved by the leaders of the work-unit. In addition to surrounding walls, there are many crucial resemblances in terms of structure and functions between the traditional extended Confucian family (*jia*) and work-units, of which one of the most important aspects is the duty of care to group members (Bray 2005). Work-units are the "public family" (*gong jia*) that operated to guarantee the livelihood and welfare of their members.

The surrounding walls of work-unit compounds also created a space that promoted the socialist collectivized lifestyle, which is achieved spatially through purposefully arranged housing and common facilities within the enclosed compounds (Bray 2005). The collectivist life within work-units compounds were arranged at different levels. At the most basic level of collectivity, every three to five families shared toilets and kitchens within each basic housing unit (*danyuan*). At the next level, each two or three buildings shared facilities like laundries, bicycle sheds, and open space, and at the work-unit level, all residents shared facilities like canteens, clinics, and kindergartens. Thus people living in these compounds were not only colleagues with intensive interactions at the workplace but also neighbors with close contacts at home. People knew each other well, and the sense of community was strong. Because of similar lifestyles, a unique micro-culture often emerged, especially in the large compounds for government ministries and departments in Beijing, forming so called "big compound culture" (*dayuan wenhua*).

Because work-units provided not only employment but also a wide range of social, political, and civil functions (He 1993), they have become the principal source of identity for urban residents. People living in the same work-unit compounds shared a strong attachment to their work-unit, and a collective identity was often formed among them. "We live in the same work-unit compound" was often used to indicate the strong connection and comradeship between members. The physical walls and gates helped to strengthen the collective identity among its residents, and project the image of collectivity and solidarity to outsiders.

In summary, with a significant resemblance to courtyard houses for Confucian families, work-unit compounds added a new layer of gating to Chinese cities, creating enclosed residential and social spaces based on occupation and industry. As a result of socialist urban planning and housing development, collectivism was transformed by the socialist ideology, and work-unit attachment became the basis for people living collectively behind walls.

Chinese cities under housing reform (1988–)

In 1988, China launched a nationwide housing reform to privatize the welfare-oriented housing system. While existing work-unit housing were sold with heavy subsidies, massive private housing has been developed especially in the suburbs in the form of *xiaoqu*, literally small districts or residential quarters. "Residential quarters" are planned neighborhoods where private housing is provided with communal facilities (Bray 2005). Ranging from the low-end affordable housing to high-end villa complexes, virtually all "residential quarters" are walled and guarded, and many are equipped with security and monitoring system such as intercoms, surveillance cameras, infrared alarm systems, and card activated entrances (Figure 8.3). "Enclosed residential quarters" (*fengbi shi zhuzai xiaoqu*) are sanctioned by national planning codes and have become the basic unit in planning and developing residential construction (Miao 2003). Even in older areas, urban renewal programs are blocking off minor streets to create enclosed communities. In 2000, the municipal government of Beijing issued an order requiring all residential quarters to be developed with gates (www.cmen.net, 2002). According to Miao (2003), 83 percent of all residential communities in Shanghai underwent some form of physical separation in the 1990s, and by year 2000 there were 54,000 gated residential communities in Guangdong province.

Scholars have identified these private residential quarters, especially those at the higher income level, as "gated communities," and attribute them to similar factors as in the US (Giroir 2005a; Miao 2003; Wu 2005). With increasing social inequality, influx of migrants, and unprecedented crime rates reported during the market transition, households are increasingly concerned about security. Gating is considered by both the government and the public as a way to reduce crimes, although there is a debate as to whether gating actually increases crime protection (Miao 2003; Su 2000; Wang, Jiang, and Gu 2001). At the same time, with the demise of the work-unit and the end of public housing provision, households affiliated with different work-units are now living in private housing estates. Public services in these private

Figure 8.3 Guarded entrance of the Purple Jade Villa, an upscale gated community in Beijing

communities, such as street cleaning, refuse disposal, gardening, and security that were provided by the government or work-units, are now provided by private developers or professional Property Management Companies (PMCs) (*wuye guanli gongsi*). Many upscale private housing estates also provide their residents exclusive facilities such as tennis courts, swimming pools, and gyms. Developers often use catch-phrases such as "24-hour professional management" and "5-star hotel management" in advertisements, emphasizing prestige and luxury to attract residents. There is no doubt that Western-based discourses – the discourse of fear, the club realm of services, and the pursuit of prestige and luxury – contribute to this new wave of gating.

Despite significant changes in housing development, these private housing estates also represent a continuity of neighborhood enclosure in the past, and the collectivist tradition continues to play an important role in this new wave of gating. With the dominance of nuclear families in Chinese cities and the end of housing provision in the form of work-unit compounds, family/clan lineage and work-unit affiliation can no longer force people to live together collectively. With increasing inequality and social-spatial sorting (Huang 2005; Wang and Murie 2000), common socio-economic status, life style and property-related interests have become the key to attracting people into different types of gated communities, and thus form territorial collectives. Despite a new socioeconomic context, walls and gates continue to be the preferred planning tools to promote collectivist living. Although being privately produced and with no connection to workplace, "residential quarters" bear a resemblance to work-unit compounds. "Residential quarters" are designed with communal facilities such as kindergartens, clinics, restaurants, convenience stores, and sports facilities, forming self-sufficient communities, similar to work-unit compounds. Designers for "residential quarters" focus particularly on the communal spaces of the compound, aiming to promote attributes like social cohesion, neighborliness, and feelings of security and belonging (Zou 2001).

In addition to the physical design, the Ministry of Civil Affairs (MCA) launched a new campaign of "*shequ* building" (*shequ jianshe*) in 2000 to foster a stronger sense of community among residents with the demise of the work-unit system, the widespread of enclosed private communities, the emergence of an increasingly liberal society, and high unemployment and crime rates. According to MCA (2000), a "*shequ*" is "a social collective (*shehui gongtongti*) formed by people who reside within a defined and bounded district." In contrast to the emphasis on interpersonal relations embodied in the concept of "community," the "*shequ*" is a clearly demarked territory. *Shequ* Committees and *Shequ* Service Centers are set up for one or several enclosed communities to deliver social, welfare, health, and administrative services (MCA 2000). In the city of Shenyang, *Shequ* Service Centers have been built as close as possible to the physical centers of the *shequ* territories that they serve (Bray 2005). This seems to be a deliberate spatial strategy to convey the message that residents now should consider *shequ*, instead of their work units, as their new collective home. Through this campaign, *shequ* has replaced the work-unit as the basic unit of urban life for everything apart from employment. While it is debatable whether this campaign has been successful, it is clear that the government wants to promote territorial collectives to serve and control its population, when the strong foundation for collectives in previous eras disappears.

Second, instead of clans/guilds and work-units, developers and PMCs now play important roles in defining and forming collectives. In addition to housing price that "sorts" households into different neighborhoods, developers often create a unique identity attached to their estates to attract households who are choosing from many gated communities with similar price ranges. Socioeconomic status, architectural design, and exclusive amenities are often used to promote unique collective identities and lifestyles. For example, the SOHO New Town in Beijing stands out among numerous estates with colorful high-rise apartment buildings, multifunctional spaces, brand-name stores and art shops, fancy restaurants, teahouses and cafes, and a fully equipped club. The developer projected an American urban lifestyle for the emerging young professionals in Beijing by creating neighborhoods similar to the SoHo area in Manhattan (New York City) (http://www.sohochina.com). In contrast, the developer of Purple Jade Villas, an upscale gated community with multi-million-dollar villas in central Beijing, created a luxury but heavenly peaceful oasis amidst the chaotic urban environment. In addition to modern amenities such as swimming pools, a fitness center, spa, and an ice rink, there are two artificial lakes decorated with lotuses and weeping willows, a hiking trail along an artificial mountain with several water falls, a central grand lawn, and numerous exotic animals such as peacocks, swans, and pheasants wandering around. The Purple Jade Villa offers the richest a retreat to the nature and an idealized Chinese country life without leaving the heart of Beijing.

Third, there is a strong desire from residents to form collectives especially when their property-related interests are under threat. Most homeowners in Chinese cities are first time homeowners whose strong proprietary attitude often leads to collective actions to protect their property-related interests. For example, the developer of Fulun Homes, a gated community for medium-income households in Beijing, sold a facility building in the compound that was originally planed for a daycare center without consulting homeowners, and the new owner converted it into a spa center with the approval from the Beijing Planning Committee for commercial use (Cai 2004). Residents were very angry about the change because not only they could no longer enjoy the daycare center but also there would be heavy traffic and "undesirable" outsiders coming into the compound for the spa service. Even though the Homeowners' Association (HOA) (*yezhu weiyuanhui*) had not been established then in this neighborhood, 149 homeowners organized themselves together and sued the developer and the Beijing Planning Committee, and won. In addition to traditional meetings and phone calls, new communication tools such as internal BBS and email listserv served important roles to inform residents and organize actions. This case demonstrates that homeowners in enclosed neighborhoods can form a strong collective to protect their property interests and life style, even though they do not share a common organizational basis such as clan lineage and work-unit affiliation, they do not know each other before moving in, and sometimes there are no official grassroots organizations such as HOAs to represent them. Of course, HOAs have been established in most private communities, and most homeowners join HOA to protect their property interests (Read 2003). Housing disputes and conflicts mostly between homeowners and developers and PMCs have increased significantly in recent years. The collective nature of housing interests and the relatively autonomous space in these enclosed neighborhoods have provided an active catalyst in the formation of collective identities, which come with the right to protest (within the walled complexes) and responsibilities of self-government (Tomba 2005).

In summary, gating has become more popular since housing reform. The increasing concern about security, the privatized provision of public services, and the desire for prestige and luxury are undoubted factors. Yet, collectivism and collective identities, now based mainly on common socioeconomic status, life style, and interests in property rights, continue to bring households together to live behind walls. In addition to residents' desire to protect their own interests and life style, collectivist living is also promoted by housing providers, for marketing purposes, on the one hand and the government, for social control, on the other hand. In other words, despite dramatic changes in the housing system and significant transformation in the basis for collectives, collectivist tradition and culture continues to contribute to gating in the reform era. Thus the transition from living in work-unit compounds to private enclosed residential quarters does not denote a progression from socialist collectivity to individuated identity, but rather "a transition from one form of collectivity to another" (Bray 2005, p. 185).

Conclusions and Discussion

In recent decades, there has been a boom of gated housing development in Chinese cities, which seems to echo the "global spread" of gated communities. Yet, there are different types of gated housing complexes in China that have very different social constructions, and the concept of enclosed neighborhoods is more appropriate in the Chinese context. Using a historical perspective, we examined different types of enclosed neighborhoods built in different historical periods in Chinese cities, and argued that the collectivist tradition and culture deeply embedded in Chinese society contributes to the prevalence of gating in China, despite the fact that the concept of collectives has experienced significant transformations over time. This is not to downplay other factors in gating; rather we hope to contribute to the literature by adding the cultural and historical dimensions to the understanding of gating.

Gating is another example demonstrating the uniqueness of Chinese cities in the sense that they are different from other Third World cities and Western cities. While none of four theories outlined in the introduction chapter can fully explain the gating in Chinese cities, the market transition theory can shed some light to the latest wave of gating, which shares many characteristics with gated communities in the West; yet, it also represents continuities with the past. With the rapid privatization in housing and service provision and neighborhood governance, the increasing social inequality, and rising crime and unemployment rates in the reform era, gating is pursued by both developers for marketing purposes (supply side), and by residents for security and better services (demand side). At the same time, as with other aspects of the reform, gating shows reminiscences of the socialist era. The new unit of housing development – residential quarters – bears resemblances with the socialist work-unit compounds as they are also enclosed housing compounds with integrated communal services to promote collectivist living. Furthermore, the government continues to organize its population into territorial collectives through "*shequ* building," and new institutions such as *Shequ* Committees are not very different from Residents' Committees in the socialist era. Yet, the current gating and neighborhood enclosure in Chinese cities are much more complex than a result of market transition. It has a much deeper historical and cultural root, in addition to political and economic forces in the specific historical period. The collectivism embodied in gating and neighborhood enclosure goes much further back beyond the socialist era such that a historical approach is needed to understand better the gating phenomenon in China.

It is clear from our findings that gating and neighborhood enclosure in China differs in both social construction and social implication from gated communities in the US based on profound differences in cultural, social, and architectural history. This comparison has enabled us to conclude that the dominant discourses drawn from the US experience, the club realm of service provision, fear of crime and others, and the pursuit of prestige and luxury, cannot fully explain the complexity of enclosed neighborhoods in the Chinese context. However, there is evidence suggesting that they are becoming increasingly important in explaining newly constructed gated communities in Chinese cities. While stressing the uniqueness, long history, and changing dynamics of gating in the Chinese context, we acknowledge the fact that there is a certain degree of imitation of housing development in China, and Western style gated communities, with a twist of Chinese flavor, have probably reached walled Chinese cities with the rapid globalization. Yet, the relationship between American gated communities and Chinese enclosed neighborhoods is probably a two-way street, with each borrowing something from the other to serve local needs. For example, gates, walls and enclosed neighborhoods have always existed in China, but not on a massive scale in the US until recently; gating has been used in the US to promote privacy and reinforce social segregation, while it has been used in China to define collectives and foster social solidarity, and only recently is there evidence that walls begin to separate people in different classes and result in nascent residential segregation. In other words, in the US, a mostly wall-free country to begin with, walls are adopted to solve the social problems emerged with increasing social inequality and ethnic diversity, the withdrawal of the welfare state, and the transformation of its political economy. In China where cities began with walls and gates, the social construction and implication of walls have been very different from those in the US, and have changed over time. Yet now class and lifestyles similar to those in the States are becoming part of the social construction of walls and gating. Thus, with rapid globalization and market transition in China, a certain degree of convergence in housing development and use of walls and gates is taking place between China and the US; yet, with the persisting collectivist tradition and culture, gating and neighborhood enclosure in China are also local in meaning and production.

REFERENCES

Beito, D. (2002) "The Private Places of St. Louis." In D. Beito, P. Gordon, and A. Tabarrok (eds.), *The Voluntary City*. Ann Arbor: University of Michigan Press.
Bejing Bureau of Civil Affairs (BBCA). (2003) *Beijing shi xingzhen quhua* [The Administrative Division in Beijing]. Beijing: China Society Press.
Belsky, R. (2000) The Urban Ecology of Late Imperial Beijing Reconsidered: The Transformation of Social Space in China's Late Imperial Capital City." *Journal of Urban History* 27: 54–74.
Bian, Y., J.R. Logan, H. Lu, Y. Pan, and Y. Guan. (1997) "'Working Units' and the Commodification of Housing: Observations on the Transition to a Market Economy with Chinese Characteristics." *Social Sciences in China* 18, 4: 28–35.
Blakely, E.J., and M.G. Snyder. (1997) *Fortress America: Gated Communities in the United States*. Washington DC: Brooking Institute.
Blandy, S., D. Lister, R. Atkinson, and J. Flint. (2003) "Gated Communities: A Systematic Review of the Research Evidence." Unpublished report presented at Gated Communities: Building Social Division or Safer Communities? University of Glasgow, September, 18–19.
Boyd, A. (1962) *Chinese Architecture and Town Planning: 1500 BC–AD 1911*. London: Alec Tiranti.
Bray, D. (2005) *Social Space and Governance in Urban China: The Danwei System from Origins to Reform*. Stanford, CA: Stanford University Press.
Brodie, J. (2000) "Imagining Democratic Urban Citizenship." In E.F. Isin (ed.), *Democracy, Citizenship and the Global City*. London: Routledge.
Cai, F. (2004) "Finally There is a Constraint to the Widespread Disease of Change in Plan (Biangen jihua de liuxinbin zhongyu youle kexing)." Editorial, *Beijing Youth Daily* (Beijing Qinnian Bao), Sunday, July 11: A4.
Caldeira, T. (1996) "Fortified Enclaves: The New Urban Segregation." *Public Culture* 8: 303–28.
Caldeira, T. (2000) *City of Walls: Crime, Segregation, and Citizenship in São Paulo*. Berkeley: University of California Press.
Davis, M. (1990) *City of Quartz: Excavating the Future in Los Angeles*. London: Verso.
Devine, J. (1996) *Maximum Security*. Chicago: University of Chicago Press.
Foldvary, F. (1994) *Public Goods and Private Communities*. Cheltenham: Edward Elgar.
Giroir, G. (2005a) "The Purple Jade Villas (Beijing), a Golden Ghetto in Red China", in K. Frantz, G. Glasze, and W. Webster (eds.), *Private City: the Global and Local Perspective*. London: Routledge.
Giroir, G. (2005b) "The Fountainebleau Villas (Shanghai), A Gloden Ghetto in a Chinese Garden." In F. Wu (ed.), *Globalization and the Chinese City*. London: Routledge.
Hayden, D. (2003) *Building American Suburbia*. New York: Pantheon.
He, X. (1993) "People in the Work-unit [Ren zai danwei zhong]." In Yanxiang Shao and Xianzhi Lin (eds.), *People and Prose* [*sanwen yu ren*]. Guangzhou: Huachen Press: 157–66. (Translation in Dutton, M. (1999) *Streetlife China*. Cambridge: Cambridge University Press: pp. 42–53.)
Hu, X., and Kaplan, D.H. (2001) "The Emergence of Affluence in Beijing: Residential, Social Stratification in China's Capital City", *Urban Geography* 22, 1: 54–77.

Huang, Y. (2005) "From Work-unit Compounds to Gated Communities: Housing Inequality and Residential Segregation in Transitional Beijing." In L. Ma and F. Wu (eds.), *Restructuring the Chinese Cities: Changing Society, Economy and Space*. London and New York: Routledge: pp. 192–221.

Judd, D. (1995) "The Rise of New Walled Cities." in H. Ligget and D.C. Perry (eds.), *Spatial Practices*. Thousand Oaks, CA: Sage.

Knapp, R.G. (1990) *The Chinese House: Craft, Symbol, and the Folk Tradition*. Hong Kong: Oxford University Press.

Knapp, R.G. (2000) *China's Walled Cities*. New York: Oxford University Press.

Liu, Z. (1990) "A Brief History of Chinese Residential Architecture [*zhongguo zhuzai jianzhu jianshi*]." Beijing: China Architecture Industry Press [*zhongguo jianzhu gongye chubanshe*].

Low, S. (1997) "Urban Fear: Building Fortress America." *City and Society Annual Review*, pp. 52–72.

Low, S. (2003) *Behind the Gates: Life, Security and the Pursuit of Happiness in Fortress America*. New York: Routledge.

Lu, H. (1999) *Beyond the Neon Lights: Everyday Shanghai in the Early Twentieth Century*. Berkeley: University of California Press.

Ma, L.J.C. (1981) "Introduction." in L. Ma and E.W. Hanten (eds.), *Urban Development in Modern China*. Boulder, CO: Westview Press: pp. 1–18.

McKenzie, E. (1994) *Privatopia*. New Haven: Yale University Press.

McKenzie, E. (1998) "Homeowner Association and California Politics." *Urban Affairs Review* 34, 1: 52–75.

Merry, S. (1993) "Mending Walls and Building Fences: Constructing the Private Neighborhood." *Journal of Legal Pluralism* 33: 71–90.

Miao, P. (2003) "Deserted Streets in a Jammed Town: The Gated Community in Chinese Cities and Its Solution." *Journal of Urban Design* 8, 1: 45–66.

Ministry of Civil Affair (2000) A *Suggestion on Promoting Nationwide Community Building*, document no. 23.

Ministry of Construction (1994) *Methods for Managing New Urban Residential Neighborhoods*, document no. 33.

Read, B.L. (2003) "Democratizing the Neighborhood? New Private Housing and Home-Owner Self-Organization in Urban China." *The China Journal* 49: 31–59.

Romig, K. (2005) "The Upper Sonoran Lifestyle: Gated Communities in Scottsdale, Arizona." *City and Community* 4, 1: 67–89.

Sit, V.F.S. (1995) *Beijing: The Nature and Planning of a Chinese Capital City*. John Wiley & Sons: Chichester.

Skinner, G.W. (1977) "Introduction: Urban Social Structure in Ch'ing China." In W.W. Skinner (ed.), *The City in Late Imperial China*. Sandford: Sandford University Press.

South, S., and K.D. Crowder. (1997) "Residential Mobility between Cities and Suburbs: Race, Suburbanization, and Back-to-the-City Moves." *Demography* 34, 4: 525–38.

State Council (2003) *Wuye guanli tiaoli* [*A Regulation on Property Management*], no. 379. Beijing: Law Publishing House.

Su, N. (2000) "Rang Baixing You Yige Anquan De Jia [Let residents have safe homes: report on the project of creating safe neighborhoods in Shanghai]." *People's Daily*, June 14.

Tomba, L. (2005) "Residential Space and Collective Interest Formation in Beijing's Housing Disputes." *The China Quarterly* 184: 934–51.

Wang, Y.P., and A. Murie. (1999) *Housing Policy and Practice in China*. London/New York: Macmillan.

Wang, Y.P., and A. Murie. (2000) "Social and Spatial Implications of Housing Reform in China." *International Journal of Urban and Regional Research*, 24: 397–417.

Wang X., L. Jiang, and Y, Gu. (2001) "Shihua xiaoqu zhian [Security at Shihua Residential Quarter]." Online at: http://www.sdrfz.jsol.net/ssxz/shxq.html (accessed March, 17 2007).

Webster, C. (2001) "Gated Cities of Tomorrow." *Town Planning Review* 72: pp. 149–69.

Webster, C., G. Glasze, and K. Frantz (2002) "The Global Spread of Gated Communities." *Environment and Planning B* 29: 315–20.

Wilson-Doenges, G. (2000) "An Exploration of Sense of Community and Fear of Crime in Gated Communities." *Environment and Behavior* 32, 5: 597–611.

Wu, F. (2005) "Rediscovering the 'Gate' under Market Transition: From Work-unit Compounds to Commodity Housing Enclaves." *Housing Studies* 20, 2: 235–54.

Wu, F. and K. Webber (2004) "The Rise of 'Foreign Gated Communities' in Beijing: Between Economic Globalization and Local Institutions." *Cities*, 21, 3: 203–13.

www.cment.net, (2002) "Wele Anquan, youguan Bumen Yaoqiu Jumin Xiaoqu Ying Fengbi [The Government Demands Residential Quarters to be Enclosed for Safety]." Online at: http://www.cment.net/docc/news-034.htm (accessed Feb. 22, 2007).

Zhang, X.Q. (1998) *Privatization: A Study of Housing Policy in Urban China*. New York: Nova Science Publishers.

Zhang, Y. (2002) *Zhongguo chengchi shi* [*The history of cities in China*]. Tianjin: Baihua Literature and Art Publishing House.

Zou, D. (2001) *A History of Modern Architecture in China* [zhongguo xiandai jianzhu shi]. Tianjin: Tiangjin Science and Technology Press [Tianjin kexue jishu chuban she].

Part III
Impacts of Migration

9

Urbanization in China in the 1990s: Patterns and Regional Variations

Zai Liang, Hy Van Luong, and Yiu Por (Vincent) Chen

Introduction

Demographically speaking, the 1980s and the 1990s were the best of times for migrants in China. After several decades of control and retrenchment, Chinese cities are experiencing an unprecedented increase in their migrant populations. From the most desirable migrant destinations of Beijing, Shanghai, and Guangdong to remote areas of Xinjiang and Heilongjiang, migrants are everywhere and changing the face of Chinese urban landscape. In response to the large volume of migrant population, students of migration in China have been studying different aspects of this great wave of migration (Goldstein, Goldstein, and Guo 1991; Hare 1999; Liang and Ma 2004; Roberts 1997; Yang 1994). Since a large volume of China's recent migrant population is characterized by rural to urban migration, migration clearly has consequences for China's urbanization patterns. Indeed, in the words of Rogers and Williamson (1984, p. 262), "urbanization is attributable to *mobility revolution*, the transformation experienced by societies with low migration rates as they advance to a condition of high migration rates."

Similar to the study of migration, for a long time, research on urbanization in China was hampered by the lack of systematic and adequate data. With the release of results from China's 1982 and 1990 Population Censuses (and more recently the 2000 census), along with other data on China's urban statistics, studies of Chinese urbanization have gained a significant momentum. Previous studies of Chinese urbanization cover a wide-range of issues and clarified much confusion regarding data and definitions of Chinese urbanization and provided many benchmarks from which future researchers can draw. Chan (1994), Ge (1999), and Kirby (1985) published in-depth historical overviews of urbanization patterns and dynamics dating back to 1949. Goldstein (1985) provided a detailed analysis of characteristics of rural and urban population and levels and patterns of urbanization. Other important scholarly work includes analysis of regional patterns of urbanization focusing on the Pearl River Delta region (Guildin 1992; Wang 1996; Xu 1990); a demographic accounting of China's urban growth (Chan 1994; Chan and Hu 2003; Chen, Valente, and Zlotnik 1998); as well as strategic discussions and planning of China's future roads of urbanization (Ge 1999; Wang and Sun 1997).

In this chapter, capitalizing on the recent data from the 1990s, we provide an overview of regional patterns of urbanization and focus on the demographic sources of urbanization in China. The central question we address is the extent to which China's recent urbanization can be explained by migration/boundary reclassification vs. urban natural increase. Our study is expected to contribute to the current literature on urbanization in China in several aspects. First, in some of the most influential international studies of demographic sources of urbanization, China is often excluded because of lack of data. For example, in Preston's 1979 study, China was not included. This changed as more data from China became available. The recent study by Chen et al. (1998) used data from the 1982 and 1990 China Population Censuses, but not after. Our study will update Chen et al.'s (1998) work by using data from the 1990s.

Second, most studies of demographic sources of urban growth by and large use a country as the unit of analysis. This makes sense if a country is relatively homogeneous. However, given the large regional variation in socioeconomic development, migration, and fertility patterns in China, we plan to carry out this exercise one step further, i.e. to examine urbanization patterns at the province level. Zhou and Ma (2003) suggested that data from different provinces may not be comparable because provinces may have different ways of adopting urban administrative changes. If so, it would give more justification for conducting this exercise at the province level.

Third, the current exercise also has theoretical significance because of its potential contribution to theorizing the experience of urbanization in different parts of the world, especially the developing world. For example, recent work by Lee and Feng (1999) challenges the traditional Malthusian notion of Chinese demographic behaviors (universal marriage, high fertility, and high mortality), which has been taken for granted until recently. By the same token, Preston's 1979 article has been widely cited as accepted wisdom of describing urbanization experience of developing countries (also see Brockerhoff, 2000). However, for lack of data, the article does not include China. Furthermore, the article was written 25 years ago during which time fertility dropped significantly in many parts of the world especially in Asian countries. Moreover, like other chapters in this volume, we also plan to address the extent to which following four theoretical perspectives are relevant to explain the experience of urbanization process in China: modernization theory, dependency and world system theory, theory of the developmentalist state, and transition from socialism.

Finally, in light of this new thinking on patterns of urbanization in less developed countries, we will use some recent data from Southeast Asia and especially Vietnam to demonstrate that the migration-dominated pattern of urbanization is by no means unique to China, it is happening in other parts of Asian countries as well.

Regional Patterns of Urbanization in China, 1990 and 1995

Unlike many other developing countries, there are tremendous regional variations in China. To begin with, China's population is unevenly distributed, which has been noted by many scholars (Goldstein 1985). Aside from population, there is also important variation in socioeconomic development. With transition to a market-oriented economy, regions with easy access to markets and capital are able to progress quickly whereas other regions either remain static or fall further behind. The well-known story is the inequality between coastal and non-coastal regions. This inter-regional inequality has clear implications for urbanization. One major pattern of migration that characterizes the 1990s is the large concentration of migrants in the coastal regions, especially coastal cities. For example, since 1987, migration to the coastal region has accelerated, especially to cities (Liang and Ma 2004). Thus, we expect that the urbanization in the coastal region is likely to experience a more rapid growth than other regions. Before we present detailed results of our analysis, let us define what is considered urban population in this chapter. Although a full explication of the historical evolution of various definitions of urban population/urbanization is beyond the scope of this chapter, we provide a working definition that is used in the current exercise. In our analysis, we follow the definition of urban population as used in China's 1990 Population Census (SSB 1991). The National Bureau of Statistics of China adopted two sets of criteria for enumeration of population of cities, towns, and counties for 1990. It is the second set of criteria that we use here. Without getting into too much detail, urban population mainly refers to people who reside in residential committees in cities and towns. This is a more restrictive definition than the definition using the first set of criteria.[1]

We take two steps to examine regional variations in urbanization. First, we describe the urbanization level in 1995 by province and examine changes from 1990 to 1995. Next, we will examine variations of the impact of migration on urbanization. For China as a whole, the urbanization level experienced a modest increase of 2.34 percent over the period of 1990 to 1995. The annual rate of growth during this time was 1.72 percent, which is consistent with corresponding rate of 1.7 for other developing countries during 1975 to 1995 (Chen et al., 1998).[2] However, there is tremendous regional variation in the pace of urbanization. The biggest increase in the level of urbanization is found in Shanghai, from 66.14 percent in 1990 to 83.87 percent in 1995, an increase of 17.6 percent![3] We suggest that this dramatic increase in urbanization in Shanghai is largely driven by Shanghai's recent development of "Pudong" (across the Huangpu River), an enterprise zone mainly developed by large transnational companies. Today's Pudong looks like New York's Manhattan on a smaller scale. A 1993 survey of the floating population in Shanghai found that more than one-third of Shanghai's floating population work in development zones, including Pudong district (Wang 1995). Migrant workers are employed in many occupations related to Pudong's development, including construction, transnational corporations, and service-related jobs. Table 9.1 summarizes the urbanization level and growth rate by province.

Given the fact that the coastal region has attracted so many migrants, we now look at changes in urbanization between the coastal region and non-coastal region.[4] For example, mean annual growth rate for the coastal region as a whole was 1.54 percent while mean annual growth rate for non-coastal region was 1.27 percent. However, this coastal vs. non-coastal difference is actually dominated by Shanghai, whose urbanization level increased dramatically in the first part of the 1990s. If we exclude Shanghai from the coastal region, the mean annual growth rate would be only 1.24 percent, even slightly lower than that of non-coastal regions. This suggests two things: first, that even within coastal provinces, there is much heterogeneity, which makes generalizations very difficult; and second, with a few exceptions, the migrant population still constitutes only a small proportion of a province's population and most of the migrants went to cities. It should be noted that urbanization level is not only determined by the population living in the urban areas, but is also determined by the size of rural population in a locale.

Also of interest is the case of Guangdong, located in the Pearl River Delta region. Due to its close proximity to Hong Kong, Guangdong has experienced a very solid economic growth during the economic reform era. It is the most attractive place for migrants, receiving the largest migrant population volume in China (Liang and Ma 2004). Therefore, it may surprise some readers to find that urbanization levels in Guangdong only increased slightly from 38.92 percent in 1990 to 39.28 percent in 1995. We argue that this is largely attributable to Guangdong's relatively high level of fertility, especially in rural areas. In 1990, for example, Guangdong's crude birth rates in rural and urban areas were 24.99 and 17.98 respectively. The corresponding rates for China as a whole in 1990 were 22.73 and 16.73.[5] Our analysis of the source of urban

Table 9.1 Urbanization level and growth rate by province in China, 1990 and 1995

Region	*1990 [1]*	*1995 [2]*	*Net change [2]-[1]*	*Annual growth rate*
North				
Beijing	73.22	69.75	–3.46	–0.96
Tianjin	68.20	69.20	1.01	0.29
Hebei	17.91	16.58	–1.33	–1.53
Shanxi	26.68	29.89	3.22	2.30
Inner Mongolia	35.53	37.69	2.16	1.19
Northeast				
Liaoning	51.40	46.04	–5.37	–2.18
Jilin	42.83	49.16	6.32	2.79
Heilongjiang	49.28	46.68	–2.60	–1.08
East				
Shanghai	66.14	83.74	17.60	4.83
Jiangsu	22.59	27.30	4.71	3.86
Zhejiang	30.58	32.58	2.00	1.28
Anhui	17.66	19.09	1.43	1.57
Fujian	22.67	22.68	0.00	0.00
Jiangxi	20.87	17.84	–3.03	–3.09
Shangdong	26.80	31.94	5.14	3.57
Central and South				
Henan	15.15	14.91	–0.24	–0.32
Hubei	29.64	31.20	1.56	1.03
Hunan	17.36	23.08	5.72	5.86
Guangdong	38.92	39.28	0.36	0.18
Guangxi	15.42	18.45	3.03	3.65
Hainan	20.59	26.93	6.35	5.52
Southwest				
Sichuan	19.74	26.73	6.99	6.25
Guizhou	20.05	24.02	3.98	3.68
Yunnan	14.74	13.82	–0.92	–1.28
Tibet	18.06	12.58	–5.48	–6.97
Northwest				
Shaanxi	20.65	27.81	7.16	6.13
Gansu	20.81	23.15	2.33	2.15
Qinghai	25.28	28.31	3.03	2.29
Ningxia	28.46	28.95	0.49	0.34
Xinjiang	32.98	33.15	0.17	0.10
China	26.24	28.58	2.34	1.72

population growth later in the chapter indicates that natural increase explains more than half of the urban increase in Guangdong province between 1990 and 1995.

One may also be puzzled by the fact that Beijing's level of urbanization actually declined slightly from 73.22 percent in 1990 to 69.75 percent in 1995. Since the floating population accounted for 15 percent of Beijing's total population in 1995, we suggest this decline in urbanization level is due in large part to the residential distribution of the floating population. Our calculations (not shown here) reveal that in 1990 14.7 percent of Beijing's floating population lived in rural areas. In contrast, by 1995, the corresponding percentage nearly doubled to 29 percent. With more and more migrants moving to Beijing, the housing is saturated which has forced many migrants to move to nearby rural areas. Thus, while holding on to jobs in Beijing, migrants can enjoy the low cost of living in the rural areas. We note that a situation like this will certainly underestimate the level of urbanization in China. In fact, in both the 1990 census and the 1995 China 1% Population Sample Survey (which we use), these people are not considered urban population.

While many researchers have focused on fast-developing coastal regions, our results suggest other non-coastal provinces also experienced a solid increase in the level of urbanization during this time period. The following non-coastal provinces saw their level of urbanization increased by more than 5 percent: Hunan, Sichuan, Shannxi, and Jilin. With the exception of Jilin province, the other three provinces started with a relatively low level of urbanization in 1990.

In sum, then, the 1990s were accompanied by a modest gain in the level of urbanization in China. Whereas coastal regions exhibit variations in urbanization, some non-coastal provinces experienced rather solid gains. It should be noted that urbanization level is not only affected by migration (as in the case of Beijing) but is also affected by the dynamics of fertility and mortality, as well as major economic development initiatives (as in the case of Shanghai).

Migration, Fertility, and Demographic Sources of Urbanization

The textbook version of migration and urbanization can roughly be restated in the following way. The standard economic analysis of migration and development suggests the following reasoning (Lewis 1954; Todaro 1969). The Industrial Revolution introduced new technological innovations, which increased agricultural productivity. The technological innovations, along with the enclosure movement lowered the demand for farm labor and made peasants increasingly redundant in agricultural production, therefore peasants migrated to cities in order to obtain manufacturing jobs and higher wages.

According to these views, moreover, the excess labor supply is eventually absorbed into the newly created jobs. In this process the income differential between rural and urban areas is reduced which in turn causes migration to reach equilibrium (Portes and Benton 1984; Rogers and Williamson 1984; Williamson 1988).

The urbanization experience of European countries suggested that in most cases migration was the major contributing factor to urbanization. Using data from Great Britain during the period of 1776 to 1871, Williamson (1988) provided empirical support for this argument. In the years during the initial stage of the Industrial Revolution, rural to urban migration accounted for more than 50 percent of urban growth. In fact for the periods of 1781–6 and 1801–6, rural to urban migration accounted for nearly 90 percent of the urban growth. For most of remaining years during 1776 to 1871 (roughly a century), rural to urban migration contributed about 50 percent to urban growth. Thus Williamson (1988, p. 293) concluded that "immigration was a much more important source of city growth during the First Industrial Revolution than it is in the contemporary Third World."

Underlying this pattern of urban growth is the notion that urbanization in developed countries historically happened at a time when the importance of urban natural increase gradually declined, therefore rural to urban migration became a leading factor in urban growth. In contrast, it is argued that in today's developing countries, urbanization is taking place at a time when the fertility level in urban areas remains quite high. Therefore, the demographic sources of urbanization in less developed countries are different from that of more developed countries. As early as the mid-1960s, Davis (1965) argued that urban natural growth contributed much more than rural-urban migration to the level of urbanization in the developing countries. Research by Preston (1979) provided a more systematic empirical support for this generalization. Preston (1979, p. 198) wrote: "of the 29 developing countries whose data support a decomposition of the sources of urban growth during the most intercensal periods, 24 had faster rates of urban natural increase than of net in-migration." Thus, it is clear that the major source of urbanization in most of today's less developed countries is driven by natural increase rather than migration.

Urban fertility is certainly high in some of the African countries (Brockerhoff, 2000), the story for China in the 1990s is clearly different. We argue that although China is still a developing country by economic measures, it may not follow the patterns of urbanization of other developing countries for two main reasons. First, China's family planning program and policies have played a very important role in fertility decline both in urban and rural areas. Since the 1970s China has adopted a series of family planning policies, the most well-known one being "the one child policy" which began in 1979 (Banister 1986). In the minds of policy-makers, these policies are necessary for the well-being of the Chinese population and for the smooth progress of China's economic development program (Lee and Wang 1999). To implement the one child policy, a system is also developed to reward individuals who pledge having only one child and to punish couples who decide to have more than one child. The penalty for non-compliance among cadres and family planning worker is especially severe (Hardee-Cleveland and Banister 1988). As expected, China's "One Child Policy" has had a tremendous impact on fertility decline in China, more so in urban areas than rural areas. Evidence suggests that this policy has been much more successful in cities than in the countryside (Retherford et al. 2004). As a result, the established TFRs are as follows: 1.11–1.19 for cities, 1.47 for towns, and 1.84–2.00 for rural areas for the period of 1990 to 1995 (Retherford et al. 2004). Aside from the one child policy, China has one of the highest contraceptive usages in the world (Lee and Wang 1999), 90 percent of married women used contraceptives in 1990. Therefore, we argue that fertility in urban China in the 1990s is so low that its contribution to urban growth is quite small, especially compared to the impact of migration.

Second, since the early 1980s, there has been a major increase in China's migrant population, especially the so called "floating population," i.e. migrants who do not have household registration status in places of destination. Data from censuses and survey show that the floating population (cross-county) has increased from about 22 million in 1990 and by 1995 it has more than doubled. This increase in migration is driven by several factors. One is the implementation of "household responsibility system" in the countryside that has greatly increased the efficiency of agricultural production and created a large amount of surplus labor in the countryside. Perhaps more importantly, China's transition to a market economy has led to major changes in the way in which cities operate, so much so that peasants no longer need to obtain urban household registration status to stay and work in cities. At the same time, booming joint enterprises, private businesses, and service industries, especially along the booming coastal regions, create constant demand for labor (Liang 2001; Liang and Ma 2004). As a result, any visitor to China today cannot fail to see migrants in cities. To take two extreme examples, the floating population in Shanghai and Guangdong accounted for about a quarter of their respective population in 2000. In sum, the combination of low fertility and large volume of migrant population in the 1990s prompt us to hypothesize that urban population growth in China in the early 1990s is mainly driven by migration rather than natural increase.

Data and Methods

Data for this analysis come from the 1990 China Population Census and the 1995 China 1% Population Sample Surveys, conducted by China State Statistical Bureau. Two considerations dictated our decision to use the 1990

Census and the 1995 China 1% Population Sample Survey. One is that the definitions of city, town, and rural area are identical in the two data sources. Without a consistent definition of city, town, and rural areas, a decomposition exercise would be impossible. The second advantage of using these data sources is that the period covered (1990–5) is important because the impact of high rates of migration on urbanization is likely to be significant. Although it may be desirable to conduct this exercise using the 1990 and 2000 Chinese Censuses, there are differences in terms of how urban is defined in these two censuses that is likely to introduce uncertainty to our estimates (Zhou and Ma 2003).

To conduct this demographic accounting, we follow a standard demographic balance equation which has been used extensively in studies of urbanization (Rogers 1984). To use Rogers' similar notation, for each province we have:

P_{95}: urban population in 1995
P_{90}: urban population in 1990
D_{90-95}: number of deaths of the urban population during 1990–95
B_{90-95}: number of births in urban areas during 1990–95
MIG_{90-95}: residual number indicating migration/classification.

Thus for each province, we have the following equation:

$$P_{95} = P_{90} + B_{90-95} - D_{90-95} + MIG_{90-95}$$

Therefore, the proportion of urban growth that is explained by migration/reclassification (P_{mig}) is given by the following term:

$$P_{mig} = \frac{MIG_{90-95}}{P_{95} - P_{90}}$$

For urban population in each province in 1990 and 1995, we obtained data from the 1990 Population Census (SSB 1991) and the 1995 China 1% Population Sample Survey (SSB 1997).

Although birth and death data are available for each province, no published data are available for rural and urban separately for each province. Based on available information and working under certain assumptions, we derived crude birth rates and crude death rates by rural/urban for each province. Suffice it is to say that we assume that cities and towns (which define urban areas) have the same levels of crude birth rates and crude death rates. In anticipation of this exercise, we also estimated rural and urban crude birth and death rates by province. The details of this procedure along with our estimated rural and urban crude birth and death rates by province are available from the authors on request. The next step is that for each province, we will generate P_{mig} shown in the above equation.

Impact of migration on urbanization, 1990 to 1995

Table 9.2 summarizes the results from this exercise. From 1990 to 1995, China's urban population (cities and towns) increased by 56 million, yielding a 19 percent increase. More than 70 percent of this increase in urban population can be explained by migration and reclassification. This result is consistent with the findings of Chen et al. (1998) who used data from China 1982 and 1990 Population Censuses. For the period of 1979 to 1990, Chan (1994) reported a slightly higher number of 75 percent.

Looking at the last column of Table 9.2, we see that Shanghai has the largest percentage, i.e., more than 95 percent of Shanghai's urban growth during 1990–5 can be explained by migration/reclassification. This echoes earlier results from Table 9.1 regarding Shanghai's spectacular increase in urbanization during this period. There are seven provinces where the results reveal negative values for the residual terms (migration/reclassification) and thus calculation of the percentage explained by migration/reclassification does not apply. Among the remaining valid 23 provinces, there are 19 provinces for which migration/reclassification explains well over 50 percent of the urban population growth. In fact, for 17 provinces migration explains over 70 percent of urban population growth. This suggests that in the 1990s, migration/reclassification was the dominant force contributing to China's urban population growth. Overall, Preston's statement regarding urban growth in developing countries does not apply to China's urban growth in the 1990s.

It is also interesting to examine the provinces for which migration/reclassification explains less than 50 percent of urban population growth, which includes Fujian (21.39 percent), Guangdong (46.7 percent), Ningxia (48.99 percent), and Xinjiang (35.26 percent). All four provinces share one characteristic: a higher crude birth rate than China's urban average (16.36). The crude birth rates in 1990 for each of the province are 18.78 (Fujian), 17.98 (Guangdong), 19.03 (Ningxia), and 20.96 (Xinjiang). Furthermore, among these four provinces, only Xinjiang's urban crude death rate (8.66) slightly exceeds the crude death rate of urban China (7.23). Therefore, the natural increase in population dominates urban changes during this period for these four provinces even though there has been significant migration in these provinces.

The case of Guangdong province deserves some further explanation. As is well known, Guangdong province is ranked number one province for attracting the largest number of migrats in China, as high as a quarter of China's floating population went to Guangdong (Liang 2001). On the basis of this, one would expect migration would contribute an exceptionally high percentage to

Table 9.2 Impact of migration on urban growth by province in China, 1990–95

Region	*Urban Pop. 90 (in thousands)*	*Urban Pop. 95 (in thousands)*	*Residual (in thousands)*	*Due to mig./ reclassifi. (%)*
North				
Beijing	7,958	8,999	778	74.75
Tianjin	6,019	6,715	586	84.16
Hebei	10,795	10,981	–379	N.A.
Shanxi	7,516	9,460	1,472	75.72
Inner Mongolia	7,498	8,854	931	68.70
Northeast				
Liaoning	20,552	19,398	–2,005	N.A.
Jilin	10,770	13,119	1,773	75.47
Heilongjiang	17,136	17,783	–204	N.A.
East				
Shanghai	8,933	12,212	3,128	95.42
Jiangsu	15,402	19,857	3,617	81.17
Zhejiang	12,490	14,489	1,548	77.44
Anhui	9,941	11,808	1,175	62.95
Fujian	6,941	7,549	130	21.39
Jiangxi	7,988	7,445	–1,070	N.A.
Shangdong	22,359	28,623	5,269	84.13
Central and South				
Henan	13,049	13,948	–12	N.A.
Hubei	16,233	18,518	1,362	59.61
Hunan	10,521	15,202	4,009	85.64
Guangdong	24,600	27,699	1,447	46.70
Guangxi	6,560	8,614	1,714	83.45
Hainan	1,321	2,000	584	86.06
Southwest				
Sichuan	21,003	31,100	9,164	90.76
Guizhou	65,616	86,578	1,690	80.64
Yunnan	54,161	56,579	–82	N.A.
Tibet	402	310	–117	N.A.
Northwest				
Shaanxi	6,704	10,036	2,893	86.84
Gansu	4,772	5,791	752	73.80
Qinghai	1,119	1,398	205	73.48
Ningxia	1,327	1,519	94	48.99
Xinjiang	5,070	5,655	206	35.26
China	296,958	353,397	39,647	70.25

urban growth in Guangdong. The reality is that the 47 percent contributed by migration/reclassification to urban growth is not exceptionally high. We suggest there are two reasons that contributed to this. One is that fertility in urban Guangdong is relatively high. For example, urban crude birth rate is 18 per thousand, much higher than China as a whole (around 16.36). The second factor is that like migrants in other parts of China, a high proportion of Guangdong's migrants live on the edge of cities for the sake of low rent in these locations. Very often these places are officially categorized as rural even though most migrants find work in cities.

The seven provinces that have negative values of residual (migration/reclassification) indicate a negative net migration. This negative net migration is caused by the following factors: intra-provincial rural to urban migration, intra-provincial urban to rural migration, inter-provincial out-migration from urban areas, inter-provincial in-migration to urban areas, and boundary reclassification. The volume of intra-provincial urban to rural migration is likely to be small. Since these provinces are not very attractive for inter-provincial migrants, inter-provincial in-migration to urban area should be quite modest. To examine how some of the dynamics work, let us look at following three provinces with negative migration and urban population decline from 1990 to 1995: Liaoning, Jiangxin, and Tibet. We suggest the major reason that these three provinces experienced urban population decline was that of large inter-provincial out-migration from urban areas. Unfortunately, the tabulations from the 1995 China 1% Population Sample Survey do not allow us to identify the rural/urban origin among inter-provincial migrants from these provinces. However, tabulations from the 1990 China Population Census provide some insight. For example, among inter-provincial out-migrants from Liaoning province, 62 percent came from urban areas. The corresponding percentages for Jiangxi and Tibet are 61 percent and 89 percent! We should note that for China as a whole, only 38.78 percent of inter-provincial migrants came from urban areas during this period. If we assume a similar magnitude of migration for the period of 1990 to 1995, it is no wonder these three provinces experienced urban declines, i.e. they lost their urban population through inter-provincial out-migration, and at the same time these places are not particularly attractive for inter-provincial in-migrants.

Beyond China: Recent Urbanization Patterns of Southeast Asian Countries

So far, we have focused on the experience of China in terms of demographic sources of urban growth. Recent evidence from Southeast Asian countries suggests that, like China, these Southeast Asian countries have also begun exhibiting a pattern of urban growth dominated by migration/reclassification rather than natural increase. According to Philip Guest (2004), migration/reclassification accounted for 64 percent of urban population growth during the period of 1990 to 2000. Furthermore, it is projected that during the decade of 2010 to 2020, migration/reclassification will contribute two thirds of the urban population growth for the whole Asian region. Guest (2004) cited increasing rural/urban migration as a major contributing factor.

Within Southeast Asia, we focus on Vietnam since Vietnam and China both have a state-sanctioned family planning program, as well as a shift from a command economy, with the state's control of urban population growth through a rigid household registration system (called *ho khau* in Vietnam), to a market economy with less effective state control of migration.

In Vietnam, urbanization also began accelerating in the late 1980s, due to the shift to the market economy, the considerable slowing of the urban-to-overseas exodus, and the increase in spontaneous migration in the context of the collapse of the subsidized-food-rationing system. The percentage of population *officially* living in urban areas increased from 19.4 percent (12.5 million) in 1989 to 23.7 percent (18 million) in 1999.[6] According to 1989 and 1999 census data, natural increase (2.4 million) accounted for 43.5 percent of the urban population growth in this period; migration, for 24.6 percent; and annexation, for 31.9 percent (Le Thanh Sang 2004, table 4). The average annual rate of natural increase at 1.6 percent in Vietnam from 1989 to 1999 reflects less draconian measures by the Vietnamese government than its Chinese counterpart to curb population growth. The Vietnamese government exhorted both urban and rural families to have no more than 2 children. During the command economy era, in urban contexts, as each state worker could buy rice and other foodstuffs at subsidized prices only for himself/herself and one child, a two-worker family in effect was discouraged from having more than 2 children. Since the systematic introduction of the market economy in the late 1980s, penalties have been imposed mainly on party members and bureaucrats for having more than 2 children in each of their families (normally promotion denial or demotion in the case of 4–5 children in a family). The rest of the population face little penalties for having more than 2 children per couple.

However, official census data underestimate the urban population growth in Vietnam since the late 1980s, as spontaneous migration began intensifying then and as migrants having resided for less than 6 months in a locality were not considered residents of that area for census purposes. A well-designed 1996 study of migrants in 16 sites in 8 districts in Ho Chi Minh City reports that in those sites, due to migration, the population was actually 7.4 percent higher than reported by local governments (Ho Chi Minh City-IER 1997: 27). In 2004, a census in Ho Chi Minh City with greater sensitivity to the presence of migrants with less than 6 months in the city reveals an one-year "increase" of 470,000, while the official population increase averaged only 110,000 for the preceding three years (*Tuoi Tre* [Daily newspaper] December 22, 2004; HCMC-SO 2004, p. 19). If the big increase from 2003 to 2004 reflects not a new and significant influx of migrants in 2004 but the regular underestimation of migrants in previous population trackings by about 300,000–360,000 in the 1999–2003 period, it means that the urban population in Ho Chi Minh City at the time of the 1999 census was probably 4.4–4.5 million instead of 4.1 million as officially reported. In other words, until 2004, the urban population in Ho Chi Minh City was probably underestimated by 7.3 percent to 8.7 percent due to the undercounting of migrants.[7] Adjusted for this new migration flow figure, migration would account for 42–44 percent of the population increase in Ho Chi Minh City from 1989 to 1999, instead of 28 percent as officially reported on the basis of the 1989 and 1999 census data; while annexation and natural increase would each account for 28–29 percent instead of 36 percent as officially reported (Le Thanh Sang 2004, table 5). Unfortunately, although Hanoi and many mid-sized Vietnamese cities like Hai Phong, Da Nang, and Bien Hoa probably had a stronger migration flow than officially reported for the 1989–99 period due to the major influx of rural migrants, comparable data are not available for these cities and other urban areas of Vietnam.

A regional analysis of the components of *officially recorded* urban growth from 1989 to 1999 reveals that annexation and migration accounted for a greater proportion of the urban growth in the Red River delta (around Hanoi), the Southeast (around Ho Chi Minh City) and the Central Highlands (respectively 63 percent, 61 percent, and 59 percent) (Le Thanh Sang 2004, table 7). It was no coincidence that in the 1994–9 period, the number of inter-provincial (mostly rural-to-urban) migrants above the age of 5 was highest to those three regions (.34, .75, and .34 million migrants respectively to the Red River delta, the Southeast, and the Central Highlands). The available data for 1993–8 also reveal that the Red River delta, the Southeast, and the Central Highlands enjoyed relatively higher economic growth, no matter whether measured in the growth of GDP or per capita expenditures (see also Luong 2003).[8] As per capita urban expenditures grew at twice the rate of rural ones for this period, rural-to-urban migration accelerated, leading to an increasing importance of migration and annexation in Vietnamese urban population growth. (In the past decade and a half, while needing migrants' labor for the construction boom in cities and for the labor-intensive garment and footwear industries, municipal governments have erected barriers to migration by not hiring migrants for local administrative jobs, by giving low priority to migrants' children in local school admission, and by creating red tapes for migrants' housing purchases in cities. Municipal governments have cited inadequate infrastructure and insufficient funds to support services to migrants. These local urban barriers stand in sharp contrast to the central government's affirmation of citizens' right to spatial mobility within the country.) In general, the rate of urbanization and the relative importance of migration in urban demographic growth are inextricably linked to the political economy of urban-rural inequality and the state's lesser role in controlling the migration flow after the end of the food-rationing and subsidy system in the 1980s.

Summary and Conclusion

This chapter provides an overview of patterns of urbanization and assessment of sources of urban growth in China during the first half of the 1990s. The growth rate of urbanization in China is not exceptionally high by international standards and is at similar level for other developing countries. Our approach to move this demographic accounting equation to the province level led to some surprising findings when we look at results from different provinces. The general logic that migration helps speed up the process of urbanization is quite straightforward. However, this simple logic produces quite complex dynamics in context of Chinese urbanization. Take the example of Shanghai, Beijing, and Guangdong. All three provinces (municipalities in the cases of Shanghai and Beijing) experienced major migration flows in the 1990s. In fact, Guangdong has been the "province of migration" in China for the past two decades and has received the largest migration flow in China. Migration affects urbanization in different ways in these three places, however. In the case of Shanghai, thanks to migration along with an exceptionally low fertility, urbanization level increased from 66.14 percent in 1990 to 83.74 percent in 1995. Whereas in Beijing, it is a different story. Despite a large volume of migration, the level of urbanization actually declined over time because migrants increasingly are forced to live in nearby rural areas due to increasing housing costs and its relative scarcity in urban Beijing. In the case of Guangdong, however, urbanization levels increased only slightly from 1990 to 1995 for still another different reason. Many rural and urban residents benefit from Guangdong's recent economic prosperity and average household income increased substantially. With money in hand and a strong son-preference in their heads, couples who wanted more than one child were not hesitant to have the second or third and pay the fines. Our analysis of demography of city population suggest that even cities in Guangdong provinces have a higher total fertility rate than the much less developed province of Sichuan (results are not shown here). The consequences of this relatively high fertility in Guangdong are (1) small-step increase in urbanization; and (2) natural increase became a major source of urban growth (more than 50 percent) in Guangdong.

The main body of the chapter has been focused on explaining the sources of urban growth in China. The long held view in the urbanization literature is that urban growth in more developed countries was dominated by migration whereas in the case of less developed countries urban growth has been explained mainly by urban natural increase. The main idea is that urbanization happened in more developed countries at a time when fertility was experiencing significant decline. In contrast, urbanization happens in today's less developed countries when urban fertility in these countries remains relatively high. This assessment has been in the writings of several well-known scholars such as Kinsley Davis, Jeffery Williamson, and Samuel Preston. Although this assessment is certainly true in many developing countries today, we argue and provide empirical evidence that the case of China along with other countries in Southeast Asia do not conform to this assessment.

To empirically validate our argument, we draw on detailed data on mortality and fertility for all provinces in China. We also decided to provide systematic analysis of the Vietnamese case to suggest that this new pattern of urbanization is not limited to China. This comparative exercise led us to find some parallels between China and Vietnam, which may provide some foundations for understanding patterns of urbanization for other countries which do not conform to the long- held view of urbanization patterns in developing countries. Both China and Vietnam have effective family planning programs. In China it is the "one child policy" which dramatically reduced China's urban and rural fertility. In Vietnam, the policy is more akin to a "two children policy" for both rural and urban residents which also resulted in fertility decline. Moreover, both China and Vietnam are transitional societies, namely transition from socialism to a market-oriented economy, with China's economic reform program started somewhat earlier than the Vietnamese one. Both countries have a household registration system that was designed to control rural to urban migration during the socialist era. With transition to a market-oriented economy, both countries witnessed significant increase in migrant population, especially rural to urban migration. Therefore, the combination of a reduced fertility and dramatic increase in migration makes migration the dominant factor in explaining urban growth in these countries. With the continued rise in migrant population, we anticipate this pattern of urban growth will continue in the foreseeable future in both China and Vietnam. Moreover, in light of the fertility decline in many other countries and the trend toward high rate of migration across the globe, time may be right to re-visit demographic sources of urbanization in other developing countries as well.

Finally, this may be a good point to respond to the four theoretical perspectives introduced in the Introduction of this volume: modernization theory, dependency/world system, develomentalist state, and transition from socialism. We admit that some of the theoretical perspectives are perhaps more relevant than others but overall they are complementary in explaining patterns of urbanization in China. They are complementary in the sense that each perspective by itself explains part of the urbanization experience, but they together provide a more complete and accurate picture. For example, our analysis suggests that in the context of China and Vietnam, it can be argued that the continuing role of the state in population control and the significant increase in rural to urban migration in the reform era provide support for theories of developmentlist state and transition from socialism. We note that in the case of migration, it is rather the relaxation of migration control (again by the state) and the increasing role of market (i.e. in food, housing, jobs etc) that led to the tidal wave of migrants. Having said that, we also argue that theories of modernization and dependency are relevant. In fact, demographic transition and urbanization process are key components of what modernization theory would address substantively. We find that as far as the migration component of urban growth is concerned, both modernization and dependency theories can find their relevance. Modernization theory would predict that if you remove the strong hand of the state that will be exactly what we should expect: migration (and urbanization) will be more responsive to the pace of industrialization. It also predicts less natural growth rate with higher education, and helps to explain the rural-urban fertility differences. Likewise, to the extent that dependency and world system theory are concerned with structural inequality and uneven economic growth as intrinsic to the capitalist world system, it highlights how rural-to-urban migration is embedded in the rural-urban structural inequality and how it swells the ranks of an exploited urban working class. However, it is not clear that rural-to-urban migration has led to excessive urbanization in either China or Vietnam. Furthermore, there is no evidence that the stronger integration of China (and Vietnam) into the capitalist world economy in the past two decades has led to the underdevelopment of the national economy in either case. In short, while the role of the state is clearly the dominant story in this chapter, we do find some level of relevance of the first two theories as well in the urbanization process of China and Vietnam.

ACKNOWLEDGMENTS

This research is supported, in part, by a FIRST Award from the National Institute of Child Health and Human Development (1R29HD34878-01A2). We are grateful to Grace Kao, Michael J. White, Tufuku Zuberi and an anonymous reviewer for their comments on earlier versions of the chapter.

NOTES

1 Readers can consult papers by Chan and Hu (2003) and Zhou and Ma (2003) for further details on this.

2 The annual rate of growth in urbanization r is calculated using the following formula: $\log(1 + r) = (\log P_{95} - \log P_{90})/5$.

3 We should note that our analysis of China's three municipalities (Beijing, Shanghai, and Tianjin) along with other provinces may not be entirely appropriate. This exercise is legitimate only to the extent that these three municipalities do have

rural counties attached to them and thus we can calculate level of urbanization and make comparisons with other provinces.

4 The coastal region covers the following 12 provinces or municipalities: Beijing, Tianjin, Hebei, Liaoning, Shanghai, Jiangsu, Zhejiang, Fujian, Shandong, Guangdong, Guangxi, and Hainan.

5 This high level of fertility and high level of socio-economic development in Guangdong has been previously noted by Feeny and Wang (1993).

6 In Vietnam, the urban population include the residents of urban wards (*phuong as* the smallest urban administrative unit), rural district towns *(thi xa)*, and market towns *(thi tran)* (cf. Douglass 2002: I-24–I-34). Ho Chi Minh City and Hanoi have both urban wards and communes whose residents are excluded from their urban population. More specifically, of the 5.1 and 2.7 million people in Ho Chi Minh City and Hanoi according to the 1999 population census, only 4.1 and 1.6 million people respectively are considered urban residents in Vietnamese statistics (Vietnam-GSO 2003, pp. 29–30, 35–6).

7 In 2003, when the urban population of Ho Chi Minh City was reported at 4.66 million (HCMC-SO 2004, p. 19), many city officials informally estimated the HCMC urban population at 6–7 million.

8 The Central Highlands had a relatively low growth in per capita expenditures. As the GDP growth in the Central Highlands resulted mainly from the dramatic increase in coffee export (from 90,000 tons in 1990 to 734,000 tons a decade later), and as a good part of this GDP growth flowed to plantation owners and coffee exporters residing in Ho Chi Minh City, the per capita expenditure growth in the Central Highlands was much lower than that of the GDP growth in this region.

REFERENCES

Banister, J. (1986) *China's Changing Population.* Stanford: Stanford University Press.

Bradshaw, Y.W. and E. Fraser. (1989) "City Size, Economic Development, and Quality of Life in China: New Empirical Evidence." *American Sociological Review* 54: 986–1003.

Brockerhoff, M. (2000) "An Urbanizing World." *Population Bulletin* 55, 3, Washington, DC: Population Reference Bureau.

Chan, K.W. (1994) *Cities with Invisible Walls: Reinterpreting Urbanization in Post-1949 China.* Hong Kong: Oxford University Press.

Chan, K.W, and Y. Hu. (2003) "Urbanization in China in the 1990s: New Definition, Different Series, and Revised Trends." *The China Review* 3: 49–71. (Special Issue on "Chinese Census 2000: New Opportunities and Challenges," K.W. Chan, Guest editor).

Chandler, T., and G. Fox. (1974) *3000 Years of Urban Growth.* New York and London: Academic Press.

Chen, N., P. Valente, and H. Zlotnik. (1998) "What Do We Know about Recent Trends in Urbanization?" In R.E. Billsborrow (ed.), *Migration, Urbanization, and Development: New Directions and Issues.* New York: United Nations Population Fund and Kluwer Academic Publisher: pp. 59–88.

Davis, K. (1965) "The Urbanization of the Human Population." *Scientific American* 213, 3: 41–53.
Douglass, M. et al. (2002). *The Urban Transition in Vietnam.* Hanoi: United Nations Development Programme-Vietnam.
Feeny, G., and W. Feng. (1993) "Parity Progression and Birth Intervals in China: The Influence of Policy in Hastening Fertility Decline." *Population and Development Review* 19: 61–102.
Ge, R. (ed.). (1999) *The Road to Chinese Urbanization.* Shanghai, Xuelin Press.
Goldstein, S. (1985) *Urbanization in China: New Evidence from the 1982 Census.* East-West Center Paper Series.
Goldstein, S. (1990) "Urbanization in China, 1982–1982: Effects of Migration and Reclassification." *Population and Development Review* 16: 673–701.
Goldstein, S., A. Goldstein, and S. Guo. (1991) "Temporary Migration in Shanghai Households, 1984." *Demography* 28: 275–91.
Gu, S., and X. Jian (eds.). (1994) *Population Migration and Urbanization in Contemporary China.* Wuhan, Hubei: Wuhan University Press.
Guest, P. (2004) "Urbanization and Migration in Southeast Asia." Paper presented at conference organized by US Social Science Research Council and Institute of Social Sciences in Ho Chi Minh City, Vietnam, February.
Guildin, G.E. (1992) "Urbanizing the Countryside: Guangzhou, Hong Kong, and the Pearl River Delta." in G.E. Guildin (ed.), *Urbanizing China.* New York: Greenwood Press: pp. 157–84.
Hardee-Cleveland, K., and J. Banister. (1988) "Fertility Policy and Implementation in China, 1986–1988." *Population and Development Review* 14: 245–86.
Hare, D. (1999) "'Push' versus 'Pull' Factors in Migration Outflows and Returns: Determinants of Migration Status and Spell Duration among China's Rural Population." *The Journal of Development Studies* 35, 3: 45–72.
Ho Chi Minh City, Census Steering Committee. (2000) *Dan so thanh pho Ho Chi Minh: Ket qua tong dieu tra ngay 01-04-1999* (Ho Chi Minh City Population: Data from the April 1, 1999 Census). Ho Chi Minh city.
Ho Chi Minh City, Institute of Economic Research (IER). (1997) "Bao cao ket qua dieu tra di dan tu do vao TP. Ho Chi Minh (Report on Spontaneous Migration to Ho Chi Minh City)." Unpublished report, Project VIE/95/004.
Ho Chi Minh City, Statistical Office (HCMC-SO). (2004) *Niem giam thong ke-Statistical Yearbook 2003.* Ho Chi Minh City: Statistical Office.
Kirby, R.J.R. (1985) *Urbanization in China: Town and Country in a Developing Economy 1949–2000.* New York: Columbia University Press.
Lewis, W.A. (1954) "Economic Development with Unlimited Supplies of Labour." *Manchester School of Economics and Social Studies* 22: 139–91.
Lee, Z.J., and W. Feng. (1999) "Malthusian Models and Chinese Realities: The Chinese Demographic System 1700–2000." *Population and Development Review* 25: 33–65.
Liang, Z. (1999) "Foreign Investment, Economic Growth, and Temporary Migration: The Case of Shenzhen Special Economic Zone, China." *Development and Society* 28, 1: 115–37.
Liang, Z. 2001. "The Age of Migration in China." *Population and Development Review* 27: 499–524.
Liang, Z., and Z. Ma. (2004) "China's Floating Population: New Evidence from the 2000 Census." *Population and Development Review* 30(2):467–88.

Luong, H.V. (2003) "Wealth, Power, and Inequality: Global Market, the State, and Local Sociocultural Dynamics." In H.V. Luong (ed.), *Postwar Vietnam: Dynamics of a Changing Society*. Boulder: Rowman and Littlefield: pp. 81–106.

Montgomery, M.R. (1988) "How Large is Too Large? Implications of City Size Literature for Population Policy and Research." *Economic Development and Cultural Change* 36: 691–720.

Perkins, D. (1990) "The Influence of Economic Reforms on China's Urbanization." In R.Y.-W. Kwok, W.L. Parish, A.G.-O. Yeh, with X. Xueqiang (eds.). *Chinese Urban Reform: What Model Now?* Armonk, NY: M.E. Sharpe: pp. 78–106.

Portes, A., and L. Benton. (1984) "Industrial Development and Labor Absorption: A Reinterpretation." *Population and Development Review* 10 (December): 589–611.

Preston, S.H. (1979) "Urban Growth in Developing Countries: A Demographic Appraisal." *Population and Development Review* 5: 195–216.

Retherford, R.D., M.K. Choe, J. Chen, X. Li, and H. Cui. (2004) "Fertility in China: How Much Has It Declined?" Paper presented at International Conference on the Chinese 2000 Population Census, Beijing, April 27–29.

Roberts, K. (1997) "China's Tidal Wave of Migrant Labor: What Can We Learn from Mexican Undocumented Migration to the United States?" *International Migration Review* 31: 249–93.

Rogers, A. (1984) "Urbanization in Third World Countries, 1950–2000: A Demographic Accounting." In A. Rogers (ed.), *Migration, Urbanization, and Spatial Population Dynamics*. Boulder, CO: Westview Press: pp. 281–304.

Rogers, A., and J.G. Williamson. (1984) "Migration, Urbanization, and Third World Development: An Overview." In A. Rogers (ed.), *Migration, Urbanization, and Spatial Population Dynamics*. Boulder, CO: Westview Press: pp. 261–80.

Sang, L.Th. (2004) "Urban Growth in the Pre-and Post-Reform Vietnam." Paper presented at the conference on urban poverty in Ho Chi Minh City, February.

State Statistical Bureau (SSB). (1991) *10% Tabulations of the 1990 China Population Census*. Beijing, China Statistical Publishing House.

State Statistical Bureau (SSB). (1997) *Urban Statistical Yearbook of China, 1996*. Beijing, China Statistical Publishing House.

State Statistical Bureau (SSB). (1998) *China Statistical Yearbook*. Beijing, China Statistical Publishing House.

Todaro, M. (1969) "A Model of Labor Migration and Urban Unemployment in Less Developed Countries." *American Economic Review* 59: 138–48.

Williamson, J.G. (1988) "Migrant Selectivity, Urbanization, and Industrial Revolutions." *Population and Development Review* 14: 287–314.

Vietnam, General Statistical Office. (1999) *Tu lieu kinh te-xa hoi 61 tinh va thanh pho* [Socio-Economic Data for 61 Provinces and Cities]. Hanoi: Statistical Publishing House.

Vietnam, General Statistical Office. (2003) *Niem giam thong ke-Statistical Yearbook 2002*. Hanoi: Nha xuat ban thong ke.

Wang, C., and S. Hui. (1997) *The Road to Urbanization in China*. Kunmin: Yunnan People's Press.

Wang, S. (ed.). (1996) *Regional Patterns of Chinese Urbanization*. Beijing, Higher Education Press.

Wang, W. (1995) *Shanghai's Floating Population in the 1990s*. Shanghai, East China Normal University.

Whyte, M., and W.L. Parish. (1984) *Urban Life in Contemporary China*. Chicago: The University of Chicago Press.

Xu, X. (1990) "Urban Development Issues in the Pearl River Delta." In R.Y.-W. Kwok, W.L. Parish, A.G.-O. Yeh, with X. Xueqiang (eds.), *Chinese Urban Reform: What Model Now?*. Armonk, NY: M.E. Sharpe: pp. 183–96.

Yang, X. (1994). "Urban Temporary Out-migration under Economic Reforms: Who Moves and for What Reasons?" *Population Research and Policy Review* 13: 83–100.

Yeh, A.G.-O., and X. Xueqiang. (1990) "Changes in City Size and Regional Distribution, 1953–1986." In R.Y.-W. Kwok, W.L. Parish, A.G.-O. Yeh, with X. Xueqiang (eds.), *Chinese Urban Reform: What Model Now?*. Armonk, NY: M.E. Sharpe: pp. 45–61.

Zha, R., Z. Yi, and G. Zhigang (eds.). (1996) *Analyses of the Fourth Census of China: Volume I.* Beijing: China Statistics Publishing House.

Zhou, Y. and L.J.C. Ma. (2003) "China's Urbanization Levels: Reconstructing a Baseline from the Fifth Population Census." *The China Quarterly* 173: 176–96.

10

Trapped in Neglected Corners of a Booming Metropolis: Residential Patterns and Marginalization of Migrant Workers in Guangzhou

Min Zhou and Guoxuan Cai

Introduction

Internal migrants in China arrive in cities in search of better employment, higher earnings, and, for some, better opportunities for social mobility opened up by market reform. While a lucky few have squeezed into the emerging urban middle class, a great majority have been trapped in a marginalized outsider status, simultaneously treated as much needed labor for the fast-growing urban economy and an indispensable scapegoat to be blamed for crime, disorder, over-crowdedness, squatter living or itinerant homelessness, traffic congestion, and other urban ills, as well as job competition and wage depression (Xiang 1999; Zhang 2001).

Migrants' experiences in China may be shared by migrant workers in other post-socialist countries or other newly industrialized countries in Asia and Latin America, but present a unique case of new mechanisms of stratification emerging from the processes of market transition and the developmental state (Castells 2000; Chen and Parish 1996; Evans 1995; Murray and Szelényi 1984; Nee 1989; Walton 1976). In this chapter, we aim to document new mechanisms of social inclusion and exclusion that have emerged from urban transformation. Specifically, we examine how migrants' residential patterns are structured by non-market forces, which in turn constrain their adjustment to the new urban environment. While long-term effects of migrant vulnerability in China have yet to be identified, existing research has shown that internal migrants tend to fall victim to their initial disadvantages as non-citizen outsiders (Chan 1994; Solinger 1999; Xiang 1999). We argue that internal migrants are not only culturally but also institutionally blocked from incorporation into urban life. To illustrate our argument, we first describe the prevailing residential patterns of migrant workers in a state-designated industrial development zone.[1] We then offer an analysis of how residential patterns are associated with social marginalization. Finally, we discuss the theoretical and policy implications of our case study to speculate on whether migrant workers will remain apart from, or become part of, the city in which they work.

Migrant Workers in the Guangzhou Economic and Technological Development District

In China, as in many other developing countries, economic development tends to rely on export-oriented production as firms from advanced industrialized countries seek outsourcing and subcontracting to reduce production costs at home and maximize profit (Castells 2000). Prime examples are *special economic zones* (SEZ), such as Shenzhen and Zhuhai initially created by the state in the early 1980s. SEZs are designated by the central or local governments. Located mostly on the fringes of urban centers and equipped with state-built infrastructure – roads, power supply, water, and sewer systems – these SEZs aim to attract foreign and domestic capital investment. A latent purpose is for the state to control effectively, or "manage," the anticipated influx of migrant labor.

In the early 1980s, the municipal government of Guangzhou set aside a large plot of farmland to the eastern edge of the city (some 50 kilometers away from the urban core) and zoned it for use as an economic development district, named Guangzhou Economic and Technological Development District (GETDD). The GETDD is an administrative district directly under the jurisdiction of the municipal government.[2] By the end of the 1990s, the GETDD had become home to a variety of manufacturing establishments, mostly in export-oriented, labor-intensive industries such as electronics and electric devices, automobile and auto parts, steel, chemicals, wrapping material, and packaged foods and drinks, but also mixed with some high-tech industries such as telecommunications and pharmaceuticals. As of 2003, the GETDD had about 600 foreign or joint-venture companies, employing more than 100,000 workers, 85 percent of whom are migrant workers without Guangzhou *hukou*.

Like those in other major metropolises such as Beijing, Shanghai, and Shenzhen, the migrants in the GETDD have come from rural areas of inland provinces of Guangxi, Hunan, Sichuan, and Anhui, as well as from poor counties in Guangdong Province beyond the Pearl River Delta region

(Fan 1996; Lang and Smart 2002). Our 2003 survey data show that migrants working in the GETDD share several characteristics with other internal migrants elsewhere in China: they are predominantly youthful and of rural backgrounds. Of the 85,000 migrant workers, 90 percent were under 35 years of age, close to 70 percent of the migrants were female, and the majority from rural areas or small townships outside Guangzhou Province. Although most of these migrants lacked industrial skills, they were relatively better-educated compared to their counterparts in their places of origin. Most had middle school or high school diplomas, a level of educational attainment considered quite extraordinary in rural China. There were also a small number of migrants who held urban *hukou* from smaller cities or county towns outside Guangzhou. This less noticeable group of urban migrants was better-educated, more goal-oriented, and more aggressively seeking out opportunities for upward social mobility than their rural counterparts. Diverse backgrounds aside, working in the GETDD was the first industrial wage job for the majority of migrant workers.

Migrants in the GETDD were primarily sojourners. They left behind their native places to seek better means of livelihood for themselves and their families with a clear intention to return. Predictably, most of these migrants came to Guangzhou to work for wages in order to support their families back home. But others intended to acquire new knowledge, modern technologies, and know-how skills that would facilitate their prospects of upward social mobility in their native places in the near future. Still others were simply driven by mere curiosities to see "what's out there" and "what can one possibly do away from home." A young woman from an electronic joint-venture company recalled:

> I still remember this day – two years ago, I and 30 of my former schoolmates came to Guangzhou with a beautiful dream. We were considered extraordinary students recommended by our school and employed by the company because we were better than others. Now, two years have passed, we have grown to be adults rather than young kids … in our journey to a meaningful career; I began to feel that there are many things one can do in life – things that make you happy and things that make you sad …

Migrant workers in the GETDD were often referred to as *waidi-ren* (meaning migrant from a different place) by *bendi-ren* (local people with Guangzhou *hukou*). While their sojourning orientation and accented language might serve as markers of outsiders, their residential patterns reinforced the *waidi-bendi* divide and constrained their incorporation into the local insider group. In the GETDD, a migrant's right to decent housing or even to living on the GETDD premise was tied to his or her job. In our study, we found that migrant workers in the GETDD were most concerned about job security among a list of issues such as wages, benefits, and working conditions, because losing a job meant not only breaking a "rice-bowl" but also breaking a "bed." Whenever their factories or companies made small changes to their duties and schedules of their current jobs, migrant workers would feel extremely anxious. A worker talked about her job change within the company:

> Any job change should be regarded as a new challenge, but for me a slight switch of my post would make me look bad and feel depressed. I was an operator before I was switched to be a quality inspector in another department. Many of my co-workers thought I got switched because my performance was not satisfactory. It gave me a sense of shame. I walked to my new post every day with my head low and voice down. No one told me that this was a normal job change … for a few weeks I was really down, not in the mood for any thing but stuck with that boring work. I felt quite depressed and had no self-confidence …

Market transition and economic development have allowed for unprecedented migrant flows into Chinese cities. However, because of the rigidity of China's *hukou* system, most migrant workers, including those holding urban *hukou*, were unable to transfer their *waidi hukou* to *bendi* (Guangzhou) *hukou*. This institutional constraint was further reinforced by managed housing arrangements in the GETDD.

Residential Patterns of Migrant Workers in the GETDD

In the GETDD, the majority (more than 85 percent) of the workers were *waidi-ren* as opposed to *bendi-ren*. *Bendi* workers are mostly native Cantonese. Almost all of them lived in the city and commuted to the GETDD daily by bus or car. Their computing time would take up to two hours each way on a daily basis. Aside from the very long and tiresome commuting time spent off work, *bendi* workers were rarely seen hanging out and socializing with coworkers in the residential areas of the GETDD in the evenings or during weekends. In contrast, almost all *waidi* workers lived in the GETDD, within walking distance of their place of work, with a small number living within a short bike ride nearby.

The difference in residential patterns between *bendi* and *waidi* workers were mainly due to the government's biased policies that favored economic development over community development. Land in the GETDD was almost entirely zoned for industrial use and little for residential use. Residential development in the district did little to attract *bendi* workers to relocate out of the city proper, but focused instead on the management and control of *waidi* workers. In the GETDD, *waidi* workers basically spent all their waking hours in the same place as they work. Three main residential patterns were readily observable from our 2003 study: factory/company dormitories, union housing compounds, and rental housing owned by local peasants whose land has given way to urban development.

Factory/company dormitories (zhigong sushe)

In pre-reform China, housing was treated as a welfare provision to which everybody was entitled, but in reality such entitlement was not granted to everyone as ideally intended. Nonetheless, many large state-owned factories and work units (*danwei*) in the productive sector with high administrative ranking cultivated a long-standing tradition of providing housing to their employees as an important part of socialist work management and welfare provision (Bian et al. 1997; Zhou and Logan 1996).

As housing became increasingly commercialized and state enterprises privatized, social housing gradually declined. Interestingly, large foreign or joint-venture factories and companies (new forms of *danwei*) that were located in industrial development districts seemed to continue this "socialist" tradition by building dormitories, or offering housing stipends, for their workers. From the perspective of management, housing provision was an effective means to cut cost, to control or manage workers, and to make them more productive. From a practical standpoint, housing provision was also a necessity, because most of the industrial development districts were newly built from farmlands on the fringes of a city quite far from the urban core with little access to public transportation.

In the GETDD, approximately two-thirds of *waidi* workers lived in factory/company dormitories (see Figure 10.1a). From the outside, newer factory/company dormitories were much like regular apartment buildings. Some of these buildings were owned by the factories/companies while others were leased from the market by the factories/companies for their workers. The quality of factory dorms varied but dorm rooms were almost always barely furnished with nothing more than just a few bunk-beds and desks. There was neither a telephone nor a bathroom within each room. Public bathrooms (with 4–6 toilets and showers) were situated on each floor, often shared by more than 60 dorm inhabitants. Each building had a custodian and was equipped with one public phone. Rooms vary in size, but the density is similar regardless of size. Smaller rooms house six to eight people in four two-story bunk-beds while bigger rooms can house as many as 14–16 in eight bunk-beds (see Figure 10.1b). There is little accommodation for married couples. All workers, single or married, are treated as single and packed into single-sex dormitories.

Figure 10.1a GETDD housing compound building

Figure 10.1b Inside a bedroom of GETDD union housing compound

Union housing compounds (yuangong lou)

Union housing compounds are quasi-governmental housing projects sponsored, built, and managed by the Labor Union Federation of the GETDD. Labor unions in China are independent non-governmental agencies, same as the Women's Federation, and the Youth League, but are functionally incorporated into the administrative body of the government. Almost all state-owned factories, companies, and non-profit institutions such as schools have unions. Private companies can establish labor unions depending on management and the organization of the work force; once established, these unions are under the jurisdiction and leadership of the Labor Union Federation of the local or municipal government. Since the GETDD is an administrative district of Guangzhou, the Labor Union Federation of the GETDD is a quasi-government umbrella agency with jurisdiction over all labor unions in the district.

The GETDD's Labor Union Confederation conducted a survey about *waidi* workers in 1993 and found that their living conditions were extremely problematic. First, the environment in and around their living places, including factory/company dormitories, was dirty. Some workers rented abandoned factory buildings or warehouses and made them into temporary living quarters while others built makeshift sheds right beside stinky sewage. Inside some of those makeshift factory buildings, more than 100 young men and women would be packed together with only clapboard partitions. Second, the living places were hazardous. Usually there were no fire extinguishers and no fireproof measures – an obvious threat to personal safety. Third, the living places were cramped. A small room would be shared by 10 to 14 people. Inside the room the two-story bunk beds would be turned into three-story ones, and one bed would be shared by two people who worked in different shifts. Fourth, the workers themselves had little access to leisure and popular entertainment. The only way for them to kill time was to play card games, which easily led them to illegal gambling. Last but not least, those who lived in rental housing faced frequent break-ins and often fell victim to crime.

Designed to improve the housing conditions of migrant workers and to help migrants adjust to urban life, the governing board of the GETDD raised funds in 1994 to build two identical 9-story housing compounds (Figure 10.2). Together, they accommodated 10,000 workers. Recently, the GETDD added another compound to the east side, housing about 5,000 more workers. These three housing compounds were financed by the district's government and managed by the Labor Union Confederation and were referred in the GETDD to as union housing compounds.

Rooms in union housing compounds were much like those in factory/company housing illustrated in Figure 10.1b, furnished with bunk-beds and desks. There was no accommodation for married couples. But housing was of much better quality. The most marked improvement in union housing was that every

Figure 10.2 GETDD housing compound building built in 1994

room in the union housing compounds had its own bathroom and was equipped with a telephone. There were 4–6 double bunk-beds that accommodate 8–12 people in each room along with a desk and a luggage rack for each person. There were other public facilities inside the compound, too, for the daily-life convenience of the workers. Each floor of the union housing compound had a least one commons room with a television set and couches. There was a cafeteria, a barbershop, guest rooms for short visits of families and relatives, hot water dispensers, and indoor bicycle stands. There were also entertainment and recreational facilities.

However, only 15 percent of *waidi* workers in the GETDD were able to secure a space at union housing compounds. In general, room spaces were allocated to participating factories and companies in the GETDD, who supposedly paid for room rental and all related expenses, including utilities, commons room usage, activities fees, and management fees (around 2,000 *yuan* per room). Then the factory or company assigned the rooms to its workers based on seniority and merit. Some companies might charge their workers a nominal housing fee (around 50 *yuan*), but the fee cannot exceed 10 percent of their wages according the union rule.

Peasant-owned rental housing (nongmin wu)

The GETDD is located on the urban fringe areas of what used to be largely farmland and villages. Government-sponsored development of industrial zones and districts in rural areas surrounding urban centers has created a new phenomenon referred to as *chengzhongcun*, or villages encircled by an expanded city (Li et al. 2003). A *chengzhongcun* retains the physical structure of the original village with old housing inhabited by original owners who have become landless peasants as well as by newly arrived renters who are primarily *waidi-ren*. As more and more rural farmland is converted for industrial use or residential housing development, many peasants have either moved into the city or in newer housing built to compensate for their lost farmland. As a result, there has existed a transitional zone between the fast-encroaching urbanized land and rural land, in which local peasants share space with migrant workers. Because of the stigma attached to *bendi* peasants and *waidi* workers, these new urban pockets are socially isolated and are often perceived by the government and the public as problematic neighborhoods plagued with concentrated poverty, crime, and social vice such as drugs, gambling, prostitution, and other urban ills.

Chengzhongcuns exist in the transitional areas in and surrounding the GETDD. In and around these transitional neighborhoods, a rental housing market owned and managed by *bendi* (native) peasants has emerged and taken shape to meet the growing demand of migrant workers. This new phenomenon is often referred to as *geng-wu* (figuratively, to "farm" housing, which means to live on rental incomes) as opposed to *geng-tian* (to "farm" land). The peasant-owned rental housing market has evolved in several notable stages. The first stage refers to a budding rental market that contains original houses vacated by peasants. When villagers moved into the city or simply left for better housing, they usually rented out their old houses. These vacant houses were typically rundown with few utilities, little furnishing, and no amenities.

The second stage refers to the rebuilding original houses with add-ons on top or by the side to create more housing space for rental use. In the beginning, local peasants tore down their original one- or two-storied houses and rebuilt them into three- or four-storied houses with multiple rental units to maximize incomes. As demand grows, some owners even add on temporary structures by the side or on the top of their buildings to create more housing space for rent (see Figure 10.3a). Renters may be individuals or couples who may in turn sublet to more people. This type of housing, though owner-occupied, also lacked basic utilities and proper maintenance for renters.

The more recent stage involves housing cooperatives that build larger apartment structures (see Figure 10.3b). Most of these new apartment buildings consist of two-, three-, or four-bedroom units with a bathroom and a kitchen as well as single rooms with no private kitchens or bathrooms, just like factory/companies dormitories. The first floor of some apartment buildings

Figure 10.3a Structure added to the top of building to increase rental value

Figure 10.3b Housing co-op building

consists of space for commercial use for retail or storage. Viewed from the outside, these peasant-cooperative apartment buildings appear no different from other high density urban housing development elsewhere in the city, except that apartment buildings are usually built irregularly and haphazardly with no coordinated urban planning.

Initially, peasant-owned housing was mostly rented out to *waidi* entrepreneurs who came to Guangzhou for business and were used either for temporary accommodation or for storage of goods. When the GETDD was established there, rental housing in this transitional zone became an important source of housing for migrant workers. About 20 percent of the migrant workers in the GETDD lived in peasant-owned rental housing.

Rents were not particularly high, averaging about 600–700 *yuan* per two- or three-bedroom unit in 2003. Since many migrants worked all day in the district, they often found themselves crowding in to these apartments for the sake of saving money. It was not uncommon to see eight or nine people cram in a small room of just 30 m^2 or in a two-bedroom unit. Most peasant-owned rental housing was overcrowded and lacked utilities.

Peasant owners generally provided little maintenance to their rental properties. Local authorities also showed little interest in establishing or reinforce codes to monitor the conditions of rental housing. As a result, areas where rental housing was concentrated showed a steady trend of deterioration and decline over time and would rapidly turn into socially identifiable and stigmatized *chengzhongcuns*. However, in recent years, peasant housing cooperatives have also begun to build luxurious apartments to cater to the growing needs of upwardly-mobile migrants working in the GETDD who are now in managerial and professional occupations and have higher earnings.

Analysis: Adaptation or Marginalization

Migrants' residential patterns in the GETDD serve reinforce the *bendi-waidi* division. The three prevailing housing patterns in the GETDD, which we have just described, have varied effects on how well these workers adapt to urban life and how well they feel about themselves and about the city in which they strive to make it – whatever "it" means. From our own observations and interviews, we find some clear association between housing patterns and discernible outcomes regarding migrant workers' quality of life, social relations, and group formation.

After-work life

Most workers in the GETDD work six days a week. Few *bendi* workers live in the district and commute some 50 km back to "town" (Guangzhou) after work.

Although their place of work is separated by some geographical distances from their place of residence, their urban life remains relatively intact. *Waidi* workers, in contrast, have little contact with *bendi* workers and their social life is largely isolated from that of *bendi* workers. What do *waidi* workers do after they get off work during the week and on Sundays? "Nothing much" is perhaps the modal answer. Adjectives such as "boring," "monotonous," "tasteless," and "pointless" were used repeatedly from our interviewees to describe their after-work life in the district. Regardless of where they live, the feeling of "*hun rizi*" (killing time) appears quite prevalent. Many stay where they are on Sundays and rarely go into "town" since downtown Guangzhou is far away and public transportation is not convenient and not easily accessible. Their after-work life appears characteristically repetitive and lacks cosmopolitan color and flavor. However, there are some notable variations among workers living in different types of housing.

The workers who live in factory/company dormitories and those in peasant-owned rental apartments do not spend much time in them except for sleeping. Living in factory/companies dormitories means a worker must share his or her room with many others. Living in peasant-owned rental housing isn't much better because many workers have to share their apartments also with many others in order to save money. In such crowded conditions, many workers take up overtime work in their factories or companies, or part-time work in or nearby the district. When they are back "home" (to their dormitories), they lead a typical bachelor's life because they are all treated as singles regardless of their marital status. Some would gather together to watch TV or videos or play cards, while others would do their own chores or just simply sleep. Mr Zhang, a young man who works in a small shop as a part-time job when he gets off his full-time regular job from an electronics factory, said:

> I would feel restless and bored if I am sitting here [in his rental apartment shared with 10 other people] doing nothing, because I have no friends, relatives, or *laoxiang* here. I like to find extra work to do so that I can kill time and make more money. Recently, our factory's business is not going too well. Fortunately, I found another part-time job through a *laoxiang*; the salary is 300 *yuan* per month … enough to pay for two months' rent. But I have no time to rest and play.

In contrast, after-work life in the union housing compounds is more colorful. Even though residents there share their rooms with many others, and their living space is just as limited as those in dorm rooms or rental apartments, the public space within each housing compound is much more available and better resourced. Workers living in these compounds have been able to conduct and organize their social life after work. They have formed various social clubs. The GETDD's Labor Union Confederation and union housing compound management have also routinely offered classes and seminars for job training and general knowledge enrichment. The union also publishes a newsletter and has a radio station to cover life and work issues concerning migrant workers. Indeed, these three housing compounds have become the center of a migrant community. A migrant worker who has lived in the compound for more than two years told us in an interview:

> I have to work 8 hours every day on the assembly line. It's just too dead boring. It's only when I return to the dorm that I feel being myself again. Here I can take part in activities organized by the union and the Youth League and have lots of fun with roommates and other friends.

The quality of life affects productivity. According to our survey result, foreign companies are more willing than local companies to place their workers in union housing compounds, even when such an arrangement is more costly, especially when they have their own dormitories. For example, Mitsubishi once decided to place more than 200 workers in a union housing compound, which is a 15-minute bus ride away. Rent and transportation cost combined are twice as much as what it costs to put up workers in the company's dorms. But the company did it anyway because, as the general manager pointed out, it actually helped in reducing the cost of worker training.

Apparently absent is family life among workers in the GETDD. Neither factory/company housing nor union housing is designed and built for married couples. Like many migrant workers in other cities, most of the *waidi* workers in the GETDD are young and single. Among the "single" workers, some may be married leaving their spouses and children in their native villages and towns while others may be dating with the intention of getting married. In fact, some of the organized leisure activities in union housing compounds, such as dance and karaoke, encourage workers to make friends and get to know one another. But there is no tangible institutional support for workers to establish families and have children and no public policy intention to encourage workers to settle permanently.

Social interaction and networking

The process of moving to a new urban environment often results in social disruption. Original social networks are disrupted and new ones are built or rebuilt. Often times, *waidi* workers migrate into the city by themselves or in small groups. While some may have friends and families who have arrived there earlier to receive them and help them get settled, many others arrive on their own without knowing anybody initially. Facing unfamiliar people, things, and the environment, they naturally seek out the obviously familiar lot – those *laoxiang* from the same village, township, or region, who come from similar socioeconomic backgrounds, speak the same dialects, and share similar beliefs, values, behaviors, and food ways. In order to survive the strange land, they have to form social groupings and rebuild social relations for self-help. The newly built or rebuilt networks are predominantly based on *laoxiang* connections, or connections to the common place of origin. These *laoxiang*-center networks or social groups are usually formed in a relative short time upon arrival and they function as an important, and sometimes the single most important, source of social support, since the migrants have very little formal institutional support and informal social support from the host city. Mr Du, an assistant to a small shop keeper told us in an interview: "When you are in a new place like this, you need the help from your own friends and *laoxiang*. We can trust them and find true friendship in them because we are from the same place and on the same boat."

In the GETDD, the social life of most migrant workers is rather narrow and isolated. But particular living arrangements give rise to different patterns of interpersonal interaction and social networking, which has implications for migrant adaptation to urban life. In general, factory/company dormitories are an extension of the workplace. Dorm occupants tend to socialize among themselves working in the same factory or company and thus know one another very well both at and off work. Since their work and social life after work are intertwined, they tend to develop tightly knit social networks that may cut across places of origin and occupations but are largely confined to the same workplace. Also, when co-workers had a small fight at work, their bitter feelings would likely get worse and intensified after work. Likewise, roommates having disputes over even the smallest things can carry those unpleasant feelings into the workplace, thus disrupting cooperative work on the assembly line. While these workplace-based networks are strong and intense, they can easily be broken once a worker is detached from the workplace.

Workers living in union housing compounds tend to have broader circles of friends and develop social groupings that cut across place of origin, occupation, and workplace. Even though roommates may be from the same factory or company, they do not have to be bound by the same type of workplace-based relations as their counterparts living in factory/company dormitories. Moreover, social networks built among residents tend to be more loosely knit and more stable than workplace-based ones. Over time, these residents tend to become more sophisticated, cosmopolitan, sociable, open-minded, collectively oriented, and socially attached. They display a visibly stronger sense of self and community than other migrant workers.

Residents living in peasant-owned rental housing are generally as diverse as those in union housing compounds in terms of place of origin, occupation, and workplace. Unlike those living in factory and union housing that are collectively organized, though to a varying degree, residents living in rental housing tend to be less social, less collectively oriented, more individualistic, and more "on their own." After-work life there has no observable structures; rather, it is largely organized haphazardly and arbitrarily. Interpersonal relations that have developed from overcrowding living are more depersonalized, anonymous, and transient than they are personal, familiar, and permanent. It is interesting to note that these areas where peasant-owned housing is concentrated resemble some of the most typical characteristics of urbanism – heterogeneity, density, and social disorganization – as described by Wirth (1938). However, unlike rural-to-urban migration in which peasants gradually become urbanized and cosmopolitan as they make contacts with urbanites, learn new ways, shed old habits, and become acculturated into the city, what we find here is something abnormal. Since the areas are originally rural villages, *waidi* workers are actually transplanted from remote villages to *chengzhongcuns* and their contacts are either among themselves or with natives who are peasants. So the rural way of life gets reproduced and reinforced here among migrant workers who lead bachelors' lives rather than family lives. Nonetheless, social networks do get formed, more on the basis of *laoxiang* relations than on workplace-based relations, and these networks would facilitate the flow of information about job and housing opportunities. As a result, renters move in and out of their rental housing a lot more frequently than residents in other forms of housing and they also tend to change jobs more frequently (mostly within the district or in nearby towns).[3] Their changes in housing or job appear to be horizontal without any significant improvement.

However, the emerging luxurious housing market in the GETDD allows a migrant middle class to take shape. Renters there tend to be a more selective lot. Many of them are professionals, managers, and skilled workers working in the district. They usually do not socialized with other migrant workers living in crowded rental housing. The rise of this luxurious housing market for the emergent white-collar middle-class among *waidi* workers presents a paradox that requires further investigation.

Social group formation

As discussed earlier, segregated living in the GETDD exacerbates the social divide between *waidi* and *bendi* workers. *Waidi* workers are looked down upon by local residents as a backward, narrow-minded, unsophisticated, and unassimilable lot, and their group distinction is often heightened by their obvious ways of speech, dress, and mannerism. *Waidi* workers have little contact with *bendi* workers after work and have little chance of incorporating into close friendship networks of *bendi* workers. They are stigmatized and thus prejudiced and excluded by local residents while being also ignored by various levels of government, public social services agencies, and non-governmental organizations. As a result, they are increasingly marginalized into a disadvantaged "non-citizen outsider" group, being blamed for urban pains and ills – social disorganization, disruption of routine rhythms of urban life, traffic jams, and crime – that are more associated with rapid growth and market transition than with their being there.

In the GETDD, even though union housing appears to serve best the adaptational needs of migrant workers compared to the other two prevailing housing modes, it still leads to a similar outcome – social isolation. As the emerging stereotype dictates, *waidi-ren* live in the GETDD and *bendi-ren* in the city.

The GETDD creates a strong impression among locals that people living inside are a homogenous *waidi* group who just want to make money in Guangzhou, and the district is a stigmatized *waidi-ren*'s enclave or village. One *bendi* worker remarked in an interview: "No matter how long they [migrants without Guangzhou *hukou*] have been in Guangzhou and no matter what they do, they are *waidi-ren*, and we can tell that they are …"

Some migrant workers grow resentful while others feel depressed. One of the migrant workers said:

> After all, who can tell how many *waidi* workers have left their sweat and youth with Guangzhou? Who can tell how much hardship we have gone through and how many tears we have shed behind all the splendid lights of night-time Guangzhou? If it were not for our work, Guangzhou would never be what it is now.

The structured and controlled lives in the industrial development district are very different from migrants' former lives in the rural area. Having to face many laws and regulations, they had to change their own living styles, which resulted in them being in a city but not part of it. Mr Zhong, worked in Guangzhou for three years, told us in an interview:

> When living in the midst of this metropolitan city which is full of energy, excitement, and competition, you would develop some kind of inferiority complex. Here, you are a just an average worker giving out your labor. You see city folks live different lives and act differently toward you. They surely have different values. Sometimes I would make mistakes, get upset about things, and get annoyed with a colleague, and I would be scolded or criticized for things I didn't quite understand. Before, we were used to the comfortable life at home. It is difficult to adapt to the new life here … Everything is so new and uncertain.

Another worker at an electronic company talked about his feelings at work: "Work occupies all my life here; I feel bored when I am working on these electronic boards piece by piece and following the same routine like a robot day in and day out here."

Also because of the nature of work that does not allow for much interpersonal interaction, coworkers feel like strangers. Ms Fan, an office secretary, who works in a soft drink company, said:

> I don't feel this city is welcoming me. I don't feel I can find a place in such a big city. It has been so difficult for me to stick to one job. I first worked for a food packing company, but I couldn't adapt to the boring work there. So I quit and found another one; but same thing happened, and I changed again. So what I can do is to change jobs frequently, but when I go to another company, I would feel people in management are unkind, always wear a scornful look at us and consider us unworthy, which would make us feel a chill deep in our hearts.

Waidi workers' outsider status in turn effectively reinforces their sojourning orientation. Most of the workers whom we have interviewed express a clear intention to "return" to their native places in the near future. However, their notion of "return" suggests the resettlement in towns or cities close to their native villages rather than the return to the villages per se. They would aim to develop new careers rather than to go back to farming, and most would want to open up their own businesses. An antithetical couplet at the gate of a company's dormitory building has the following Chinese phrases written on it:

> Guangzhou's builders today,
> Hometown's entrepreneurs tomorrow.

Discussion: Reflection on Theory

Migrant workers in the GETDD have remained residentially segregated and socially isolated. As a distinguishable social group with little social contact with *bendi-ren* (except with *bendi* peasants), they are treated by the state, the industries, and the public as necessary labor, but not as blood-and-flesh human beings, and are pushed into the corners of the booming metropolis as non-citizen outsiders. Some of the causes of their residential segregation may arguably be voluntary and self-selective. For example, many migrant workers arrived in the GETDD as sojourners, and they came to find work and had no intention to settle in Guangzhou permanently. However, if the host city were more receptive and welcoming, they would possibly shift their sojourning orientation to settlement.

Social marginality associated with internal migration in China shows certain uniqueness in comparison with that associated with urbanization in developing countries as well as with international migration to developed countries. In the Latin American urbanization literature, the marginality perspective posits that the isolation of the poor is not simply a matter of individual incomes; but more importantly, it is part of the spatial and physical organizations of cities where economic growth and housing construction as well as the provision of public services are severely mismatched. The lack of adequate housing and social services leads to squatter settlements, or the formation of shanty towns, which in turn exacerbates the social marginality of the poor as evident in many fast-growing Latin American cities (Cornelius 1973; Morse 1971a, 1971b; Roberts 1973). In the case of the GETDD, however, a large proportion of migrant workers' housing need is met by industries as mandated by the state with a smaller proportion provided directly by the state or by the emerging market. Although state-intervention effectively prevents squatter or slum settlements and shanty towns, it intentionally and unintentionally reinforces social separation and marginality.

However, there are different ways marginality is viewed by the public and acted upon by the government. Some treat marginality as a social problem intrinsic to rapid urban growth and economic development and the marginalized as an unfortunate lot (Perlman 1976). Others see marginality as the poor's own making as illustrated by the culture of poverty thesis. The culture of poverty thesis predicts that the lifestyle of the poor, arising from low-income jobs, poor living conditions, and lack of education, perpetuates fatalistic orientation and the passing of disadvantages from one generation to the next (Lewis 1961). These two views of marginality are criticized as elitist in conception, regarding the poor as victims either of structural circumstances or of their own cultural deficiency (Roberts 1978).

Another long-standing theoretical paradigm explaining social marginality has been that of internal colonialism in the development literature (Blauner 1969; Hechter 1976; Walton 1976). Internal colonialism is defined as "a process that produces certain intranational forms of patterned socioeconomic inequality directly traceable to the exploitative policies through which national and international institutions are linked in the interests of surplus extraction of capital accumulation" (Walton 1976, p. 58). Underprivileged groups, referred to as internally colonized groups, may be differentiated by race and ethnicity, religion, language or some other cultural characteristics and are excluded from participating in mainstream social, economic, and political institutions based primarily on these group markers. In the US, internal colonialism has been applied primarily to the study of race and ethnicity, where involuntary immigrants are treated as "colonized" minorities and voluntary immigrants, including white ethnic, as "immigrant" minorities (Blauner 1969; Ogbu 1974).

In Latin America, internal colonialism has been applied to the study of socioeconomic inequality among culturally distinct groups (Walton 1976). Scholars from this perspective liken the marginalization of rural-to-urban migrants to foreign colonialism (Gonzales Casanova 1969), suggesting that its causes not only lie in the acute cultural differences between city and the countryside but also in the encroachment of urban-based capitalism on the periphery and with it the emerging machinery of increased political control and economic exploitation (Casanova 1969; Walton 1976). From the perspective of internal colonialism, the state plays an important role in establishing and reinforcing neocolonial relationships of domination and subordination.

In many newly urbanized Chinese cities, however, the mechanisms of social marginality suggested by existing literature – mismatch, culture of poverty, internal colonialism – do not fully account for the current state of migrant workers. Our contemporary case of the GETDD shows that migrant workers are perceived as *waidi-ren* by the locals while assumed to be so by the state and social institutions that serve them. Various levels of the government privilege economic development over human development. The concern over migrant workers has been primarily with how to keep them productive at work and how to keep them from becoming *bendi-ren* who can lay claim on social service and welfare benefits accorded to *bendi hukou* status. In fact, state and local development polices have never considered facilitating migrant workers' adaptation to urban life. Of social policies concerning migrant workers in cities, most have targeted "problem"-solving and none resettlement. Residential patterns in the GETDD show a minimum involvement of either the municipal or the district government. Even in union housing where government intervention is substantial, the goal is to manage workers better, to make them more productive on the work front with no consideration about housing for married couples. Industries provide housing for workers for the same purpose and show little interest in doing more than necessary to maximize profit. The emerging peasant housing market, with little intervention from the state, functions to maximize profit and do little for improving worker's housing conditions and social environment. If that housing market expands and dominates in the near future, it remains an empirical question whether similar problems associated with Latin American urbanization would be reproduced.

Nonetheless, the GETDD case is not entirely unique to China's transitional economy. It illustrates some of the similar lessons found in the development literature on urbanization in Latin America and immigrant adaptation in the US. First, contemporary migrants arrive in urban centers or host countries where economic restructuring and globalization have created bifurcated labor markets with numerous labor-intensive and low-paying, part-time/temporary jobs on the one end and a growing sector of knowledge-intensive and well-paying jobs on the other (Sassen 2000). This bifurcated labor market renders many migrants, even those with education and skills, underemployed in substandard wages or occupational over qualification, which, in turn, limits their social mobility (Zhou 1993). In China, the kinds of investments attached to the SEZs tend to be labor-intensive industries that absorb a vast surplus labor from the countryside. Migrant workers attracted to fast-growing cities tend to concentrate in the lower end of the labor market with few opportunities for upward social mobility.

Second, contemporary migrants come into contact with the existing systems of stratification in the host society, which are often interacting closely with class, race, and gender hierarchies, but often find it difficult to resist it. In the US, non-white immigrants are often confronted with racism along with racial discrimination in the rigid system of racial hierarchy, and this hierarchy is in many ways intertwined with the system of class stratification (Massey and Denton 1993). In Latin America, however, class is a more prominent determinant than race to structure inequality (Walton 1976). As for internal migrants in China, non-*hukou* status connotes deprivation of citizenship rights and structural disadvantages that institutionally sort them into an outsider group, despite similarities in racial and cultural characteristics and even in class status.

Third, contemporary migrants, especially the undocumented, are often constrained by the state – regulations restricting entry and settlement (e.g., backlog in visa processing, work permits, driver licenses, residency requirement, language requirement, and citizenship rights) and public policies (e.g., public assistance, public housing, education, healthcare, and government loans) – in their struggle to combat disadvantages associated with migrant status. Even in situations where the state promotes immigrant assimilation and multiculturalism, it may inadvertently reinforce barriers inhibiting the migrants' chance of mobility. In the Chinese case, the *hukou* system and state policies, including housing policies, that aim to manage or control internal migration reinforce migrants' outsider status.

Moreover, some migrants are involuntarily brought to the host country (for example, slaves, indentured labor, and refugees) while others intend to migrate only to sojourn for a given period of time and eventually return home (e.g., seasonal labor migrants and some undocumented immigrants). Typically, migrants who initially intend to work in the labor market of a host city (or host country) for an extended period of time in order to make money to improve and enrich their future lives in their places of origin are referred to as "sojourners." To sojourn is to work in a place away from home with no intention of eventually settling in that place. As time goes by some of the sojourners may become settlers. However, rigid state policies make it difficult to switch from sojourning to settlement. For example, America's Chinatowns had continuously been a bachelors' society until the 1970s because of legal exclusion (Zhou 1992). Internal migrants in China arrive in cities initially to sojourn. Their lived experiences as outsiders uphold their sojourning goals, which simultaneously trap them in their marginalized outsider status.

Conclusion

In China, the causes of social inclusion and exclusion do not merely stem from culture, place of origin, and the rural-urban divide, but also emerge from the processes of market transition that is intimately involved by the state. New mechanisms of migrants' urban adaptation are being formed in ways that are market driven on the one hand while being constrained by state policies on the other. As suggested by the theories of market transition and social stratification, market reform has opened up new revenues for social mobility, allowing some from the rank-and-file to capitalize on their human capital and individual motivation to gain privileged social positions, but continued to maintain and even reinforce the advantages of the state elite (Bian 2002; Bian and Logan 1996; Nee 1989; Nee and Matthews 1996; Walder 1989). However, market reform has promoted particular forms of capital investment and development that disproportionately utilize export-oriented production as the means for economic expansion. Consequently, rapid growth of labor-intensive industries has unleashed unusually high volumes of internal migration, which has in turn compelled the state to exert concerted effort to control and manage migrants, a phenomenon that may be better captured by theories of the developmental state.

Research on internal migration and urban transformation in China has highlighted the importance of the state in initiating market reform, maneuvering its directions, and lauding its positive outcomes to the neglect of unintended social consequences. Our case study of the GETDD shows that residential segregation and social marginalization of migrants are not inherently the effect of migration but rather a result of the way migration is being handled by public policies at the state and local levels. The marginalization of migrant workers may be explained by the theories of market transition. Its root causes, however, result from deliberate policies of the developmentalist state which, on the one hand, encourage internal migration to fuel economic growth but, on the other, control it so as to avoid over-urbanization.

The implications of migrants' residential segregation and social marginalization may be far-reaching beyond the GETDD. Witnessing rapid economic development and tremendous wealth that have arisen from it and yet being trapped in the neglected corners of the booming metropolis, aspiring migrants will not passively react to their structural disadvantages. One possibility is that migrant workers are likely to utilize their newly acquired industrial skills and personal savings for their own betterment, but to do so by returning to their places of origin where they have *bendi-ren* privileges and advantages. Indeed, the trend of return migration is already underway, which helps boost economic development in the migrant-sending inland regions, redirects the courses of migrant flows, and causes labor shortages in migrant-receiving coastal regions. Another possibility is that migrant workers would carve out their own economic niches and social space in their host cities, creating urban enclaves much more like *Zhejiangcun* in Beijing than *chengzhongcuns* (or rural villages) within the walls of a metropolis. Still another possibility is that migrant workers in their host cities are likely to become a politically organized social group resisting marginalization, pressing their own issues onto the agenda of public policy and discourse.

ACKNOWLEDGMENTS

The original draft of this chapter was presented at the Conference on Urban China, Four Point Sheraton Hotel, Santa Monica, May 1–2, 2004 and a revised version at the 2nd Conference on Urban China, Radisson Hotel New Orleans, January 15–16, 2005. The authors wish to thank John Logan, Susan Fainstein, Alejandro Portes, Neil Brenner, and anonymous reviewers for their helpful comments and suggestions. They also thank Ying Na Deng, Ly Lum, and Angela Sung for their research assistance.

NOTES

1 Our analysis was based on a 2003 qualitative study of migrant workers in an industrial development zone in Guangzhou. The study involved a survey of non-*hukou* migrant workers and follow-up face-to-face interviews, conducted in the summer of 2003, under the sponsorship of the Labor Union of the Guangzhou Economic and Technological Development District (GETDD). Through face-to-face interviews and focus group discussions, we obtained information about migrants' work and job satisfaction, quality of life, concerns and demands, and views about their current status and others' perceptions of them.

2 Guangzhou is the capital of Guangdong Province. It boasts the most economically potent and prosperous coastal city on the forefront of China's reforms. It has been called the "Hong Kong on the Mainland" or as the "Gold Mountain by the Pearl River." The Guangzhou metropolitan region grew to more than 10 million in population (from 3 million in 1989) and its total area grew to 5,000 km^2 (from 55 km^2 in 1989) as of 2003. The region now encompasses 10 administrative districts, five of which were formerly neighboring counties with an agriculture-based economy.

3 The development of the GETDD has stimulated industrial and urban development in nearby areas. Ample labor-intensive construction work and numerous low-paying service jobs become available as a result. Many migrant workers find jobs in these newly developed areas.

REFERENCES

Bian, Y. (2002) "Chinese Social Stratification and Social Mobility." *Annual Review of Sociology* 28: 91–116.

Bian, Y., and J.R. Logan (1996) "Market Transition and the Persistence of Power: The Changing Stratification System in Urban China." *American Sociological Review* 61: 739–58.

Bian, Y., J.R. Logan, H. Lu, Y. Pan, and Y. Guan (1997) "Work Units and the Commodification of Housing: Observations on the Transition to a Market Economy with Chinese Characteristics." *Social Science in China* 18, 4: 28–35.

Blauner, R. (1969) "Internal Colonialism and Ghetto Revolt." *Social Problems* 16: 393–408.

Casanova, G.P. (1969) "Internal Colonialism and National Development." In I.L. Horowitz, J. de Castro, and J. Gerassi (eds.), *Latin American Radicalism*. New York: Vintage Books.

Castells, M. (2000) *End of Millenium*. Oxford: Blackwell Publishers.

Chan, K.W. (1994) *Cities with Invisible Walls: Reinterpreting Urbanization in Post-1949 China*. Oxford: Oxford University Press.

Chen, X. and W. Parish (1996) "Urbanization in China: Reassessing an Evolving Model." in J. Gugler (ed.), *The Urban Transformation of the Developing World*. New York: Oxford University Press: pp. 61–90.

Cornelius, W. (1973) *Political Learning among the Migrant Poor: The Impact of Residential Context*. Beverly Hills: Sage.Evans, P. (1995) *Embedded Autonomy: States and Industrial Transformation*. Princeton: Princeton University Press.

Evans, P. (1995) *Embedded Autonomy: States and Industrial Transformation*. Princeton: Princeton University Press.

Fan, C.C. (1996) "Economic Opportunities and Internal Migration: A Case Study of Guangdong Province, China." *Professional Geographer* 48, 1: 28–45.

Hechter, M. (1976) "Ethnicity and Industrialization: The Proliferation of the Cultural Division of Labor." *Ethnicity* 3, 3: 214–24.

Lang, G., and J. Smart (2002) "Migration and the 'Second Wife' in South China: Toward Cross-Border Polygyny." *International Migration Review* 36, 2: 546–69.

Lewis, O. (1961) *The Children of Sanchez*. New York: Random House.

Li, L., S. X. B. Zhao and J. P. Tian (2003) "Self-Help in Housing and Chengzhongcun in China's Urbanization." *International Journal of Urban and Regional Research* 27, 4: 912–37.

Massey, D., and N. Denton (1993) *American Apartheid*. Cambridge, MA: Harvard University Press.

Morse, R.M. (1971a) "Trends and Issues in Latin American Urban Research, 1965–1970." *Latin American Research Review* 6, 1: 1–27.

Morse, R.M. (1971b) "Trends and Issues in Latin American Urban Research, 1965–1970." *Latin American Research Review* 6, 2: 1–45.

Murray, P., and I. Szelenyi (1984) "The City in the Transition to Socialism." *International Journal of Urban and Regional Research* 8: 90–108.

Nee, V. (1989) "A Theory of Market Transition: From Redistribution to Markets in State Socialism." *American Sociological Review* 54, 663–81.

Nee, V., and R. Matthews (1996) "Market Transition and Social Transformation in Reforming State Socialism." *Annual Review of Sociology* 22, 401–35.

Ogbu, J.U. (1974) *The Next Generation: An Ethnography of Education in an Urban Neighborhood*. New York: Academic Press.

Perlman, J.E. (1976) *The Myth of Marginality*. Berkeley: University of California Press.

Roberts, B. (1973) *Organizing Strangers*. Austin: University of Texas Press.

Roberts, B. (1978) *Cities of Peasants: The Political Economy of Urbanization in the Third World*. Beverly Hills and London: Sage.

Sassen, S. (2000) *Cities in a World Economy*. Thousand Oaks: Pine Forge Press.

Solinger, D.J. (1999) *Contesting Citizenship in Urban China: Peasant Migrants, the State, and the Logic of the Market*. Berkeley, CA: University of California Press.

Walder, A.G. (1989) "Social Change in Post-Revolution China." *Annual Review of Sociology* 15, 405–24.

Walton, J. (1976) "Urban Hierarchies and Patterns of Dependence in Latin America: Theoretical Bases for a New Research Agenda." In A. Portes and H.L. Browning (eds.), *Current Perspectives in Latin American Urban Research.* Austin: The University of Texas Press: pp. 43–69.

Wirth, L. (1938) "Urbanism as a Way of Life." *American Journal of Sociology* 44, 8–20.

Xiang, B. (1999) "Zhejiang Village in Beijing: Creating a Visible Non-State Space through Migration and Marketized Traditional Networks." In F. Pieke and H. Mallee (eds.), *Internal and International Migration: Chinese Perspectives.* Richmond, Surrey: Curzon Press: pp. 215–50.

Zhang, L. (2001) *Strangers in the City: Reconfigurations of Space, Power, and Social Networks within China's Floating Population.* Stanford, CA: Stanford University Press.

Zhang, L., X.B. Simon, and J.P. Tian (2003) "Self-Help in Housing and Chengzhongcun in China's Urbanization." *International Journal of Urban and Regional Research* 27, 4: 912–37.

Zhou, M. (1992) *Chinatown: The Socioeconomic Potential of an Urban Enclave.* Philadelphia: Temple University Press.

Zhou, M. (1993) "Underemployment and Economic Disparities among Minority Groups." *Population Research and Policy Review* 12, 2: 139–57.

Zhou, M. and J.R. Logan (1996) "Market Transition and the Commodification of Housing in Urban China." *International Journal of Urban and Regional Research* 20, 3: 400–21.

11

Migration and Housing: Comparing China with the United States

Weiping Wu and Emily Rosenbaum

Introduction

China's recent migratory flow is perhaps the largest tide of migration in human history and reflects a rapidly urbanizing society undergoing a transition from a command to a market economy. Primarily from rural to urban areas, much of the migratory flow (officially estimated at 120–150 million) involves circular movements of rural labor in search of work to augment agricultural income. Regarded as temporary migrants by government authorities, they tend to be concentrated among the most economically active group (between the age of 15 and 34). The increasing level of mobility is accompanied, however, by an institutional structure at urban destinations that still separates migrants from local residents through a household registration system (*hukou*). Migrants without local *hukou* have very limited access to local public schools, citywide welfare programs, state sector jobs, and the housing distribution system. These and other restrictions inevitably increase the costs and hardship borne by migrants. It was not until 2003 that a national circular began to call for local authorities to abolish discriminatory measures (*Migrant News*, April 2003).

Of the four theoretical approaches to China's urban transformation that are proposed in the Introduction of this volume, two – transition from socialism and the developmentalist state – are most relevant to this chapter on migration and housing. The concept of "citizenship," which we use to frame our analysis, was developed under the socialist state in the form of the *hukou* system. It is this system that determines the range of opportunities afforded migrants, including those in the housing sector. Although some small cities and towns recently have experienced a weakening of the *hukou* system, it continues to be enforced in the largest Chinese cities. The *hukou* system is one socialist institution that is slow to change in the course of transition, and remains a way for the state to exclude many peasant migrants from acquiring urban citizenship and its attendant social rights. In addition, Solinger (1999) identifies another source of migrant exclusion: low political place of labor in general. The dominance of the Communist party over the whole labor movement also means that there is hardly a question of any genuine representation of labor's interests. Where the power of permanent, resident labor is strong and its place more or less secure, the fate of migrants is less promising. Both the political system and its continuing strategy of effectively sacrificing peasants' interests for urban modernization have undercut any influence by groups that are outsiders to the decision-making process.

However, the onset of changes in this institutional structure (and the potential for future changes) reflects the rise of a new developmentalist state intent on being fully engaged in the global market. The *hukou* system has, over time, helped to create a large reservoir of surplus rural labor by making migration difficult if not impossible. Because of its low cost, this source of labor is highly desired for the full functioning of the urban economy, and thus for China's ability to compete strongly in the global economy. Hence, temporary migration (without change of *hukou*) has become an alternative to permanent migration (with *hukou* change) in meeting urban labor force needs. Migrant workers and entrepreneurs provide substantial human impetus for the rapid modernization of China's cities. The willingness to break with previous migration policy of strict control reflects the formation of a new developmentalist state that is increasingly more interested in economic growth than ideological pursuits. Under such changing development strategies during a time of transition, migrants are placed in a position where their labor is desired but their presence unwanted in cities.

Our work on migration and housing focuses on the factors that allocate different sets of housing opportunities to migrants from those available to local residents. Our approach will be to analyze migrant experience in the Chinese context by comparing and contrasting it to the situations of immigrants prevailing in the United States (US). The heterogeneity of the immigrant population of the US complicates the comparison with Chinese migrants in general and specifically with respect to housing. Yet one group of immigrants, the undocumented, share a number of commonalities with Chinese migrants, including the ambiguous position of having their labor greatly desired but their presence generally unwanted (Roberts 1997). Estimated at nearly 11 million (nearly 60 percent of whom originate in Mexico) nationwide as of March 2005, the undocumented population grows by an average of almost half a million per year (Passel 2005).

We choose to compare China's internal migrants with undocumented immigrants in the US, primarily based on the notion of exclusion. Much unlike their counterparts in other developing countries who migrate to cities with the intension to settle, migrants in China are forced into a temporary or circular pattern of mobility. The *hukou* system ensures that Chinese migrants are excluded from urban society, just as the illegal status forces the undocumented to the margins of US society, its institutions, and markets. Such exclusion shapes their housing choices, mobility patterns, and living arrangements in similar ways. Our elaboration of this theme is guided by a conceptual framework that is based on a theoretical issue: citizenship rights. In particular we highlight the important role of social networks in channeling Chinese migrants and the undocumented to certain locations in their destinations, and how the resulting residential configurations may facilitate or inhibit the creation of social institutions that can aid the adjustment of newcomers, facilitate social mobility and integration, and allow a small group of entrepreneurs to earn profits from servicing the needs of their co-ethnics.

Citizenship rights in a comparative context

Citizenship has long been an important concern of social science. Most fundamentally, as framed by Bryan S. Turner (cited in Solinger 1999), citizenship is structured by two issues. First is social membership or belonging to a community. Second is the right to an allocation of resources. Therefore, the hallmark of citizenship is exclusivity as it confers rights and privileges just to those legally living within specific borders. Members typically have a broader range of rights than do non-members, which, in combination with the distinction between these two classes of individuals, makes citizenship a factor underlying patterns of inequality (Glenn 2000). Membership may be defined in terms of civic activities (such as working and paying taxes) or on ethno-cultural terms, such as speaking the dominant language or belonging to the dominant nationality.

Underlying the ethno-cultural model of citizenship are such key questions as: How can the "Other" become a citizen when subjected to a gamut of exclusionary practices? How are ethnic minorities to be incorporated as citizens with equal rights (Castles and Davidson 2000)? The nature of citizenship has become even more complex as a result of intense globalization, transnationalism, and the associated mass movement of people and goods. Countries vary greatly in their notions of membership, as well as in their generosity with respect to the access to key institutions, such as the housing market, that they afford to non-members.

One may question whether it is appropriate to employ the term citizenship in dealing with internal migrants. But Solinger (1999) explains convincingly that citizenship, especially in terms of civil and social privileges, applies well in China. A number of factors make China's rural-urban migrants more like immigrants from developing to developed counties than internal migrants within developing countries. For example, a large gap in the income, cultural values, and living standards exists between areas of origin and destination. More importantly, there are restrictions preventing or obstructing migrant settlements in destinations, ranging from labor market discrimination to China's *hukou* system (Roberts 1997; Solinger 1999). In fact, the *hukou* system segregating migrants from the urban population may be a much more important factor than cultural barriers in accounting for migrant marginality and denial of their citizenship rights.

Solinger (1999) argues that unlike what is predicted by previous social theorists, the coming of the market in China does not guarantee a full-fledged citizenship for peasants, as the state continues to use its rules and bureaucracies to prevent peasant migrants from becoming urban citizens. The two kinds of citizenship, urban and rural, that were created during the era of state socialism now form the basis of the broadest kind of social inequality in China. Whereas urban Chinese citizenship comes with full provision of social welfare, rural citizenship essentially entails self-responsibility in food supply, housing, employment/income, and lacks most of the welfare benefits enjoyed by urban Chinese; as such, rural citizenship may be characterized as half-citizenship or second-class citizenship. With deepening privatization and commercialization, citizenship rights no longer neatly correlate with the rural-urban dichotomy (Zhang 2001). Inevitably there are variations among migrants. Some have gained access to limited benefits (e.g., dormitory housing) by signing employment contracts with urban enterprises through official sanctions. On the other hand, some with capital and skills have resorted to self-employment and paradoxically been living on the fringe, often literally, in the cities. But, in general, the citizenship of migrant workers in Chinese cities can be conceptualized as falling under the ethno-cultural rubric – being subject to various exclusionary practices and without equal rights despite their nativity status. Specifically their rights to housing are restricted by the unique institutional barrier of the *hukou* system, causing them to live in poorer-quality housing than non-migrants (Wu 2002a).

In the US, there are far fewer legal structures differentiating citizens from non-citizens in terms of their rights and their access to important institutions, such as the housing market. In general, legal permanent residents have access to most of the civil rights (except voting) afforded citizens, whether naturalized or born there, and a number of Supreme Court cases have, over the years, recognized the rights of even undocumented individuals (Alienikoff 2003; Glenn 2000; Plascencia, Freeman, and Seltzer 2003; Ueda 2003). The broad overlap between the civil rights of non-citizens and citizens likely underlies the often-repeated finding that immigrants' locational outcomes remain largely a function of the spatial assimilation process (Alba and Nee 2003; Rosenbaum and Friedman 2007). In other words, the quality and location of immigrants' housing in the US tends to be determined more by levels of socioeconomic status and acculturation-related variables than by whether they have acquired legal citizenship.[1] In short, once immigrants arrive in the US, there are few, in any, legal structures that shape their access to housing in a way that is comparable in strength and significance to China's *hukou* system. Indeed, on some dimensions, immigrants actually occupy housing of higher quality than that occupied by native-born households (Schill, Friedman, and Rosenbaum 1998).

However, the general absence of legal restrictions on immigrants' housing opportunities does not always translate into complete equality of outcomes. The housing opportunities available to immigrant households of color are often of the same poor quality as those available to native-born households of color, indicating that race/ethnicity is a far more potent predictor of locational outcomes than is nativity status (Friedman and Rosenbaum 2004a, 2004b; Rosenbaum and Friedman 2001, 2007). Moreover, the persistence of racial/ethnic inequalities in housing and neighborhood conditions underscores the notion that having legal protection of one's civil rights does not always translate into the *actual* protection of those rights (cf. Basok 2004; Glenn 2000). The findings from the most recent Housing Discrimination Study (HDS) revealed that while the incidence of inadequate treatment of home-seeking minority households has declined over time, blacks and Hispanics continue to encounter obstacles in their housing searches, especially in the area of steering[2] (Charles 2003; Turner et al. 2002). Thus, in the US, while there are no legally recognized or constructed barriers to housing choice that separate members from non-members and ensure the lower quality of non-members' housing environments, the segmentation of the housing market along the lines of race/ethnicity – regardless of citizenship or nativity status – ensures the physical separation of whites from racial minorities and disproportionately places racial minorities (and especially blacks and non-white Hispanics) in housing environments that are not just of low quality, but that are devoid of many of the place-based resources that help to facilitate upward social and economic mobility. Membership and access to rights, then, in the US context are based on ascribed characteristics but do not rest upon a legally recognized foundation.

Confronting the "Double Divide" in China's Housing System

There are two layers of barriers that most migrants encounter in the urban housing sector as a result of being from rural China (rural-urban divide) and being outsiders (local v. non-local divide). While major barriers are institutionalized in legal restrictions, others are social norms shaped in time. The rural-urban divide has its historical roots. By the early twentieth century, the old Western cliché of urban superiority became a new trend in Chinese society. Treaty ports and the concentration of modern industries and commerce in urban areas brought job opportunities and material comfort as well as progressive ideas and advanced education to the city (Rudolph and Lu, Chapter 8). Despite efforts to reduce the distinction between city and countryside after the Communist Party took power in 1949, the rural-urban distinction continued to widen. The pursuit of a development strategy that promoted heavy industries during the 1949–78 period was a key root cause. This strategy aimed at rapid industrialization by extracting agricultural surplus for capital accumulation in industries and for supporting urban-based subsidies. The main enforcement mechanisms were a trinity of institutions, including the unified procurement and sale of agricultural products, the commune system, and *hukou* (Yang and Cai 2000). The *hukou* system has long been used to restrict migration, especially from rural to urban areas. Bound to collective farming, peasants were completely cut off from many urban privileges.

The rural-urban divide in housing provision dated back to the early socialist period as housing had long been a form of social welfare to urban residents only until recent housing reforms. The dominant route, prior to 1999, was through a system of low-rent housing distributed by either work units or municipal governments (such provision of welfare housing ended in 1999). This urban welfare housing system, however, did not apply to local residents with rural *hukou* or peasants in the countryside, who did not have access to either municipal or work-unit public housing. Traditional family houses and private housing constructed on land allotted by production brigades were the norm for them, even in rural pockets within cities (Wu 2002a). Even with housing reforms, these residents still do not have access to the housing provident funds established in the cities and related low-interest mortgages, as they have not been treated officially as urban population since 1964 in many cases.

The local v. non-local divide also has historical roots, as migrants throughout China's history have encountered varied forms of mistreatment. One of the best known examples is the low socioeconomic status in Shanghai held by migrants from northern Jiangsu in the republican period (Honig 1992). Distained by local residents, they were given a common depreciating nickname and confined to undesirable jobs and overcrowded shack settlements. Discrimination against migrants also was not limited to urban areas as shown in the case of Hakkas people throughout southeastern China. A migrant group distinguished primarily by dialect, they were often the objects of prejudice, described as uncivilized and poor (Honig 1992).

The system of *hukou* has institutionalized this local-non-local divide and exerted a profound impact on China's migrants. *Hukou* defines whether a migrant is permanent or temporary (also see Chan 1996). The distinction between permanent and temporary migration is important, since permanent migration with official change of *hukou* has continued to be strictly controlled. To permanent migrants, access to urban amenities and state-sector jobs is fully guaranteed. These are often not available to migrants without local *hukou*. Recent *hukou* reforms have begun to address the double divide, but changes tend to focus on towns and small cities.[3] The reforms enable rural migrants with stable jobs and residences to register as urban residents and to obtain social services, primarily their children's education. Almost all the large cities, however, continue to place significant limits on eligibility for urban registration.

Housing reforms gradually implemented during the past two decades seem largely to overlook the needs of the migrant population in cities even though they have broadened housing choices for urban residents (see Table 11.1). A local urban *hukou* continues to be an important qualification for accessing several types of urban housing, particularly those that are more affordable. Migrants cannot acquire either the use right or ownership of municipal and work-unit public housing directly because only sitting tenants (local urban residents) can do so. Both the Economic and Comfortable Housing and affordable rental units also are reserved for local urban residents only. On the secondary housing market where older housing units are traded, participation generally requires a local *hukou* although theoretically migrants can purchase housing there after completing a lengthy process of official approval. Commodity housing, the only real property sector open for migrant ownership, is not affordable for most migrants. Even local residents report difficulty in affording new commodity housing (Rosen and Ross 2000). In addition, a local urban *hukou* is required to qualify for bank mortgages for new commodity housing (Bi 2000). As a result of these restrictions, the rural-urban divide in housing continues even after rural migrants move to cities. Migrant status has brought a new dimension (local v. non-local) to this divide, as migrants from urban origins do not enjoy the same access as local urban residents.

Given this larger context, migrants display different housing behaviors from not only local residents but also migrants in other developing countries. Home ownership is yet to become an attainable goal and, therefore, the security offered by housing tenure is less relevant as a motivation for migrants in making housing decisions. China's migrants, however, do share some behaviors with their counterparts elsewhere in that they all tend to invest little income to improve housing conditions. The critical factor often lies in the intention of migrants and their commitment to cities. Informal settlements are not a viable option for China's migrants, unlike in many other developing countries (particularly in Latin America), largely due to municipal authorities' intolerance of migrant congregation and squatting. However, a number of large migrant settlements or communities have existed in Beijing for over a

Table 11.1 Types of urban housing and their availability to migrants in urban China

Type of housing	*Qualification*	*Availability to migrants*	
		Own	*Rent*
Commodity housing	Anyone, but only those with local urban *hukou* can qualify for bank mortgage loans	Yes	Yes
Economic and comfortable housing	Local urban residents with low or medium income can purchase at subsidized price		Yes
Municipal public housing	Sitting local urban tenants can purchase and trade units on secondary housing market		Yes
Work-unit public housing	Sitting local urban tenants can purchase and transfer on secondary housing market		Yes
Low-rent housing	For rental to local urban residents with the lowest income		
Resettlement housing	For local urban residents relocated from areas undergoing redevelopment		Yes
Private housing	Pre-1949 urban housing units passed on within family and housing in rural areas	Yes	Yes
Dormitory	Housing managed by local enterprises or institutions		Yes
Migrant complex	Housing managed by local government agencies for migrants		Yes

Rental is directly from sitting residents with rental permits in all types of housing except dormitory housing and migrant housing complex.
Source: Based on Wu (2004).

decade (Ma and Xiang 1998). Located mostly in suburban areas where rental housing is readily available in contrast to the city's downtown districts, these migrant enclaves or villages are formed by migrants from the same province or region. Unlike most migrant squatter settlements in other developing countries, migrants rent from local residents or live in market areas constructed by local governments or private businesses.

Evidence from Shanghai and Beijing substantiates these patterns. Home ownership is minimal (about 1 percent) for migrants. Renting represents the best choice for migrants and more than half of them in both cities are renters. Private rental housing accommodates the largest number of migrants (about

43 percent), especially in suburban areas that used to be or still are agricultural. Another 14 percent of migrants rent public housing. In addition, when migrants find jobs in state and some collective enterprises, many of them (33 percent) also obtain access to dormitory housing provided by the enterprises (Wu 2004). A small percentage of migrants (less than 10 percent) have no adequate housing and live in less stable arrangements such as on boats and in public spaces.

Migrants' growing demand for housing and their exclusion from the mainstream urban housing distribution system contribute to the chaotic situation of the urban rental market. As cities scramble to develop effective rental regulations, an increasing amount of deleterious building and rental activity continues, largely in the form of unauthorized construction and leasing of unsafe dwellings. This problem is particularly serious in urban-rural transitional areas where land is more readily available, the migrant population is more concentrated, and local residents have more incentive to rent out rooms due to the loss of agricultural income. Even when regulations about rental housing take shape in some cities, concerns for adequate housing conditions and rental rights tend to be secondary.

It is no exaggeration to say that once in the city, migrants continue to be on the move. With substantially higher mobility rates than local residents, they experience much more residential instability. Research on migrant housing in Latin American cities reveals that new migrants (labeled as "bridgeheaders") initially seek deteriorating rental shelter, primarily in the central city but sometimes scattered across town for good access to jobs. Over time, migrants generally occupy better housing – from rented rooms to self-built shanties or houses often closer to the urban periphery. Once this transition is made, migrants become consolidators (Klak and Holtzclaw 1993;Turner 1968). Few migrants in urban China, however, make the transition from bridgeheaders to consolidators even after years of living in the city. Instead, most remain trapped in the private rental sector or staying in dormitory housing. One possible explanation may be migrants' intention to maintain ties to places of origin, since many see migrating to cities as a seasonal pursuit to augment agricultural income. But the main explanation would lie with local controls, which force migrants (even those with families in tow) into more of a bridgeheader existence than they may otherwise prefer. Specifically, the system of granting only temporary urban residence permits to migrants discourages them from making the tenure transition into self-help housing or ownership.

Finding a Place to Stay: Undocumented Immigrants in the US

As alluded to above, the long-term process of settlement for a migrant or immigrant group can be viewed as a continuous progression from the sojourner stage to the final stage of settlement and integration. The initial sojourning stage is most relevant to the present discussion, as the majority of group members at this stage are undocumented and male, move back and forth between origin and destination, see themselves as members of the origin community, and remit most of what they earn from their unstable and often seasonal jobs (Massey 1986).

For some at this stage of the settlement process, the search for accommodations is simplified, or really nullified, by living in the housing provided by their employers. This is often the case for migrant farm workers, who may live in trailers, shacks, or other forms of shelter on site, and naturally for live-in domestics. Yet for those who cannot rely on employer-provided housing, the search for a place to stay is shaped on the one hand by preferences and needs, and on the other by the availability of housing options that suit these preferences and needs. Traditional approaches to residential mobility (e.g., Rossi 1955) would add that the pool of potential housing options is further defined by the migrant's ability to pay for the housing attributes he prefers or needs. Yet for the undocumented, cost acts more profoundly as a preference or need, given the goal of earning as much as possible in a short period to supplement their families' budgets at home. Therefore, the need to minimize costs while in the destination takes priority over other housing considerations, such as habitability, privacy, or even safety (Burgers 1998; Hagan 1994; Mahler 1995; Massey 1986). The emphasis on minimizing costs is facilitated by the fact that the typical migrant at this stage is an unattached, young male (Mahler 1995).

Another set of factors influencing the housing options the undocumented consider relate to their jobs. Proximity to work ranks high on the list of preferences and needs, given the long hours most work at what are almost inevitably physically exhausting and dirty jobs. Accommodations are used mainly as places to sleep and prepare for the next day's labor. Recognizing that their illegal status means that their jobs may be unstable, and that they may need to move on with little notice, many undocumented migrants in the US have strong preferences for flexible rental arrangements. In other words, this means that few migrants in such a position want to be encumbered by a lease, instead choosing places where transience is common and accepted behavior.

In the US, these housing needs are often difficult to meet, especially in the areas in which immigrants concentrate. In the metropolitan areas that attract a disproportionate share of immigrants to the US (e.g., New York and Los Angeles), vacancy rates are low and costs are high. Apart from hotels, most rental opportunities in the formal housing market involve signing a lease and assuming legal responsibility for the tenancy of the unit. Thus, few options suiting the needs and preferences of the undocumented exist in the formal housing market.

The absence of housing options in the formal market that fit the preferences and needs of migrants require that migrants use informal means of finding housing. Prominent here is the use of information garnered from social ties. Most new arrivals stay at least initially with friends and family members who preceded them to the destination. For example, using data from the Mexican Migration Project, Espinosa (1998) found that the majority of male household heads received help with finding a place to stay on their first trip from a relative or friend. Similar findings of the use of social capital at this stage of the settlement process are reported for undocumented Central Americans on Long Island, New York (Mahler 1995) and undocumented Maya in Houston, Texas (Hagan 1994). Stemming from relationships based in the village of origin, at this early stage of the settlement process, social ties for the undocumented are almost universally with co-ethnics (Massey 1986).

The undocumented newcomer's reliance on social capital to locate housing takes on greater significance in light of his illegal status. Compounding the difficulties stemming from the absence of knowledge about housing options in the destination, being illegal makes it essential to avoid detection and its associated risk of identification and deportation. Thus, persons in this position do not have the kind of independence or freedom to search openly for housing (or jobs; cf. Aguilera and Massey 2003), as do persons with legal documents. They are left dependent on their social ties, and the information they contain, to establish themselves in the destination.

The use of social networks to find housing tends to concentrate migrants in certain areas and neighborhoods. As more migrants arrive and their presence becomes noticeable, some long-term residents may begin to move out of the area, opening up vacancies for migrants to occupy. This process of neighborhood transition has long roots in those US cities, like New York, which have historically attracted immigrants and native-born minority groups. Eventually, offshoots of the origin community become established at the destination. As this process of neighborhood transition gathers speed, the migrant group's presence becomes more evident as some group members become small entrepreneurs offering services and goods that infuse the new neighborhood with the familiarity of home. Spatial concentration of the group, therefore, enables the growth of a local, informal infrastructure that provides many of the goods and services that are necessary for daily life but remain unavailable (at least in familiar forms) within the formal market.

One area that small entrepreneurs turn to is the provision of housing, in the form of an informal housing market that parallels the formal market but provides options that fit the needs of migrants. Mahler (1995) describes the informal housing market for the undocumented Central Americans she studied as the "encargado system." In this system, older, more settled migrants with some capital rent a house or apartment, and sublease space within the unit to co-ethnics. The encargado may also reside in the unit, or may operate it as a boarding house and live elsewhere. Encargados do not view their activities as a form of service to their compatriots, but as constituting an investment that will enable them to either cover their own housing costs, or provide them with a significant profit. In many instances, the housing created by the encargado may be illegal, as in the case of apartments illegally carved out of basements or garages. The spread of illegally converted basements and garages, moreover, often becomes a flashpoint for long-term area residents. One option for long-term residents is to move away to avoid what they see as deteriorating conditions and overuse of the local infrastructure, which hastens the process of neighborhood transition, while others become vocal protesters against both the illegal conversions and immigration in general (e.g., Sanjek 1998).

The quality of the accommodations the undocumented acquire within the informal market may be quite low, especially in the case of illegally converted units. The emphasis on cost over quality leads many migrants to rent a bed-sized space, or even just a bed in a room occupied by other migrants. The frequent churning of migrants through such accommodations (as some leave to pursue jobs elsewhere, to be replaced by new arrivals) wears down the physical amenities in the house or apartment. Similarly, the overuse of facilities, especially when this overuse accumulates over time, adds to the deterioration of the structure. Mahler (1995) provides vivid descriptions of the grim conditions her undocumented Central American informants lived in, including extensive pest infestation, bath and kitchen facilities (if present) shared by many and so forth. Moreover, most of her informants paid higher rents than the quality of the accommodations could justify. Thus, as a captive market, many undocumented migrants endure low quality living conditions while paying high rents.

There is, however, variation along these lines. Because the informal market parallels the larger formal market, the quality and cost of housing available in the former are strongly influenced by the state of the latter. Mahler's (1995) informants lived on Long Island where the segregated housing market was made up of a housing stock that was decades old and contained few rental units. Illegal conversions were therefore commonly done to provide additional units to rent out.[4] The limited supply of housing, combined with high and rising demand, created a situation of high rents and low quality. In contrast, the Maya studied by Hagan (1994) arrived in Houston after the collapse of the oil industry, where they found a housing market glutted with recently constructed and vacant units. Despite their illegal status, the Maya were greeted by landlords happy to lease otherwise empty apartments at fractions of the original rents. Hagan's Maya informants, therefore, had few problems finding affordable housing that was also in good condition.

Is the Case of Migrants in Chinese Cities Similar?

To a very large degree, the shared experience of illegality and exclusion from the formal market, combined with the need to maximize earnings, create a very similar context for Chinese migrants. As discussed above, some Chinese migrants, like the undocumented, occupy housing provided by their employers. This is sometimes the case for construction workers who occupy various (and often insubstantially built) types of shelters on site, and for single females who work as live-in domestics or in factories providing dormitory-style accommodations (Zhang, Zhao, and Tian 2003). However, others must find their own accommodations, and the absence of a formal market with affordable options to which they have a legitimate claim, forces migrants to rely on their social ties. Just as social ties for the undocumented in the US are based on common ethnicity, the social ties for Chinese migrants are based on village of birth, which acts as an ethnic marker (Logan 2002; Wu 2002b).

Chinese migrants, by virtue of needing to minimize their costs while in the city, share with the undocumented a similar set of housing needs and preferences. Thus, low cost and proximity to the workplace are higher priorities than physical quality and space. New arrivals to the city may stay with members of their social networks, or rely on the information controlled by the network to find a place to stay. Just as is the case for the undocumented, satellite or "daughter" communities of migrants from a single village develop in destination cities (Wu 2002b; Zhang et al. 2003). In both cases then, the social networks that sustain migration flows also lead to spatial concentration of co-ethnics (Fan and Taubman 2002).

The complete absence of a formal housing market available to Chinese migrants leaves them even more heavily reliant on their social networks for information than is true for the undocumented. This situation, moreover, also requires the development of an informal housing market to fill the gap left by the state's lack of involvement (Zhang et al. 2003). The parallel to the encargado system is perhaps the *chengzhongcun* (or urban village), neighborhoods created as cities expanded outward enveloping villages that were formally on the outskirts within their borders. Although located physically within the city, the local peasant residents of *chengzhongcun* have rural *hukou* status. By virtue of this status, though, the local peasants have land rights and are exempt from the building codes and regulations that are in force in the rest of the city. Many local peasants, therefore, expand their homes or build additional structures on their land to rent out to migrant peasants. The money they are able to earn this way is essential to supplement the little they are able to earn from their land (Fan and Taubman 2002; Wu 2002b). Thus, as in the encargado system, the system prevailing in the *chengzhongcun* can be said to benefit both groups. The local peasants earn needed supplemental income and the migrant peasants are provided with affordable housing options that suit their other housing needs (Zhang et al. 2003). Instead of sharing a common ethnicity (as is the case in the encargado system), renters and owners share a common peasant heritage.

The role of social networks is best exemplified in the ongoing, intense struggle over housing and the use of urban space by the Wenzhou migrants in Beijing's Zhejiang Village – a migrant community formed on the basis of a shared place identity. They first rented from local residents in this urban-rural transitional area. Then a group of migrants with more economic and social capital invested in the development of large, private housing compounds. They gained access to land for housing construction and obtained limited infrastructure resources by buying off local village and township cadres and forming informal economic alliances with them. Built on extended kinship ties, clientilist networks with local cadres, and voluntary gang-like groups, a shadow migrant community and leadership structure emerged (Xiang 2000; Zhang 2001, 2002). Within the community, allocation of housing, production. and marketing space, and policing and social services proceeded through networks centered on the migrant bosses of housing compounds and market sites. This explosive growth outside party-state structures worried authorities and led them to order the demolition of many housing compounds several times in the past decade or so. The migrants, desperate to stay in the lucrative urban market at the center of the national transportation network, continued to rebuild their community within months of each raid. This example shows that rather than demanding specific welfare benefits from the state, migrant entrepreneurs aspire to urban citizen status mainly to gain a secure space of their own for business and living quarters. Housing space, therefore, figures as the central element in their bid for urban citizenship (Zhang 2002).

Exceptions to the "Rule" of Commonality and Future Prospects

Despite the similarities linking undocumented immigrants in the US and Chinese rural-to-urban migrants, the larger social, economic, and political contexts in which they live inevitably cause their experiences and their potential for integration to diverge in certain ways. One relates to the legal status of children, and thus the family's right to certain social benefits. US-born children of undocumented migrants are granted US citizenship, while the children of Chinese migrants are limited to their parents' *hukou* status regardless of their place of birth. As a result of the US' *jus soli* system of citizenship, barriers to social rights and their associated benefits disappear for the second generation, increasing the chances for the immigrant group's eventual integration into US society.

A second key difference is that all children of undocumented immigrants are entitled to such universal resources like public elementary and secondary education, and their entitlement to other programs (such as Medicaid) varies by state. In contrast, education is one of the many things persons with rural *hukou* are barred from while in the city. Thus, children of the undocumented, regardless of their citizenship status, enjoy broader opportunities than do the children of Chinese migrants.

A third difference relates to the recent shift in US policy towards stricter exclusionary rules aimed at immigrants, which have, by definition, also affected the undocumented. Prior to the legislative changes of the 1990s that limited immigrants' eligibility for a range of social welfare programs and required that governmental agencies affirm an applicant's legal status, the undocumented could gain access to a number of programs, including rental housing assistance.[5] With the shift in policy, many households headed by undocumented individuals withdrew from such programs, fearing detection and deportation. In situations where the respective government agencies refused to identify the legal status of both applicant and within-system households, some stayed in subsidized housing (and perhaps in other programs as well). Yet recent decisions by some agencies to reverse their position on verifying legal status has either uprooted such households or have lead to rent increases for those that contain citizen members (Stewart 2004). In contrast, the exclusion of Chinese migrants from any form of state-sponsored housing in cities, and the impermeability of this barrier, has a far longer history.[6] As China further undergoes the plan-to-market transition, housing commodification and socioeconomic stratification are likely leading to urban spatial restructuring. Increasing migrant concentration, given that most migrant housing is in much worse conditions than local housing, may aggravate existing residential differentiation. Settlement patterns will also influence the future socioeconomic standing of migrants. Experience of other developing countries shows that severe shortages of affordable housing tend to force migrants to live in squatter settlements and slums, many of whom eventually become the urban underclass.

Given the enormous magnitude of migration and its potential impact on China's cities, it is important to explore ways in which migrant access may be broadened, especially in light of the declining role of the state and work units as service and housing providers. Further housing reform, for instance, should aim to help improve migrants' quality of life and prevent slum formation. Specifically, migrants can be allowed to participate in the secondary housing market, where older apartment units are more affordable. Of course, these measures do not fully address the costs and hardship migrants bear as a result of the long-standing double divide.

To respond properly to the need of migrants and their quest for citizenship rights entails that the linkage between *hukou* and the provision of urban services be discontinued. With more tolerant migration policies in China, over time urban ties will surpass rural ties and many migrants may choose to settle permanently at urban destinations. It is conceivable that migrants will make a different range of decisions regarding housing choice and tenure, probably breaking with some of the patterns seen among undocumented immigrants in the US.

NOTES

1 In fact, although traditional approaches to assimilation tend to view the acquisition of citizenship as an indicator of complete integration within US society (Glenn 2000), citizenship status is rarely, if ever, considered as a predictor of immigrants' locational outcomes.
2 "Steering" occurs when real estate actors channel home seekers toward or away from specific neighborhoods based on some aspect of the home seeker's identity, such as race/ethnicity.
3 It was reported that in 2001, approximately 600,000 rural residents acquired urban *hukou* in small urban areas (http://www.usembassy-china.org.cn/econ/hukou.html (accessed May 17, 2005).
4 To remain in their increasingly expensive homes, some long-term area residents participated in the informal housing market, by subleasing illegally converted space in their homes to members of the Central American community (Mahler 1995).
5 However, because the waiting lists for public housing units and housing vouchers are extremely long, immigrant households are far less likely than native-born households to receive such benefits (Friedman, Schill, and Rosenbaum 1999).
6 That the undocumented in the US are ineligible for publicly assisted housing could be said to parallel the ineligibility of Chinese migrants for any form of state-subsidized or state-produced housing in cities. This shared experience of exclusion from state-sponsored housing reflects, then, a common barrier to the acquisition of *affordable* housing.

REFERENCES

Aguilera, M. and D. Massey. (2003) "Social Capital and the Wages of Mexican Migrants: New Hypotheses and Tests." *Social Forces* 82, 2: 671–701.

Alba, R., and V. Nee. (2003) *Rethinking the Mainstream: Assimilation and Contemporary Immigration*. Cambridge: Harvard University Press.

Alienikoff, T.A. (2003) "Policing Boundaries: Migration, Citizenship, and the State." In G. Gerstle and J. Mollenkopf (eds.), *E Pluribus Unum: Contemporary and Historical Perspectives on Immigrant Political Incorporation*, New York: Russell Sage Foundation: pp. 267–91.

Basok, T. (2004) "Post-national Citizenship, Social Exclusion, and Migrants' Rights: Mexican Seasonal Workers in Canada." *Citizenship Studies* 8, 1: 47–64.

Bi, B. (2000) *Housing: Purchase, Sale, Transfer, Rental, and Use [zhufang: mai mai huan zu yong]*. Beijing: China Development Press.

Burgers, J. (1998) "In the Margin of the Welfare State: Labour Market Position and Housing Conditions of Undocumented Immigrants in Rotterdam." *Urban Studies* 35, 10: 1855–68.

Castles, S., and A. Davidson. (2000) *Citizenship and Migration: Globalization and the Politics of Belonging*. Basingstoke: Macmillan.

Chan, K.W. (1996) "Post-Mao China: A Two-Class Urban Society in the Making." *International Journal of Urban and Regional Research* 20, 1 (March): 134–50.

Charles, C. (2003) "The Dynamics of Racial Residential Segregation." *Annual Review of Sociology* 29: 167–207.

Espinosa, K. (1998) "Helping Hands: Social Capital and the Undocumented Migration of Mexican Men to the United States." PhD Dissertation, University of Chicago.

Fan, J., and W. Taubman. (2002) "Migrant Enclaves in Large Chinese Cities." In J. Logan (ed.), *The New Chinese City: Globalization and Market Reform*. Oxford: Blackwell: pp. 183–97.

Friedman, S., and E. Rosenbaum. (2004a) "Nativity-Status Differentials in Access to Quality Housing: Does Home Ownership Bring Greater Parity?" *Housing Policy Debate* 15, 4: 865–901.

Friedman, S., and E. Rosenbaum. (2004b) "Does Suburban Residence Mean Better Neighborhood Conditions for All Households? Assessing the Influence of Nativity Status and Race/Ethnicity." Paper presented at the 2004 annual meeting of the Population Association of America.

Friedman, S., M. Schill, and E. Rosenbaum. (1999) "The Utilization of Housing Assistance by Immigrants in the United States and New York City." *Housing Policy Debate* 10, 2: 443–75.

Glenn, E. (2000) "Citizenship and Inequality: Historical and Global Perspectives." *Social Problems* 47, 1: 1–20.

Hagan, J. (1994) *Deciding to be Legal: A Maya Community in Houston*. Philadelphia: Temple University Press.

Honig, E. (1992) *Creating Chinese Ethnicity: Subei People in Shanghai, 1850–1980*. New Haven: Yale University Press.

Klak, T., and M. Holtzclaw. (1993) "The Housing, Geography, and Mobility of Latin American Urban Poor: The Prevailing Model and the Case of Quito, Ecuador." *Growth and Change* 24, 2 (Spring): 247–76.

Lee, J. (2004) "Lost Fruit in Central Florida Means Lost Jobs for Migrants." *The New York Times*, October 10: A22.

Logan, J. (2002) *The New Chinese City: Globalization and Market Reform*. Oxford: Blackwell.

Ma, L.J.C., and B. Xiang. (1998) "Native Place, Migration and the Emergence of Peasant Enclaves in Beijing." *The China Quarterly*, 155 (September): 546–81.

Mahler, S. (1995) *American Dreaming: Immigrant Life on the Margins*. Princeton: Princeton University Press.

Massey, D. (1986) "The Settlement Process Among Mexican Migrants to the United States." *American Sociological Review* 51: 670–84.

Massey, D., and N. Denton. (1994) *American Apartheid: Segregation and the Making of the Underclass*. Cambridge: Harvard University Press.

Passel, J. (2005) *Estimates of the Size and Characteristics of the Undocumented Population*. Washington, DC: The Pew Hispanic Center.

Plascencia, L., G. Freeman, and M. Seltzer. (2003) "The Decline in Barriers to Immigrant Economic and Political Rights in the American States 1977–2001." *International Migration Review* 37, 1: 5–23.

Roberts, K.D. (1997) "China's 'Tidal Wave' of Migrant Labor: What Can We Learn from Mexican Undocumented Migration to the United States?" *International Migration Review* 31, 2 (Summer): 249–93.

Rosen, K.T., and M.C. Ross. (2000) "Increasing Home Ownership in Urban China: Notes on the Problem of Affordability." *Housing Studies* 15, 1: 77–88.

Rosenbaum, E., and S. Friedman. (2001) "Differences in the Locational Attainment of Immigrant and Native-born Households with Children in New York City." *Demography* 38, 3: 337–48.

Rosenbaum, E., and S. Friedman. (2007) *The Housing Divide: How Generations of Immigrants Fare in New York's Housing Market.* New York: New York University Press.

Rossi, P. (1955) *Why Families Move: A Study in the Social Psychology of Urban Residential Mobility*. Glencoe, IL: Free Press.

Rudolph, J., and H. Lu. (2004) "The Legacy of (Semi-)Colonialism: Urban Identity in Shanghai and Taipei." Paper presented at the Santa Monica workshop of the Urban China Research Network.

Sanjek, R. (1998) *The Future of Us All: Race and Neighborhood Politics in New York City*. Ithaca: Cornell University Press.

Schill, M., S. Friedman, and E. Rosenbaum. (1998– "The Housing Conditions of Immigrants in New York City." *Journal of Housing Research* 9, 2: 237–70.

Solinger, D.J. (1999) "Citizenship Issues in China's Internal Migration: Comparisons with Germany and Japan." *Political Science Quarterly* 114, 3 (Fall): 455–70.

Stewart, J. (2004) "Officials Get Tough on Rent Subsidies." *The Los Angeles Times* June 11: B1.

Turner, J.F.C. (1968) "Housing Patterns, Settlement Patterns, and Urban Development in Modernizing Countries." *Journal of the American Planning Association* 34, 6 (November): 354–63.

Turner, M.A., S. Ross, G. Galster, and J. Yinger. (2002) *Discrimination in Metropolitan Housing Markets: National Results from Phase I HDS 2000.* Washington, DC: US Department of Housing and Urban Development.

Ueda, R. (2003) "Historical Patterns of Immigrant Status and Incorporation in the United States." In G. Gerstle and J. Mollenkopf (eds.), *E Pluribus Unum: Contemporary and Historical Perspectives on Immigrant Political Incorporation.* New York: Russell Sage Foundation: pp. 292–330.

Wu, W. (2002a) "Migrant Housing in Urban China: Choices and Constraints." *Urban Affairs Review* 38, 1: 90–119.

Wu, W. (2002b) "Temporary Migrants in Shanghai: Housing and Settlement Patterns." In John Logan (ed.), *The New Chinese City: Globalization and Market Reform.* Oxford: Blackwell: 212–26.

Wu, W. (2004) "Sources of Migrant Housing Disadvantage in Urban China." *Environment and Planning A* 36, 7 (July): 1285–1304.

Xiang, B. (2000) *A Community that Crosses Boundaries: The Living History of Beijing's Zhejiang Village* [kuayue bianjie de sheqiu: Beijing zhejiangcun de shenghuo shi]. Beijing: Sanlian Publishing House.

Yang, D.T. and F. Cai. (2000) "The Political Economy of China's Rural-Urban Divide." Working Paper no. 62, Center for Research on Economic Development and Policy Reform, Stanford University.

Zhang, L. (2001) *Strangers in the City: Space, Power, and Identity in China's Floating Population.* Stanford: Stanford University Press.

Zhang, L. 2002. "Spatiality and Urban Citizenship in Late Socialist China." *Public Culture* 14, 2: 311–334.

Zhang, L., S.X.B. Zhao, and J.P. Tian. (2003) "Self-Help in Housing and Chengzhongcun in China's Urbanization." *International Journal of Urban and Regional Research* 27, 4: 912–37.

Part IV

Social Control in the New Chinese City

12

Economic Reform and Crime in Contemporary Urban China: Paradoxes of a Planned Transition

Steven F. Messner, Jianhong Liu, and Susanne Karstedt

Introduction

As noted throughout this volume, China has experienced profound social change as a result of its economic reform. Indeed, the changes in Chinese society have been of such magnitude and rapidity that it is hard to imagine that crime rates would not be affected. Although the data are limited in many aspects, the best available evidence indicates that crime rates have risen significantly in China over recent decades (Bakken 2005a; Liu and Messner 2001).

Contemporary China might thus be seen at first glance as simply another case providing support for the criminological variant of modernization theory. On the basis of a time-series analysis of official statistics, Liu and Messner (2001) have argued that the classic modernization thesis in criminology is broadly applicable to China. At the same time, the authors caution that a full understanding of the Chinese crime situation requires appreciation for the "distinctiveness of the Chinese context" (2001, p.18).

The purpose of this chapter is to examine the ways in which the Chinese experience with crime conforms to that observed in other nations during the transition to modernity and post-modernity, and the ways in which the Chinese experience deviates from that of other nations. We begin by briefly reviewing the major paradigms on societal development and crime that have informed criminological research. These include variants of modernization theory, world systems/dependency theory, and globalization/marketization theory. However,

China is also a case to which theories of market transition from socialism apply. We then use the insights from these paradigms to understand the crime situation in urban China during the recent period of economic reform, anticipating that each will add a new facet to our understanding.

Our basic argument is that Chinese cities exhibit both commonalities and uniqueness when considered in comparative contexts. The commonalities can be usefully attributed to the operation of certain "master forces" associated with the general process of modernization, while the unique features can be understood with reference to "pathway forces" that are unfolding within the context of a planned and "controlled" transition, which is a defining and unique feature of the Chinese transition process (Bakken 2005b, p. 5). In our concluding remarks, we summarize how each of the four theoretical paradigms discussed in the introductory chapter – modernization, dependency/world systems theory, theories of the developmentalist state, and theories of market transition from socialism – offer insights for understanding different aspects of the changing crime situation in China.

Criminological Approaches to Societal Development and Crime

The origin of the research on crime and societal development goes back to the classic work of Emile Durkheim. Durkheim argued that in times of rapid social change, consensus on societal values breaks down, resulting in a weakening of the normative order – anomie. For Durkheim, anomie is a principal source of increased crime and other social pathologies in societies undergoing profound social transition (Durkheim 1933, 1950; Hinkle 1976).

Several modern variants of the Durkheimian theory have been proposed in the criminological literature, but the most influential by far is the modernization theory formulated by Louise Shelley (1981). Shelley posits an evolutionary model that assigns causal priority to urbanization and industrialization.[1] In some ways, urban growth and industrial development inhibit crime. The greater anonymity associated with city life decreases the likelihood of the internecine feuds that characterize rural settings. Similarly, urban development reduces the salience of the extended family, thereby eliminating the source of much interpersonal violence. Industrial development also inhibits crime by improving the quality of life for many and by providing resources for a modern criminal justice system (Shelley 1981, pp. 20–1). At the same time, urbanization and industrialization unleash forces that are destabilizing and that promote crime. Traditional social structures are undermined, leading to social disorganization and anomie. Urban growth is associated with massive migration, which contributes to crime as rural migrants bring with them the traditions of violence associated with rural life and encounter problems of adjustment in their new environments. In addition, while industrial development improves standards of living, it also creates inequalities and inflames desires for the fruits of prosperity that are not available to all (Shelley 1981, pp. 31–6).

Shelley proposes that, on balance, modernization tends to increase crime, but that the changes in levels of violent and property crime vary depending on the stage of modernization and the offense under consideration. At earlier stages of modernization, both violent and property crimes rise. At later stages, property crimes continue to rise, becoming the most prevalent type of criminal activity. In contrast, violent crime subsides as rural migrants become adjusted to urban conditions. The nature of violent crime also changes. Increasingly, criminal violence occurs in the context of the commission of property crimes.

Classic modernization theory emphasizes weakened controls and elevated criminal motivations to explain the rising crime rates that typically accompany development. "Situational" or "opportunity" factors have also been cited to account for both the higher overall levels of crime in developed societies and the greater prevalence of property crimes relative to violent crimes. Drawing upon insights from "routine activities" theory (Cohen and Felson 1979), this perspective focuses on how the developmental process changes the opportunities and incentives for different types of crime. Economic development increases the volume of valuable, often portable, property, and modern merchandising makes this property readily accessible to the general public (Neapolitan 1997, pp. 69–70; see also LaFree and Kick 1986). Developed societies, and in particular their cities, have higher levels of property crime in large part because there are more goods that can be easily stolen.

While increasing opportunities for theft, societal development also decreases social interactions that are likely to lead to violent crimes. Modern activity patterns draw people away from households and neighborhoods and thus reduce the frequency of contacts among intimates and close acquaintances – the prime setting for interpersonal conflict. Instead, the routine activities in developed societies are more likely to involve distant relationships lacking emotional intensity (see Karstedt 2003a). Opportunity theory, like classic modernization theory, thus implies that property crimes will increase and become the dominant type of criminal offense as societies develop.

Notwithstanding the fact that classic modernization theory seems to give an accurate account of the historical process in the European context (Eisner 2001), it fails to account for the impact of urbanization on crime in a number of Third World countries. The influx of migrants to cities, the development of shantytowns, and the resulting social conflicts and disorganization have increased levels of violent crime and have contributed to high levels of insecurity in many cities in the Third World. Both world systems/dependency theory and globalization/marketization theory specifically address this problem.

These two perspectives on development direct attention to the interrelationships among nations. The core claim of world systems/dependency theory is that the criminogenic conditions within lesser-developed nations arise from their subordinate position within the larger world economy (Chen 1992). Economic dependency, reflected in transnational corporate penetration and trade imbalances, promotes distorted economic and political development as manifested in extreme poverty, inequality, and political oppression. High levels of criminal violence can be understood as a response to the "frustration and misdirected anger created by deprivation and demoralizing living conditions" (Neapolitan 1997, p. 76), in particular among the impoverished urban population. In essence, the distinctive claim associated with world systems/dependency theory is that the national conditions that serve as the proximate causes of violent crime reflect an inherently exploitive international division of labor that affects the process of urbanization, and makes the cities of these countries seedbeds of violent crime. The process of urbanization no longer – as in modernization theory – reduces violence in the urban environment, but to the contrary, elevates it to extraordinary levels.

Scholarship on globalization also emphasizes the interconnections among nations. The basic claim is that the expansion of markets and the accompanying triumph of capitalism have unleashed criminogenic forces in developed and developing countries alike. These forces include the spread of rampant individualism (Currie 1991), the decline of social reforms that previously mitigated the more deleterious consequences of capitalist society (Teeple 1995), and an expansion in the opportunities for organized criminal activities at a transnational level, activities that are often accompanied by violence (Karstedt 2000; Mittelman 2000). This more comprehensive approach of globalization/marketization theory provides a global perspective on the relationship between urban development and crime, and tries to explain the simultaneous increase of violence in the cities in developed and less developed nations since the mid-1960s (Taylor, Evans and Fraser 1996).

To summarize, criminological inquiry on development and crime has been guided by several paradigms. While differing in many specific claims, these paradigms share the common logic of the general theory of modernization as well as its weak points. Societal development allegedly entails basic social structural changes or "master forces." The most important of these include urban growth, industrialization and economic prosperity, high levels of population migration, increased inequality, and more extensive ties among nations that are embodied in transnational economic exchanges (some of which may be inherently exploitive) and cultural diffusion. These master forces exert pressures for higher rates of crime through a combination of enhanced criminal motivations, weakened traditional social controls, and expanded opportunities for selected forms of criminal activity (Neapolitan 1997, p. 70).

The general model of modernization has undergone major revisions during the past decades, which have not gone unnoticed by criminologists. Seminal to these revisions has been critical historical and comparative research on the nature of the modernization process in specific nations. This research has demonstrated convincingly that very different pathways towards modernization are not only possible but in some cases highly successful. The master forces are viewed as taking variable shapes and forms according to cultural traditions, the social and political institutions that are available at the start and are built along the road, and the specific mechanisms that emerge in coping with the crises, conflicts, and imbalances that inevitably accompany the distinctive path toward modernization. Each modernizing society develops its specific features on the path that leads through these crises.

These revisions of modernization theory have been predicated on three major insights. First, tradition and modernity are not seen as a dichotomy. To the contrary, modern institutions are built on traditions and derive their legitimacy from them (North 1996). Second, the central role of social institutions in the process of modernization has been acknowledged (North 1995; Polanyi 1957), in particular the rule of law and property rights. Finally, the role of political regimes, political systems, and their institutions (North 1994), as well as the impact of governments on the course of the modernization process, has been emphasized (Tilly 1992). These revisions have resulted in a more flexible historical and comparative approach that can accommodate a variety of historical experiences and contemporary routes without imposing a Western model. In particular, the emphasis that theories of the developmentalist state put on political regimes and institutions seems to match China's pathway of transition and modernization.

Theories of market transition have stressed the role of anomic pressures as criminogenic factors in order to explain rising rates of violence and property crime during this process. A recent example of such an approach is institutional-anomie theory, proposed by Messner and Rosenfeld (2007). This theory was mainly developed to grasp the consequences of neo-liberal market policies in Western democratic countries, but it has proven equally pertinent to the analysis of the transitional process from a state-run toward a market economy (Karstedt 1999, 2003b; see also Kim and Pridemore 2005). The authors argue that in the process of modernization and transformation, societies need to balance social justice and self-interest, collective and individual interests, welfare institutions and market forces. Imbalances between institutions, the dominance of market forces and the erosion of other institutions like the family or the educational system, produce anomic forces that are conducive to crime, both property and violent crime. The imbalances that emerge from shifts within the institutional regimes of transitional societies reflect the tensions in their cultural and structural spheres that cause particular types and patterns of crimes.

In the sections that follow, we consider the ways in which the insights from the criminological paradigms on societal development can help explain the impact of social reforms on crime in contemporary urban China, and the ways in which the Chinese situation must be understood with reference to distinctive social processes, institutions, and cultural traditions. We explore the role of "master forces" as well as the impact of "pathway forces" that constitute the specific transitional route that China has taken.

Social Change and Crime in China

Trends in Chinese crime rates and cross-national comparisons

The consequences of Chinese economic reform for crime have attracted a good deal of scholarly attention both inside and outside China (for example, Bakken 2000, 2005c; Liu et al. 2001). Although differing in specific details, certain general patterns have emerged in this research. The evidence concerning trends suggests that Chinese crime rates have risen significantly over recent decades. Along with many impressionistic accounts, Liu and Messner (2001) have conducted formal time series analyses of official statistics. The results of their analyses are consistent with the widespread consensus among scholars that rates of both violent and property crimes have increased over the past decades.

The research also indicates that the pace of these increases in crime rates has varied by offense type. Similar to other countries undergoing social transition, property crimes in China have grown more rapidly than violent crimes (Liu 2006). According to official statistics, from 1978 to 2003, homicide rates have more than doubled; rape has increased 30 percent; and assault has increased 7.5 times. These increases in violent crimes are substantial, as are those for property crimes. Robbery grew by 4.5 times, and larceny increased 8 times. Fraud, only reported up to 2002, has gone up almost 14 times. Grand larceny (3,000 yuan or more), for which data were only published up to 1999, increased 90 times (*China Law Yearbook Press*, various years). These official statistics on crime must be interpreted very cautiously, and some of the increases are likely to reflect reporting and recording practices. In a very careful assessment of the available data, Bakken (2005a) concludes that it is likely that robbery and rape have been on the rise in China, while the homicide rate has been increasing slightly.

In addition to the rise in common forms of crime, corruption has become one of the most important concerns of the Chinese public and the government (Chen et al. 2004; Li 2004). It is reasonable to assume that with building booms and new industries concentrated in urban areas, this crime has particularly strong effects on the urban population, both as victims and offenders. Comparable data are available between 1999 and 2001, when the number of corruption cases filed for prosecution showed a 10 percent increase from 32,911 in 1999 to 36,447 in 2001. Other indicators also show corruption on the rise: "Big" cases, which are defined as a monetary amount embezzled or bribed above 50,000 yuan increased from 13,095 in 1999 to 16,637 in 2001. The number of higher-ranking officials involved rose from 2,019 in 1999 to 2,670 in 2001 (*China Law Yearbook Press* 2000–2002). Consistent with the official Chinese data, both the Corruption Perception Index and the Bribe Payer Index of Transparency International suggest that, in international comparisons, China ranks among the countries most affected by corruption. In the period from 2000 to 2005, China mostly ranks in the lower half of the assessed countries but seems to have improved slightly, from a relative rank of 70 in 2000 to a relative rank of 48 in 2004 (www.transparency.org/cpi/), where the relative rank indicates the equivalent position within 100 countries. Throughout this period, China ranks close to post-Soviet states like Belarus and post-communist countries like Poland, whereas other Asian states or regions like Singapore, Hong Kong or Taiwan rank in the upper 20 percent far ahead of China, with Thailand in its vicinity. It seems that corruption is pervading all institutional realms, including the judiciary (Zhou 2005). The Chinese press reported several cases of members of the judiciary who took advantage of their position to collect bribes, enjoy sexual favors or exploit women (Xiaohang 2005), and cases of large numbers of corrupt officials who tried to flee to other counties that have not signed extradition treaties with China (www.sina.com.cn 2004).

We emphasize, however, that this evidence on the crime situation in China – quantitative and qualitative – needs to be considered in a comparative context. The combination of rising rates of street crime as indicated especially by the increase in robberies, and the growth of white-collar crime, including corruption, is by no means unique to the developmental process in China. Rather, this seems to be a common characteristic of post-communist transitional societies (Hagan and Radoeva 1998; Karstedt 2003b). Moreover, current levels of crime in China are not exceptional. On the basis of statistics compiled by the United Nations, Bakken (2005a, p. 60) concludes that "China's officially reported crime rates range from fairly low to very low when compared internationally." Comparisons of victimization and self-report surveys yield a similar conclusion (Bakken 2005a, pp. 63–6).[2] Thus, despite the upward trends in crime rates over recent decades, crime in China still appears to be less pervasive than in most other nations of the modern world with the exception of corruption, and in particular compares favorably to other transitional societies and less developed nations with respect to violent crime.

A central question for criminologists, then, is how to explain this complex set of facts about crime in China: a general rise in crime rates, with a disproportionate increase in property crimes relative to violent crimes, yet with levels that remain low in comparative perspective, accompanied by growing alarm about rampant corruption. To what extent can they be accounted for by general criminological paradigms that emphasize the "master forces" of the developmental process, and how might they reflect special features of social transition in China and its "pathway forces"?

Applying the criminological paradigms to China: "master forces"

Several of the "master forces" reviewed above that allegedly accompany modernization and that exert criminogenic pressures appear to be at work in contemporary China. Urbanization has proceeded steadily, with the large cities expanding in size and with mid-level cities growing as well (Chan and Hu 2003). A considerable proportion of this growth is due to the influx of non-registered migrant workers from the countryside – "the floating people" who live at the margins of urban society. Industrial production has increased greatly, enhancing the economic prosperity of the nation. This overall increase in national wealth, moreover, has been accompanied by more pronounced social inequality (Cao and Dai 2001). The introduction of market mechanisms and the associated need for the mobility of labor were driving factors of migration from the countryside to the cities (Liang and Ma 2004), while contacts with the outside world, both economic and cultural, have grown steadily (Lardy 2001). In other words, the structural developments commonly associated with modernization – urbanization, industrialization and prosperity, growing economic inequality and global interconnectedness – have been stimulated or accelerated by the economic reforms in China, particularly in its expanding urban areas.

Scholars have related many of these "master forces" to crime in China with reference to intervening processes similar to those described in the variants of criminological modernization theory. With respect to criminal motivations, researchers have pointed to the destabilizing effects of inequality. For example, Cao and Dai (2001) propose that rising crime in China is due in part to relative deprivation accompanying increased inequality (see Bakken 2005a, p. 80).[3] Rapid increases in economic inequality also are a common characteristic of societies that undergo the transformation from a state-run to a market economy (Karstedt 2003b, p. 306).

Growing cultural contacts with the West have also been cited as a source of the strengthening of criminal motivations, especially for property crime. As market reform and the open door policy have proceeded, the so-called "modern culture" has begun to permeate society. Exposure to materialistic Western lifestyles and consumer goods, through Hollywood, television, and the Internet, has increased over recent years. Some evidence suggests that an individualistic consumerism has begun to take root in China, which is likely to contribute to property crime (Deng and Cordilia 1999). This, however, is a general complaint throughout Asia that the "advanced nations' disease" and its graphic symptoms of individualism and hedonism are at the roots of rising crime rates in the region.

The "master forces" of social change in China have also been linked with breakdowns in traditional mechanisms of social control, although these mechanisms are in some ways distinctive. In the Chinese situation, economic reform has evidently weakened neighborhood committees, which in the past practiced "reintegrative shaming" to regulate misbehavior (Rojek 2001; see also Chapter 14 in this volume). The role of work units as agencies of social control has also been reduced as new forms of employment have developed. In addition, social control institutions similar to those in the West, such as the family and schools, have greater difficulty in performing their socialization functions (for illustrations of moral education in school, see Yian 1994).

Similar to other nations undergoing development, the high levels of population migration in China have also contributed to the weakening of social control. Chinese scholars have observed that migrants are "confronted with a different value system and lifestyle when they reach the cities" (Ma 2001, p. 65). The resulting value conflict is conducive to social disorganization and tenuous informal control. The weakening of control in particular affects young people and impacts on youth crime, predominantly in the group of the floating people. Presently, street children and the formation of youth gangs with links to organized crime are seen as a major problem in the urban areas of Hongkong and Shanghai (Lo and Su 2005).

Finally, the social changes accompanying economic reforms in China have created expanded opportunities for certain types of crime. Economic productivity in China has created more consumer goods that are readily available to the general public. This makes property offenses much more attractive. In addition, the reduced international isolation of China opens up new opportunities for organized, transnational crime, such as human smuggling by so-called "snakeheads" (Zhang and Chin 2002).

In sum, the Chinese case conforms to expectations derived from the criminological variants of modernization theory in several important respects. Societal development in China has unleashed many of the "master forces" commonly associated with modernization, and these forces have stimulated criminal motivations, weakened traditional forms of social control, and generated expanded opportunities for crime, especially property and transnational crime. The general pattern of rising crime rates in China, and property crimes in particular, is thus not surprising from the perspective of the modernization paradigm in criminology.

What is anomalous about the Chinese crime situation from the vantage point of criminological theory is the fact that Chinese crime rates remain relatively low in comparative perspective, with the important exception of corruption. Given the truly extraordinary pace of the economic reforms and the fundamental nature of the social changes associated with them, the "costs" of modernization in terms of crime appear to be much less than might be expected for a country that is developing and simultaneously transforming into a market economy. We suggest that this "anomaly" can be attributed to what is perhaps the most distinctive feature of the developmental process in China: the management of the social transition by a strong central government. As Bakken

(2000, p. 4) observes: "The Chinese programme of modernization can be described as an attempt at taming or binding, of completely controlling the path and the pace of the Juggernaut of modernity. The Chinese regime wants to put brakes on the runaway engine, although it also wants to let it run."

In the following sections, we describe how the planned transition has mitigated some of the criminogenic pressures commonly associated with modernization, while at the same time creating unique opportunities for special types of crimes.

Implications of the governmentally managed transition for social control: "pathway forces"

To understand the consequences of reform for crime in China, it is important to appreciate the sociohistorical context. China has a longstanding tradition of strong control by the state over the populace. Scholars have pointed out that a distinctive feature of state control in China has been its pervasiveness, achieved through the combined forces of a paternalistic organization of society and a bureaucratic regime. Law and law enforcement were deemed as less important than ethical and moral guidance through traditional values, embodied by the head of the family and the bureaucrat, and exemplified in "model behavior" (see Bakken 2000). Consequently, in contrast to Western European societies, the rule of law did not develop in China, and law remained a "secondary" mechanism to achieve domestic order, in contrast to the "primary" mechanism of traditional values and morality. Educating and controlling the thoughts and minds of people therefore remained a preeminent task of the government at all levels (Buoye 2000, p. 223).

Upon seizing power in 1949, the Communist government built upon these traditions. The pre-reform command economy that developed under communist rule placed the role of government as central to all social processes. The government planned production, allocated jobs, set prices, distributed health care, assigned housing, rationed services, and controlled marriages, divorces, and births. Migration to the cities was subject to tight controls. Government policies were normally supported and accompanied by major campaigns to educate the public. Communities and work units were the main agents of tight control.

These governmental controls have certainly been weakened compared with the pre-reform era, but it would be a mistake to underestimate the extent to which the government continues to play a critical role in the operations of contemporary Chinese society and economy. The infrastructure of the grass roots social control organizations that were created by the Communist regime is still largely intact, and it is being cultivated in newly developed urban neighborhoods. China's urban areas are administratively divided into districts;

each urban district government administers a number of "street offices," which are responsible for many political, educational, and social welfare tasks. Under each street office are residential neighborhood committees, which perform daily social service and control functions in their neighborhoods, including surveillance of the general population.

Scholars have found that the urban neighborhoods committees and rural villagers' committees have in fact provided an effective organizational framework for social control (Feng 2001; Lu and Miethe 2001; Rojek 2001; Zhang et al. 1996). For example, a study by Lu and Miethe (2001) has found that residents in old Shanghai neighborhoods, where the neighborhood committees are particularly active, perceived both informal and formal social control to be more effective then residents in new Shanghai neighborhoods. Zhang et al. (1996) also found that mediation in neighborhoods reduces the likelihood of disputes that typically lead to criminal outcomes. These governmentally linked social control organizations have not vanished along with the economic reforms. As recently as 2001, the total number of mediation cases was 4,860,695 (*China Law Yearbook Press* 2002, p. 1256). The Communist Party, moreover, has continued to strive to maintain these organizational structures.

In addition, the implications of migration from the countryside to the city have been somewhat different in China than in other nations, partially due to its specific nature in China, and partially due to the special role of the government, and its particularly strong control over rural life. After several campaigns, the government in the late 1950s finished organizing all rural residents into "People's Communes," which were local "collective" economic/governmental organizations that controlled virtually all aspects of peasants' life. Traditions of lawless criminal violence in the countryside, often found in other societies undergoing the process of modernization, were thus stifled to a large extent in China by the strong governmental control since the revolution in 1949. Throughout the course of the second half of the twentieth century, tight governmental control of migration has probably mitigated the violence-generating process that has been associated with rural/urban migration in other nation's undergoing social transition, although the common problems of adjustments for rural migrants to the cities have been noted by Chinese scholars and may indeed have been exacerbated by the household registration system.[4]

Since the economic reforms, migration to cities has increased exponentially, and to a great extent taken the form of quasi-illegal migration. However, different from many cities in developing countries, migrant residential areas are not "equivalent to concentration of poverty" (see the Introduction and Chapters 10 and 11, all in this volume). Though these non-registered and, as such illegal, migrants suffer from social exclusion like their counterparts in First and Third World countries, a considerable group among them is economically quite successful. Further, they tend to link up with family, kinship and other networks from their place of origin, thus preserving a high level of social control in the urban environment. Finally, a specific feature of migration to the new urban centers is the high number of female migrants (*Spiegel* 2005), a group with consistently low levels of violent crime.

In sum, it seems that the combination of government controls and the nature of the flow of workers into the urban centers accounts for a comparably low level of social disorganization in the course of urbanization. Migration to urban centers has neither engendered the development of shantytowns with high levels of social disorganization as in the cities of third World countries, nor have existing neighborhoods been afflicted by increasing social disorganization (see Chapters 8 and 14 in this volume). This seems to account for the rather small impact of migration on the development of violent and other types of crime during the process of urbanization. However, the future impact of the development of marginalized zones at the periphery of cities, as described by Zhou and Logan (Chapter 6 in this volume), on social disorganization has to be determined during the coming years.

Faced with the perceived challenge of a growing crime problem, the government has renewed the tradition of governing through ethical guidance, moral exhortation, and education as a primary mechanism of ensuring domestic order, at the expense of the development of the rule of law. A specific feature of its crime policies are "hard strikes," mass campaigns rooted in the Maoist idea of relying on mass movements to solve problems, and promoted by Deng Xiaoping at the start of the economic reforms back in 1983, as in the following address to the leaders of the Ministry of Public Security:

> Why can't we organize once, twice, three times battles striking hard against criminal activities? Every large and middle-size city should organize several battles within three years ... we must inspire the masses to strike hard against serious criminal activities. ... We must use capital punishment by the law on a group of offenders ..." (Deng 1983, p. 33 [translation by one of the authors]).

Deng's address laid the foundation for "strike hard" campaigns as a major criminal justice policy in China, which has continued to the present. Three nationwide strike hard campaigns have subsequently been carried out, the first from 1983 to 1986, the second from April to July in 1996, and the third from April 2001 to April 2003. Notwithstanding the fact that the effects have typically been temporary, "strike hard" campaigns remain a basic criminal justice policy of the government, which has resulted in the highest numbers of death sentences worldwide (Liu 2004).

In addition to the "strike hard" campaigns, the government has rehabilitated Confucius as a model for guiding young people, schoolteachers, and those in business (*Spiegel Special* 2004a), and has embarked upon a novel effort at "moral planning" to mobilize the traditional values of Confucianism to circumvent the destabilizing forces of modernization that would otherwise be unleashed.

Bakken (2000) has accordingly characterized China as an "exemplary society" and has described the continuous efforts of the government to provide moral education and launch moral campaigns since the early 1980s. According to Bakken, the ethical and moral culture of China has always operated on the mechanisms of "model behavior" and "exemplary norms" emphasizing a virtuous life and setting standards for "the good student," the "good worker," etc.

The prospects for the Chinese government's moral planning and its reliance on social control through "exemplary norms" require a cautious appraisal. Presently, the thriving economy offers considerable rewards and incentives to individuals and their families to conform to "exemplary norms" of thriftiness, hard work, education, and excellence on the job. In addition, exemplary behavior offers routes of upward mobility for the lower classes, as long as the government can control and reward exemplary behavior by offering positions in the economy and at all levels of government, and the expanding (private) economy provides jobs and positions which reward exemplary behavior. Upward mobility and favorable economic prospects seem to be preconditions for the exemplary society to exert its impact on crime rates during the modernization process. Consequently, increasing inequality and relative deprivation pose considerable threats as they inhibit upward mobility and engender social exclusion, and expectations of well educated young people particularly in urban centers might be disappointed when the economy tightens. Exemplary norms become meaningless to those who are excluded, while they will find support among those who mainly profit from economic growth and new opportunities, even among the poorest groups in China's cities (see Chapter 3 in this volume).

What seems to pose more problems for the rapidly modernizing society is the rudimentary development of the rule of law, the legal and criminal justice systems. At present, experts deplore the widening gap between the developing market economy and its legal institutions (*Spiegel Special* 2004b). Only in 1997 was an important step taken to eliminate the supremacy of the moral code in favor of the rule of law, and to start to "police the law," and not the moral code. The clause that allowed for the prosecution of analogous conduct, if this was deemed harmful to society even if not illegal, was abolished from the penal code (see Bakken 2005a, p. 2). Experiences from other post-communist societies demonstrate that those quickly succeeding in establishing the rule of law and reforming the criminal justice system were most successful in curbing corruption, such as Hungary and the Czech Republic (Karstedt 2003b). It is reasonable to propose that the precedence that is given to moral conduct over the rule of law by the Chinese government contributes to the observed "imbalance" of different types of crime in China, and is partially responsible for its high levels of corruption pervading all institutions of society.

Finally, throughout the recent period, governmental control has also enabled China to avoid some of the problems associated with developments that are emphasized in world systems/dependency theory. China does not have a recent history of colonial exploitation like Latin American and other Asian countries. In the pre-reform era, the Chinese government banned nearly all foreign investment and has since opted for a tight control of foreign investment. With China's entrance into the World Trade Organization in December 2001, the pressure to open markets has certainly increased. Nevertheless, to date, China has been able to stimulate impressive economic growth from within while avoiding the kind of dependency on the advanced societies that has been cited as a condition conducive to crime in other developing nations. In addition, a high proportion of the capital that bolsters growth and development in China is provided by Chinese diasporas overseas (Castells 2000, p. 317; also see the Introduction in this volume). While dependency is not a real threat to China, the capital influx from Chinese diasporas overseas can paradoxically be responsible for the amount of corrupt exchanges. In contrast to large and international financial institutions and corporations that operate with high visibility and are meticulously controlled, the extended kinship relationships and networks through which capital flows into the Chinese economy are much less controlled, and are conducive to grand corruption and corporate crime (Karstedt 2000).

Institutional incompatibilities and opportunities for corruption

The Chinese government not only continues to play a central role in maintaining social control; it remains a powerful force managing the economy. Indeed, a distinctive feature of China's economic reform is that it is government initiated, directed, and to a striking degree, government controlled (Garnaut 2001). The changes associated with these reforms should not be minimized. The policy of setting production plans for farmers has been phased out, and price controls for farm products have been gradually loosened. In addition, the government has encouraged village-owned and operated enterprises (TVEs) and family-based private enterprises while expanding the autonomy of state owned enterprises (SEOs). Stock-based ownership and modern corporations have been created. Finally, a variety of markets, including product markets, labor markets, stock markets etc, has been allowed to emerge within some limits (see Chapter 1 in this volume).

Nevertheless, although these market reforms have without question changed the basic functioning of the Chinese economy, the state still operates as the guiding force. The state directly participates in a non-trivial share of economic activities, despite the dramatic growth in the private sector. Shi (2003) estimates that about 20–30 percent of economic activities are directed by the state, and about one-fifth of industrial jobs are employed by state/public and state-owned businesses. Urban development and housing markets are subject to intense state control, notwithstanding speculative building booms in major cities. In short, selective market mechanisms have clearly been introduced into the Chinese economy to a significant degree, but these processes continue to be managed by a powerful political apparatus and elite networks in urban centers.

We suggest that this co-existence of selective market institutions along with institutional arrangements associated with a command economy leads to pronounced "institutional incompatibilities," conditions under which market institutions pose demands that are incompatible with the processes under which command economic institutions operate. Institutional incompatibilities are inevitable under a planned transition. The most important consequence of institutional incompatibilities for present purposes is that they alter the opportunity structure for crimes. Specifically, opportunities for those crimes related to the economic arena, including corruption, are greatly expanded.

For example, market institutions demand the autonomy of enterprises and require private property rights. Only recently has China acknowledged property rights in its constitution, and the government reduced control over the enterprise and granted managers some autonomy (Groves et al. 2001; Jefferson and Rawski 1999). While this has encouraged economic growth, it has also generated attractive criminal opportunities for managerial personnel. It has created a situation with multiple "principle-agent problems" (Peng 2001). Managers (the agents) use their power over state-owned firms for personal advantages at the expense of the state firms for which they bear responsibility, and strip the state company of its assets (Peng 2001, p. 1345). This might include the secret transmission of resources such as technology and clients to a company producing similar products but owned by relatives or friends while letting the state company sink into debt or bankruptcy. Other opportunities include the privatization of a previously state-owned firm; the managers and their relatives or friends purchase the company or company stock at a much-deflated price, thereby converting state assets into private property.

Institutional incompatibility is further visible in the competition between the private and public sectors. When both private firms and Town and Village Owned Enterprises (TVEs) operate in the market along with state-owned firms, they compete for limited resources, including materials, capital, technology, energy, and market shares for products (Byrd and Gelb 2001; Chang and Wang 2001), with state-owned companies traditionally having advantages over private firms and TVEs. Peng (2001) has pointed out that privately owned enterprises incur very high transaction costs including difficulty in obtaining bank loans. This invites governmental predation (debilitating taxes, fees, levies, and bribery exaction) (Peng 2001, p. 1346). Institutional incompatibility leads to a situation wherein illegal and quasi-legal methods are routine means by managers of private firms and TVEs to obtain resources necessary for production, including bribery of officials.

Another form of economic crime resulting from the mixture of state- and privately owned business is "trade by officials" (*guandao*), that was cited by protestors during the Tiananmen Square demonstrations in 1989. Officials take advantage of a dual system of fixed and market prices, and use their power to allocate the lower priced resources and goods to gain bribes, and even sell the "right to purchase" to make profit (Raby 2001). Institutional incompatibility and the ensuing increase of criminal opportunities are also reflected in the process of fiscal decentralization. To increase efficiency, a major governmental reform has been to decentralize fiscal authority (West and Wong 2001), conferring on local governments the power to tax and retain profits owned by the government, and to decide on business matters and expenditures (Oi 1996). This has created ample opportunities for officials at the local level to use their power for monetary advantages. Local revenue is used officially for personal bonuses and perks, and officials spend public funds to build luxury homes, and purchase expensive cars for private use under the name of public use (Peng 2001, p. 1347).

In sum, the institutional incompatibilities that accompany a planned transition have created an abundance of opportunities for the abuse of political position and for various forms of corruption, and are probably responsible for China's high levels of such crimes in global comparisons. Popular support for harsh punishments in particular for corruption and all sorts of illegal business (Trevaskes 2004) demonstrates that the public is well aware of illegal business practices and condemns them, notwithstanding their prevalence. Currently no unambiguous direction can be discerned in Chinese government policies. External pressures to move gradually toward a "fuller" market have increased since China joined the WTO in December 2001, however, fundamental incompatibilities remain within the core industries, the financial sectors, and the largest state-owned enterprises. These make themselves felt on the local level in housing markets and the building industries, and will affect urban dwellers. Unless these incompatibilities are resolved, opportunities and incentives for corruption are likely to continue to be abundant in the years ahead.

Summary and Conclusions

Our analysis of crime in contemporary urban China reveals both commonalities and distinctiveness when considered in comparative perspective. In a number of important respects, the Chinese case conforms to paradigms on the criminogenic consequences of modernization. The basic structural forces associated with the modernization process – the "master forces" – have exerted pressures towards higher crime in urban China in ways similar to those observed elsewhere, especially rates of violent and property crimes. At the same time, however, a striking feature of the Chinese situation is that the criminogenic pressures associated with the transition to modernity have apparently been muted with respect to violent and property crimes. Levels of these crimes have increased but much more modestly than might be expected given the profound social changes that have occurred at a remarkably rapid pace.

We suggest that the "crime cost" of modernization in China has been relatively modest because of the distinctive "pathway" that the nation has followed. The powerful role of the state in the planned transition has allowed for the persistence of strong internal (moral) controls and strong external social controls that restrain the destabilizing forces that would otherwise be associated with the social changes accompanying the economic reforms. Recently, the state has started to promote traditional Confucian values with their strong emphasis on informal social control. In addition, the strong state has enabled the nation to avoid the exploitive relationships of dependency that have characterized many other developing nations, while fostering economic growth that has been relatively inclusive and has provided incentives for law-abiding behavior for large groups of the population. We note that China is not the only country in Asia where the combination of a strong developmentalist state with economic growth has kept crime rates at relatively low levels: Singapore, Japan and Hong Kong are other prominent examples. Further, specific Asian traditions and collectivistic orientations are also likely to play a role according to the value pattern in each of these countries (Karstedt 2001).

Yet the active role of the developmentalist state that mitigates the common criminogenic processes associated with modernization has led to institutional incompatibilities with profound consequences for economic and white-collar crime. The particular distribution of white- and blue-collar crime, so to speak, is not a unique feature of the Chinese situation, but one that is shared with other transitional countries in Central and Eastern Europe. The partial implementation of market mechanisms within the context of overarching government planning produces strong incentives and opportunities for corrupt practices and predatory behavior by those in high-level positions. Moreover, the Chinese government's efforts at moral planning and control have produced institutional imbalances in the patterns of social and judicial control. While strengthening traditional forms of grass-root controls, modern forms through the rule of law have been neglected. Paradoxically, the tight control that the government still holds over the economy inhibits the development of strong legal institutions, property rights, and due process that typically accompany the process of modernization. The resulting institutional imbalances, though otherwise deeply rooted in the traditions of Chinese society, are to a large extent responsible for the high and increasing costs of economic crime on the part of elites, which according to all accounts far exceed the costs of other crime.

In sum, our analysis suggests that the four "theoretical prisms" combine into an explanatory model that can account for rising rates of violent and property crime, as well as for China's comparative advantages and disadvantages and the specific patterning of crime in China. The Chinese case in many respects reaffirms the claim of the modernization paradigm that "crime is one of the major costs of modernization" (Shelley 1981, p. 134), albeit with modestly increasing and low crime rates when compared to other Third World and transitional countries. Both dependency theory and the theory of the developmentalist state can account for this comparative advantage in terms of rates of "street crime." In the context of world system/ dependency theory, these crime figures corroborate the attenuating impact of China's independent position within the global economy on modernity's criminogenic forces. As suggested by theories of the developmentalist state, an effective, autonomous central government and bureaucracy can sustain traditional controls to a considerable degree and create an economic and social environment that mitigates the criminogenic pressures of the "Juggernaut of modernity," at least in the short run. China's specific pattern of crime combining extenuated growth of more common forms of crime and exponential growth of all types of white-collar and economic crime is not unique, but clearly caused by its specific pathway of making the transition from socialism to a market economy. Both theories of the developmentalist state and of transitional societies account for the high level of corruption and other forms of official malfeasance. Here, the developmentalist state has created those institutional incompatibilities which are made responsible for rising crime rates in transitional societies. By combining selective market mechanisms with features of a command economy, it opened up a plethora of opportunities for white-collar crime, especially in a context where the legal infrastructure is underdeveloped.

However, in the long run, the institutional imbalance within the systems of social and judicial control may erode the success of the strong traditional controls that are presently in place, and that have attenuated the growth of violent and property crime. In the absence of the development of institutional arrangements more compatible with the logic of a market economy and a modern criminal justice system, the legitimacy of the very governmental control in China that has allowed for a planned transition with muted criminogenic pressures may very well be undermined, and China's cities might suffer the same fate as that presently found in cities in many other third world countries.

NOTES

1 In subsequent work, Shelley (1986) states that her original formulation placed insufficient emphasis on cross-national variability in the impact of development on crime. Our review of modernization theory draws on Messner (2003).

2 As Bakken points out, all interpretations of crime trends in China based on official statistics and victimization surveys have to take two facts into account. First, official crime rates are heavily affected by pressure on the police to produce high clearance rates. This policy has obviously resulted in massive under-reporting during the 1990s, and its recent abandonment seems to be partially responsible for a steep rise in crime rates. Second, as a result of this policy, victim surveys show extremely low

rates of reporting to the police. Finally, and perhaps most important, official data as well as victim surveys do not include the huge number of non-registered migrants to the major cities, whose status resembles more the one of illegal migrants. They are excluded from social and health services, and from housing, and their children are not entitled to education (Qian and Zhang 2006). The estimates for the "floating people" stand between several tens of millions and one hundred million. This group, which has to be regarded as most vulnerable to both violent and property crime, will neither report to the police nor be included in victimization surveys.

3 The GINI-coefficient has increased from an extraordinarily low level of .180 in 1978 to a high level of .50 for the whole country in 2000, and similar developments have affected the urban and rural population during this period.

4 The household registration system (*hukou*) was established in the mid 1950s. This system divided citizens into urban or rural residents. Only urban residents qualified for city welfare benefits and were allocated daily necessities via rations. The major benefits that urban residents enjoyed included employment and schooling of children. In the past few years, as labor market reform has proceeded, the government has started to loosen up this system, allowing some rural residents to become citizens in small towns (see Chapters 9, 10 and 11 in this volume).

REFERENCES

Bakken, B. (2000) *The Exemplary Society: Human Improvement, Social Control, and the Dangers of Modernity in China.* New York: Oxford University Press.

Bakken, B. (2005a) "Comparative Perspectives on Crime in China." In B. Bakken (ed.), *Crime, Punishment, and Policing in China.* Oxford: Rowman and Littlefield: pp. 57–88.

Bakken, B. (2005b) "Introduction: Crime, Control and Modernity in China." In B. Bakken (ed.), *Crime, Punishment, and Policing in China.* Oxford: Rowman and Littlefield: pp. 1–22.

Bakken, B. (ed.) (2005c) *Crime, Punishment, and Policing in China.* Oxford: Rowman and Littlefield.

Buoye, T. (2000) *Manslaughter, Markets and Moral Economy.* Cambridge: Cambridge University Press.

Byrd, W., and A. Gelb. (2001) "Township, Village and Private Industry in China's Economic Reform." In R. Garnout and Y. Huang (eds.), *Growth Without Miracles: Readings on the Chinese Economy in the Era of Reform.* New York: Oxford University Press: pp. 170–80.

Cao, L., and Y. Dai. (2001) "Inequality and Crime in a Changing China." In J. Liu, L. Zhang, and S.F. Messner (eds.), *Social Control in a Changing China*, Westport, CT: Greenwood Press: pp. 73–87.

Castells, M. (2000) *The Rise of the Network Society: The Information Age.* Oxford: Blackwell Science.

Chan, K.W., and Y. Hu. (2003) "Urbanization in China in the 1990s: New Definition, Different Series, and Revised Trends." *The China Review* 3: 49–71.

Chang, C., and Y. Wang. (2001) "The Nature of the Township Village Enterprise." In R. Garnout and Y. Huang (eds.), *Growth Without Miracles: Readings on the Chinese Economy in the Era of Reform.* New York: Oxford University Press: pp. 219–43.

Chen, D. J.H. (1992) "Third World Crime in the World System: A Cross-National Study." PhD Thesis, Department of Sociology, State University of New York, Albany.

Chen, X., C. Wang, and J. Wu. (2004) "The Inside Story about the China Index for Financial Corruption: A Complete Chain Can Be Seen Behind Corruption." *China News Net* online at: http://www.sina.com.cn/

China Law Yearbook Press. (1987–2004) *Law Yearbook of China*. Beijing: China Law Yearbook Press (in Chinese).

Cohen, L.E., and M. Felson. (1979) "Social Change and Crime Rate Trends: A Routine Activity Approach." *American Sociological Review* 44: 588–608.

Currie, E. (1991) "Crime in the Market Society: From Bad to Worse in the Nineties." *Dissent* 38: 254–9.

Deng, X. (1983) "Strike Hard Against Criminals" In *Selected works of Deng Xiaoping*, vol. 3. Beijing: People's Publisher.

Deng, X., and A. Cordilia. (1999) "To Get Rich is Glorious: Rising Expectations, Declining Control, and Escalating Crime in Contemporary China." *International Journal of Offender Therapy and Comparative Criminology* 423: 211–29.

Durkheim, É. ([1893] 1933) *The Division of Labor in Society*. New York City: Free Press.

Durkheim, É. ([1897] 1950) *Suicide: A Study in Sociology*. New York City: Free Press.

Eisner, M. (2001) "Modernization, Self-Control and Lethal Violence." *British Journal of Criminology* 41: 618–38.

Feng, S. (2001) "Crime and Crime Control in a Changing China." In J. Liu, L. Zhang, and S.F. Messner (eds.), *Social Control in a Changing China*, Westport, CT: Greenwood Press: pp. 123–32.

Garnaut, R. (2001) "Twenty Years of Economic Reform and Structural Change in the Chinese Economy." In R. Garnout and Y. Huang (eds.), *Growth Without Miracles: Readings on the Chinese Economy in the Era of Reform*. New York: Oxford University Press: pp. 1–18.

Groves, T., Y. Hong, J. McMillan, and B. Naughton. (2001) "Autonomy and Incentives in Chinese State Enterprises." In R. Garnout and Y. Huang (eds.), *Growth Without Miracles: Readings on the Chinese Economy in the Era of Reform*. New York: Oxford University Press: pp. 263–80.

Hagan, J., and D. Radoeva. (1998) "Both Too Much and Too Little: From Elite to Street Crime in the Transformation of the Czech Republic." *Crime, Law & Social Change* 38, 195–211.

Hinkle, R.C. (1976) "Durkheim's Evolutionary Conception of Social Change." *Sociological Quarterly* 17, 336–46.

Jefferson, G.H., and T. Rawski. (1999) "Ownership Change in Chinese Industry." In G.H. Jefferson and I. Singh (eds.), *Enterprise Form in China: Ownership, Transition, and Performance*. Oxford, UK: Oxford University Press: pp. 23–41.

Karstedt, S. (1999) "Social Transformation and Crime: A Crisis of Deregulation?". In A. Czarnota and M. Krygier (eds.), *Rule of Law after Communism*. Aldershot: Dartmouth: pp. 308–26.

Karstedt, S. (2000) "Knights of Crime: The Success of 'Pre-Modern' Structures in the Illegal Economy." In S. Karstedt and K.D Bussman (eds.), *Social Dynamics of Crime and Control: New Theories for a World in Transition*. Portland, OR: Hart: pp. 53–68.

Karstedt, S. (2001) "Die moralische Staerke schwacher Bindungen. Individualismus und Gewalt im Kulturvergleich." *Monatsschrift fuer Kriminologie und Strafrechtsreform* 84: 226–43.

Karstedt, S. (2003a) "'Strangers on the Road': Railway Traffic and Violence in 19th Century South West Germany. A Study on the Production of Social Capital." In B. Godfrey, C. Emsley and G. Dunstall (eds.), *Comparative Histories of Crime.* Cullompton: Willan, pp. 89–109.

Karstedt, S. (2003b) "Legacies of a Culture of Inequality: The Janus Face of Crime in Post-Communist Countries." *Crime, Law & Social Change* 40, 295–320.

Kim, S.-W., and W.A. Pridemore. (2005) "Social Change, Institutional Anomie and Serious Property Crime in Transitional Russia." *British Journal of Criminology* 45: 81–97.

LaFree, G.D., and E.L. Kick. (1986) "Cross-National Effects of Development, Distributional and Demographic Variables on Crime: A Review and Analysis." *International Annals of Criminology* 24, 213–36.

Lardy, N.R. (2001) "The Role of Foreign Trade and Investment in China's Economic Tansformation." In R. Garnout and Y. Huang (eds.), *Growth Without Miracles: Readings on the Chinese Economy in the Era of Reform.* New York: Oxford University Press: pp. 385–98.

Li, Y. (2004) "Prospects for Anti-Corruption in 2004: Three Changes in Strategies." *China News Net.* Online at: http://www.sina.com.cn/

Liang, Z., and Z. Ma. (2004) "China's Floating Population: New Evidence from the 2000 Census." *Population and Development Review* 30: 467–88.

Liu, J. (2006) "Modernization and Crime Patterns in China." *Journal of Criminal Justice* 4: 119–30.

Liu, J., and S.F. Messner. (2001) "Modernization and Crime Trends in China's Reform Era." In J. Liu, L. Zhang, and S.F. Messner (eds.), *Social Control in a Changing China,* Westport, CT: Greenwood Press, pp. 3–22.

Liu, J., L. Zhang, and S.F. Messner (2001) *Crime and Social Control in a Changing China.* Westport, CT: Greenwood.

Liu, Y. (2004) "Apply Scientific Development View to Deepen the Understanding of Striking Hard Strategy." *Public Security Studies* 2004, 11: 14.

Lo, T.W.C., and S. Su (2005) *Explaining and Fighting Youth Crime in Hong Kong, Singapore and Shanghai.* Hong Kong: City University Press.

Lu, H., and T.D. Miethe. (2001) "Community Integration and the Effectiveness of Social Control." In J. Liu, L. Zhang, and S.F. Messner (eds.), *Social Control in a Changing China.* Westport, CT: Greenwood Press: pp. 105–21.

Ma, G. (2001) "Population Migration and Crime in Beijing, China." In J. Liu, L. Zhang, and S.F. Messner (eds.), *Social Control in a Changing China.* Westport, CT: Greenwood Press: pp. 65–72.

Messner, S.F. (2003) "Understanding Cross-National Variation in Criminal Violence." In W. Heitmeyer and J. Hagan (eds.), *International Handbook of Violence Research.* The Netherlands: Kluwer Academic Publishers: pp.701–16.

Messner, S.F., and R. Rosenfeld. (2007) *Crime and the American Dream.* 4th edn. Belmont, CA: Thomson.

Mittelman, J.H. (2000) *The Globalization Syndrome: Transformation and Resistance.* Princeton, NJ: Princeton University Press.

Neapolitan, J. (1997) *Cross-National Crime: A Research Review and Sourcebook.* Westport, CT: Greenwood Press.

North, D.C. (1994). "Institutions Matter." *Economic History.* Economics Working Paper Archive at WUSTL.

North, D.C. (1995) "Some Fundamental Puzzles in Economic History/ Development." *Economic History*. Economics Working Paper Archive at WUSTL.

North, D.C. (1996) " Where Have We Been And Where Are We Going?" *Economic History*. Economics Working Paper Archive at WUSTL.

Oi, J.C. (1996) "The Role of the Local State in Chinas Transitional Economy." In A.G. Walder (ed.), *China's Transitional Economy*. New York: Oxford University Press: pp. 170–87.

Peng, Y. (2001) "Chinese Villages and Townships as Industrial Corporations: Ownership, Governance, and Market Discipline." *American Journal of Sociology* 106, 1338–70.

Polanyi, K. (1957 [1944]) *The Great Transformation: The Political and Economic Origins of Our Time*. Boston: Beacon Press.

Qian, W., and Z. Zhang. (2006) "The Incorporation of Migrant Workers in the Urban Society." *Journal of Zhejiang University* 4, 25–34.

Raby, G. (2001) "The 'Neither This nor That' Economy." In R. Garnout and Y. Huang (eds.), *Growth Without Miracles: Readings on the Chinese Economy in the Era of Reform*. New York: Oxford University Press: pp. 19–35.

Rojek, D.G. (2001) "Chinese Social Control: From Shaming and Reintegration to 'Getting Rich is Glorious.'" In J. Liu, L. Zhang, and S.F. Messner (eds.), *Social Control in a Changing China*. Westport, CT: Greenwood Press: pp. 89–104.

Shelley, L.I. (1981) *Modernization, Age Structure and Regional Context: A Cross-National Study of Crime*. Carbondale, IL: Southern Illinois University Press.

Shelley, L.I. (1986) "Crime and Modernization Reexamined." *Annales Internationales de Criminologie* 24, 7–21.

Shi, X. (2003) "A Review of 23 Years of China's Economic Reform, a Non-Reversible Process." *China News Digest*. Online at: http://www.cnd.org/

Spiegel. (2005) "Chinas Migration ist weiblich." 11: 121–4.

Spiegel Special. (2004a) "Die Lehren des weisen Konfuzius". In *China. Aufstieg zur Weltmacht*. no.5: 122–4.

Spiegel Special. (2004b) "Oekonom Xu Xiaonian ueber Reformzwaenge". In *China. Aufstieg zur Weltmacht*. no.5, 123–4.

Taylor, I., K. Evans,and P. Fraser. (1996) *A Tale of Two Cities*. London and New York: Routledge.

Teeple, G. (1995) *Globalization and the Decline of Social Reform*. Atlantic Highlands, NJ: Humanities Press.

Tilly, C. (1992) *Coercion, Capital and European States AD 990–1990. Studies in Social Discontinuity*. Oxford: Blackwell.

Transparency International. Corruption Perception Index 2000–2004. Online at: www.transparency.org/cpi/ [accessed July 28, 2005].

Trevaskes, S. (2004) "Propaganda Work in Chinese Courts." *Punishment and Society* 6, 5–21.

West, L.A., and C.P.W. Wong. (2001) "Fiscal Decentralisation and Growing Regional Disparities in Rural China." In R. Garnout and Y. Huang (eds.), *Growth Without Miracles: Readings on the Chinese Economy in the Era of Reform*. New York: Oxford University Press: pp. 436–58.

Xiaohang. (2005) "Hu Emphasizes Continuing the Strike Hard Policy." Online at: http://news.sina.com.cn/c/2005-05-24.

Yian, J. (1994) "Understanding in Value Education for Changing Subjects." *Journal of Hainan University* 1, 100–2.

Zhang, L., D. Zhou, S.F. Messner, A.E. Liska, M.D. Krohn, L. Liu, and Z. Lu (1996) "Crime Prevention in a Communitarian Society: Bang-Jiao and Tiao-Jie in the People's Republic of China." *Justice Quarterly* 13, 199–222.

Zhang, S., and K. Chin. (2002) "Enter the Dragon: Inside Chinese Human Smuggling Organizations." *Criminology* 40: 737–67.

Zhou, L. (2005) "Group Corruption of An Hui Fuyang City Intermediate Court: Corrupted Judges Committee All Evils." Online at: http://news.sina.com.cn 4/30/2005.

13

Migration, Urbanization, and the Spread of Sexually Transmitted Diseases: Empirical and Theoretical Observations in China and Indonesia

Christopher J. Smith and Graeme Hugo

Introduction: Theorizing HIV/AIDS in China

> China is on the verge of a catastrophe that could result in unimaginable human suffering, economic loss and social devastation.
>
> (UNAIDS 2002b, p. 7)

This chapter focuses attention on the spread of HIV/AIDS, in China. It is argued that the spread of this disease is linked in significant ways to the dynamic urban transformations that are occurring in China and are examined in other chapters of this volume. In order to better understand these linkages we compare China's experience with respect to HIV/AIDS to that of Indonesia, another of Asia's demographic giants (2006 population 225.5 million). Indonesia shares similarities with China in terms of being a large and diverse nation but has experienced a different pace and nature of economic, social, and political transformation. Like China it has undergone rapid urbanization and has a similar pattern of incidence of HIV/AIDS. Moreover in both countries there are strong links between infection on the one hand and population mobility and urbanization. During the early years of their epidemics, the leaders in both countries refused to recognize the seriousness of their HIV/AIDS epidemics, and have only recently given their strong backing to nationwide programs of prevention and treatment. In purely statistical terms, their HIV/AIDS problem is considered to be less serious than in some parts of the world, although the speed at which new cases are occurring, and the geography of such occurrences, are causing alarm. The rate of increase in the number of infected persons in China was about 30 percent per year in the second half of the 1990s, but this appeared to have doubled by the year 2002 (Thompson 2003; Wu 2003). In Indonesia, officials initially insisted that HIV/AIDS was only a problem among small numbers of marginal population groups.[1] However, as in China, there is increasing concern because of the rapid increases in infection rates, and the understanding that official data may substantially understate the actual rates (Ministry of Health, Republic of Indonesia 2002, p. 5). Between 2000 and 2004 the number of new infections increased by 166.8 percent, with some very significant variations in the geography of incidence.

At the start of the new millennium the Chinese government admitted there may already have been between 600,000 and one million people infected with HIV (Stephenson 2001); and the United Nations predicted that by the year 2010 as many as 10 million people could be infected (UNAIDS 2001). On the basis of these projections UNAIDS (2002b) has described HIV/AIDS as China's "titanic peril." As a result China's leaders are now beginning to deal with the potentially devastating consequences of an epidemic of these proportions, and its consequences for individual health, community development, and overall social stability (National Center for HIV, STD, and TB Prevention 2001). International agencies like UNAIDS have welcomed China's new proactive stance, but there are still major concerns about institutional and infrastructural problems that threaten to derail prevention and treatment plans. The Indonesian government estimates that there are between 83,000 and 278,000 HIV/AIDS sufferers in Indonesia (UNAIDS Secretariat 2004), but the Indonesian Ministry of Health (2002) has estimated the numbers vulnerable to HIV infection at between 13 and 20 million people. As in China, some commentators feel that Indonesia is "poised on the edge of an AIDS explosion" (McBeth 2004, p. 46).

Among the theoretical perspectives discussed in the Introduction to this volume, three – modernization theory, the transition from socialism, and the role of the developmentalist state – have particular relevance to the spread of HIV/AIDS in China. At the start of its reform period China, primarily as a result of deliberate state-initiated policies, was largely able to slow-down the rate of urbanization and the accompanying negative side-effects encountered in many parts of the developing world. Within the past decade and a half, however, many of the constraints on rural to urban migration in China have been relaxed, and urbanization, fuelled by favorable state policies, has been rapid.

One of the social outcomes of this accelerated rate of urbanization in China is the increased prevalence of epidemiologically high-risk behaviors associated with three overlapping sets of forces: the dismantling of the organizational and spatial structures that helped to maintain social control in Chinese cities during the Maoist era; the increase in the overall "fluidity" of Chinese society, which has resulted in the erosion of some of the behavioral boundaries of China's recent past; and a new set of cultural values that have allowed a wide variety of behavioral changes among the Chinese population. We hypothesize that these three sets of overlapping forces might be implicated in the emergence of new behaviors and lifestyles in urban China that carry with them significantly raised risks of contracting STDs, especially HIV. There are some important parallels here with the situation in Indonesia, in spite of the obvious cultural and political differences between the two countries.[2]

A Brief Historical Geography of the STD and HIV/AIDS Epidemics in China and Indonesia

China's HIV/AIDS epidemic evolved through three phases, each of which has had a relatively specific geographic pattern. The first began as early as 1985, and was marked by geographic concentration and small numbers. Most of the newly infected individuals were in the coastal provinces, and most of them were reported to have been "overseas" Chinese (mainly from Hong Kong and Macau), as well as some foreigners from other parts of the world. The official view at the time was that STDs (and then later HIV/AIDS) were "foreigners" diseases, which had come from outside China, and which could be controlled by limiting access to potentially diseased individuals from abroad, and by rigorous testing upon entry (Farrer 2002).

A second phase, beginning to be evident by the end of the 1980's, was also a geographically limited epidemic, with infections clustered in China's borderlands, particularly in Yunnan Province and Guangxi Autonomous Region in the southwest, and in Xinjiang Autonomous Region in the far west. During this period the majority of reported HIV infections were among intravenous drug users (IDUs), and the spread of the virus was thought to be associated primarily with the cross-border travel of individuals engaged in a variety of epidemiologically risky behaviors, especially the trafficking of drugs and the sharing of infected needles. A small number of HIV infections were being reported at this time among laborers returning from abroad, patients with other STDs, and some female sex workers (UNAIDS 2000).

The third phase began in 1994, when HIV transmission began to appear in China's interior (Rosenthal 2002). In some rural areas, most notably in Henan Province, contaminated blood was circulated to both the original donors and to the recipients of blood transfusions (BBC News Online 2003; Smith and Yang 2005). Blood collection centers had opened in Henan and elsewhere in the early 1990s, encouraging peasants to donate blood and blood plasma in return for payment (J. Chan 2001). Local blood banks aggressively sought donors and many poor peasants were eager to sell their blood for money, often in less than sanitary conditions. Many of the individual donors had blood returned to them from communal vats, after the plasma had been removed and sold to hospitals (Eckholm 2001). It turns out that some of the pooled blood was contaminated with HIV, which put both the donors and the recipients at risk, and in some villages the results were disastrous, with donors being infected at rates as high as 65 percent.

As tragic as the Henan blood scandal was – and still is, in the sense that many of those infected subsequently spread the virus to others – from the perspective of public health policy, by the end of the 1990s the Chinese authorities were beginning to focus more of their attention on HIV transmissions that were occurring as a result of individual risk-taking behaviors (US Embassy 2000). UNAIDS (2000) report that the HIV infection rate among drug users increased almost tenfold in a five-year period, from 0.04 percent in 1995, to nearly 5 percent in 2000; and by the end of the decade intravenous drug use was thought to account for almost 70 percent of all new HIV infections in China.[3]

The fastest growing rate of new HIV infections in China was however from heterosexual contact, often linked to the emergence of a thriving commercial sex industry, which has been one of the unexpected and unintended consequences of China's dramatically successful economic reform project (Wu 2003). New commercial sex outlets opened in the coastal cities and some provincial capitals, and sex workers began to re-emerge on China's streets after a hiatus of more than 40 years. Sex trades have now spread to smaller cities and towns in the interior, and even to some villages (Hyde 2001; Micollier 2004). One of the primary reasons for this growth is economic, in the sense that sex is a highly marketable commodity, and as incomes rise new outlets have emerged in multiple locations to satisfy demand. More importantly, as incomes have risen in China so has inequality, and in recent years many people have been driven toward the sex trades for purely economic reasons.

The Chinese Communist Party (CCP) successfully outlawed commercial sex work in the early 1950s, and virtually eliminated the associated STDs. The prohibitions are still nominally in place – and prostitution is still a criminal activity – but in many areas they are widely ignored. It has been estimated that China may now have between four and six million female sex workers (Pan 1999), and this is likely to increase rapidly as the country moves further along the reform trajectory, and as its population grows more mobile, urban, and entrepreneurial. Large numbers of women have either chosen, or have been coerced by economic need, to sell sexual services in a variety of locations, including karaoke bars, massage parlors, barbershops, and hotels (Goodman 2003).

In geographical terms STD and HIV/AIDS was at first restricted to specific parts of China: the coastal provinces; the southwestern border areas; and parts of the rural interior. By the middle of the 1990s, however, the epidemic was spreading both spatially and socially – moving into different areas and being transmitted to different groups of people. The number of new HIV infections in 1994 was double that in 1993; in 1995 more than triple that in 1994; and with one exception, has grown by more than 40 percent every year since (UNAIDS 2002a).[4]

In Indonesia, the spread of HIV/AIDS also has gone through three phases since its introduction in 1987 (Utomo pers. comm.). In the first phase, infection was confined to small groups and spread mainly by homosexual contact. In the 1990s, numbers of notifications increased with the main method of transmission being heterosexual contact and the commercial sex sector playing an important role. Since the late 1990s, there has been a dramatic increase in transmission via shared needles by intravenous drug users. Like China, the evolution of spread of HIV/AIDS in Indonesia has had a distinctive spatial pattern with strong geographical concentrations.

Although official data on infection rates are in most instances gross understatements, they do indicate the spatial pattern of the disease. Figure 13.1 shows that in 1999 the disease was especially prevalent in the province of Irian Jaya (now Papua) where the rate was 35 times the national figure. In this province there are a number of distinctive local behavioral and cultural elements that favor the spread of infection, especially the taking of multiple sex partners and high incidence of patronage of sex workers (Hugo 2001). The largest number of reported cases and the second highest prevalence rate is in the megacity of Jakarta (2000 population 8,389,443). The third highest levels were in Bali, which is Indonesia's premier tourist destination and where there is considerable activity in the drug and sex trades. It has been suggested also that because there are higher levels of surveillance in this province, a greater

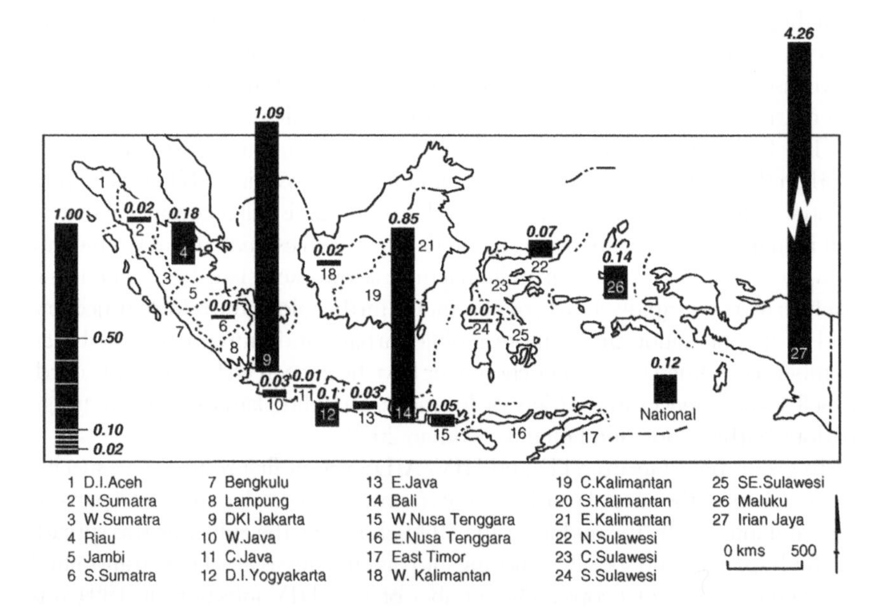

Figure 13.1 AIDS Case Rate by Province: Cumulative Cases per 100,000 People as at 31 August 1999

proportion of infected people are being detected. The fourth highest incidence is in the Riau archipelago, where a flourishing commercial sex industry provides Indonesian female sex workers, primarily serving males who travel across by ferry from neighboring Singapore. Overall there is a stronger presence in urban centers than in rural areas.

Migration and the Spread of HIV Transmission in China and Indonesia

As noted earlier, three distinctive and localized HIV/AIDS epidemics can be identified in China: 1) approximately 68 percent of those with HIV were intravenous drug users, most of whom were in China's southern and southwestern provinces; 2) about 10 percent were infected through heterosexual intercourse, which mostly involved commercial sex workers and their customers; 3) another 10 percent are thought to have been infected through unsafe blood donation practices (Rosenthal 2002). Beginning sometime in the mid-1990s, individuals in these relatively well-defined "vulnerable" groups were having regular contact with so-called "bridging" populations – individuals and groups that have the potential to spread HIV (Yang 2002a, 2002b).

Migrant activities and lifestyles

China's migrant populations – or more generally, people with transitory lifestyles or work routines – represent one of the potentially most dangerous of these "bridging" populations. The assumption made about migrants (in China and elsewhere) is that they are young and poorly educated, and although they are in the sexually most active stage of their lives, they actually have little knowledge either about STDs in general, or about HIV specifically, and almost no access to preventive education or regular health care. Many of China's rural to urban migrants are young men and women who are separated from their home places for most of the year, living in single-sex dormitory housing, and working long hours in difficult conditions.[5] It is assumed that under such circumstances they are easy prey for drug sellers, and that many of them have the time and money to pay for local sex workers. Being far from home, they are also less constrained by the norms of their local (village and small town) cultures.

A report in the Greater Mekong Sub-region (GMS), which includes much of southwestern China's Yunnan Province, concluded that population mobility has been one of the most significant factors for rapid transmission of HIV:

> HIV moves with people who, while on the move, pass through various risk situations that force or encourage them to get involved in unsafe sex or drug use … Massive population movement in the GMS has created these risk environments in many places … that are continuing to fuel the epidemic. (Chantavanich 2000)

Indonesia also has significant internal mobility, the majority of which consists of non-permanent, circular movements involving millions of workers (Hugo 1975, 1978, 1982). As in the case of China, a number of the characteristics of mobile populations have ramifications for the potential spread of HIV/AIDS. Migrant workers are selectively young adults, especially males; the movements rarely involve entire family units: and individuals are separated from their spouses and/or parents for extended periods Most of the migrants are directed to places where the commercial sex industry is already well-established, such as urban areas, plantations, remote construction sites, oil rigs, mining sites, border industrial areas, and forestry sites. Because many earn relatively good wages, and are far removed from traditional constraints on behavior, they are more likely to experiment in a range of risky activities. Last but by no means least, almost all commercial sex workers in Indonesia are themselves internal circular migrants.

The proliferation of large factories in the Jakarta-West Java region, and to a lesser extent in areas like Batam in Riau, has involved the recruitment of many young workers, especially women, to live in dormitories, barracks, and crowded boarding houses (this situation is also found in many parts of China, especially in the Special Economic Zones, where huge export-oriented factories are populated primarily by migrants from all over the countryside). Moreover the onset of the financial crisis in 1997 displaced large numbers of these urban based workers and while some returned to their rural places of origin many moved into the informal sector and also added to the numbers working in the informal commercial sex industry (Hugo 2001).

The empirical evidence

STD infection is likely to result from unprotected casual sex, the majority of which is provided commercially in China. Yang (2004) has reviewed a large number of studies conducted in different parts of the country, which have reported higher STD prevalence rates among temporary migrants than among other population groups. These results suggest that migrants are more likely to have had multiple sexual partners, and to have engaged in casual or commercial sex. Significantly, the majority of commercial sex workers in China are also migrants, and most are young and have little formal education, which implies a potentially widespread pattern of STDs across space (Hugo 2001). They are doubly-exposed to HIV infection: as risk-takers themselves, and also as victims of the risk-taking sexual behaviors of others. Because of the well-known correlation between the incidence of STDs and the (later) transmission of HIV, the migration connection is considered to be crucial in this context.

In addition temporary migrants in China may also be more likely than non-migrants to be users of illegal drugs (Qiao et al. 2001). In a southwestern province, for example, Yang shows that migrants made up only 1.8 percent of the total pool of respondents (representing 2.5 percent of the working age population in 2000); but they accounted for 13 percent of the known drug users, and 14.4 percent of injecting drug users (Yang 2002a, 2002b). A correlation analysis showed that total illegal drug use and injecting drug use were both positively correlated with the volume of temporary migration in a province. The correlations were higher for injecting drug use than the total drug use index, and both sets of correlations seemed to get stronger over time (from 1996 to 2000). The same data also pointed to a correlation between temporary migration and the prevalence of commercial sex activities, using both the number of detained commercial sex workers, and the number of entertainment establishments as indicators (Smith and Yang 2005).

There is currently no direct evidence to suggest that circular migrants in Indonesia have a higher rate of HIV infection than non-migrants, mainly because migrant workers have not been included among the sites for sentinel surveillance activities and the collection of data on HIV infection. We also must emphasize the inappropriateness of assuming that migrants *per se* have a greater vulnerability to HIV infection than is the case among the population at large. The evidence suggests that many migrants find themselves living in contexts where they are at higher risk of infection than if they had remained in their home areas. Circular mobility is occurring on a massive scale in Indonesia: In Java, for example, few rural households are not influenced in one way or anther by migration. There is a concern not just because of the high level of individual labor mobility in Indonesia but also because of the nature of that mobility and the extent to which it puts movers in potentially high risk destinations.

There is also very little hard evidence suggesting that Indonesian migrant workers have a higher rate of HIV infection than non-migrants at the destination or origin. While it is illegal in Indonesia to insist that workers be compulsorily tested for HIV before or after migration (*Asian Migration News*, December 1–15, 2002) in fact many destination countries (e.g., Singapore) and employers do insist on pre-departure testing and periodic testing while they are at the destination. One of the few sets of data publicly released relates to Brunei, which compulsorily HIV tests incoming Indonesian migrant workers: and it was found that of 33,865 tested only one was HIV positive (Parida 2000). Indonesia has an extremely high rate of premature return among its official overseas workers (Hugo 2003), although it is not known what proportion of these returnees are HIV positive. While the bulk of Indonesian commercial sex worker international labor migration is to Malaysia, there are other destinations. In 2003, for example, it was reported that more than 100 Indonesian migrants working as prostitutes in several cities in Saudi Arabia had been arrested and were being deported (*Asian Migration News*, May16–31, 2003). Another story (*Asian Migration News*, June 16–30, 2003) reported that 1,500 young Indonesian women were trapped in the sex business in Abu Dhabi after being tricked into moving there with the promise of well-paying jobs as factory workers or domestic workers.

There are several elements in the relationship between population mobility and the commercial sex industry that might be associated with the increased incidence and spread of HIV (and for the most part, the same goes for the situation in China as well). For example: almost all commercial sex workers (CSWs) in Indonesia are themselves circular migrants and hence can potentially spread the infection themselves: in addition to spreading it among their clients, many of whom are also migrants. Mobile groups are among the heaviest users of the services of the commercial sex industry: results of behavioral surveillance surveys in several Indonesian towns show that more than half of the highly mobile males paid for sexual services in the past year (Ministry of Health, Republic of Indonesia 2002: 11).

In theoretical terms, the potential for the spread of HIV infection through the strong interaction between migrant workers and the commercial sex sector is demonstrated in Figure 13.2. The infection can readily be passed from customers to CSWs or vice versa. However as the diagram shows, both CSWs and their customers then have a substantial capacity to spread the infection, by returning to their home areas and passing it on, presumably to their sexual

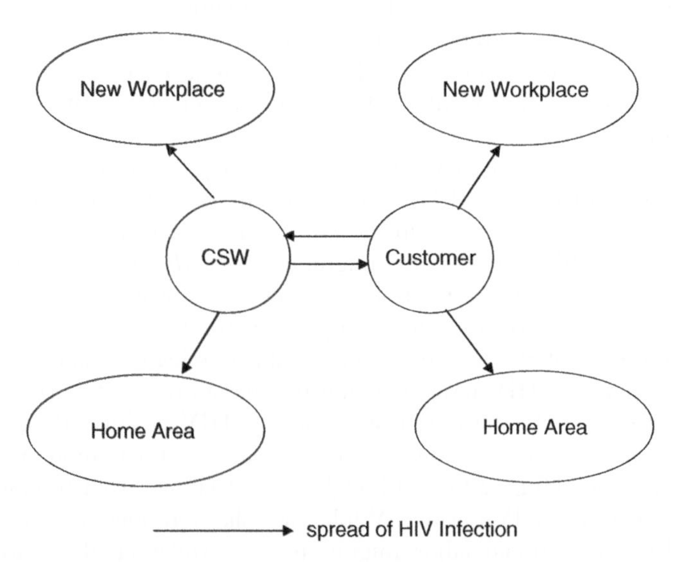

Figure 13.2 Model of Potential Spread of HIV/AIDS in Indonesia Through The Commercial Sex Industry

partners. Another important feature of the mobility of internal migrant workers and CSWs in Indonesia (and presumably in China as well) is that they do not always return to the same destination to work (Brockett 1996), implying that there are multiple diffusion effects, as both CSWs and their customers become agents of diffusion.

Urbanization and Risk-Taking Behavior in China

Over the past two decades China has been the most rapidly developing and the most rapidly urbanizing nation in the world (Solinger and Chan 2002). The economic reforms of Deng Xiaoping's modernization project transformed China dramatically, and within two decades most of the institutions and the infrastructure of Maoist socialism were transformed. China's rapid urbanization has been the subject of much new research but to date very little emphasis has been directed toward the social outcomes of urbanization. It has been in the newly expanding cities of China where the impacts of economic reform and the process of "opening-up" have been most evident, and where we might expect to find some significant impacts on human activity.

In the discussion that follows we link three overlapping dimensions of China's urbanization to new risk-taking behaviors among city dwellers. It is hypothesized that these overlapping forces – which are geographical, socioeconomic, and cultural in nature – are interwoven with, and thoroughly implicated in, the emergence of new behaviors and lifestyles in urban China (Smith 2005).

Urban restructuring and the demise of the socialist 'golden age'

During the Maoist era (1949–76) China's cities achieved an unlikely blend of high social stability matched by low prosperity. Through a combination of political manipulation and social/spatial engineering, China's leaders managed to create a new urban society that some scholars have described as the "golden age" of Chinese socialism (Solinger 2000; Whyte and Parish 1984). In the cities, remarkably few instances of social disorder were visible: mobility was highly constrained, unemployment was virtually nonexistent, and crime and other social problems were almost unheard of. The Communist Party launched significant campaigns to clean the streets and rid China's cities of all evidence of social pathology after the 1949 revolution, and by the 1960s and 1970s China's cities had been totally transformed. Almost nobody was wealthy, but neither was anyone visibly destitute, and intraurban inequality was probably as low as it has ever been measured anywhere in the world.[6] The household registration system (the *hukou*) restricted peasant entry into the cities, and controlled unwanted migration out of the countryside. By limiting the competition from rural migrants, the state virtually guaranteed city dwellers access to jobs and attendant state benefits. The workers – and usually their entire families – were covered by national labor insurance, which provided pensions, medical care, and maternity leave.

All of this changed dramatically with the reforms that began in the early 1980s, and within two decades most of the partitions and boundaries that had been assembled during the Maoist era had been dismantled (Solinger 2000). China began to urbanize very rapidly, both as result of massive population transfers from the rural areas to the cities, and as a result of the reclassification of lower order places into higher order towns and cities (Lin 2002). Market forces have reshaped the landscapes of China's cities, while at the same time restructuring the lives of workers. What had once been the norm – having a permanent, relatively well-paid and secure state-sector job, complete with a range of social benefits – is now rapidly becoming a luxury: in fact, by the end of the 1990s unemployment was probably the most serious problem in China's cities, with at least 13 million laid off in 1997 alone, and several million more in subsequent years (Solinger 2002). Wages in the state sector slipped down to (and even below) the level of those outside it; and with job security sacrificed to competition, fears about the rise of a new urban underclass population became a reality (Parish and Chen 1996). By the end of the 1990s it was apparent that the social problems that had effectively been eradicated from the Maoist city had returned.

China's cities today look prosperous; but this outward appearance obscures rising inequality, extensive poverty, and multiple signs of social instability (Pei 2001). Behavioral norms were changing rapidly at either end of the social spectrum: the new rich were starting to consume and behave in ways they had never even dreamed about just ten years earlier; while at the other end increasingly large numbers of city dwellers were facing a desperate struggle just to make ends meet (Conachy 2001; Chan 2003; Forney 2003; Schauble 2002).

The new urban mobility and the erosion of behavioral boundaries

In addition to the process of internal migration that has been reshaping China's urban landscapes, one of the recurring themes in debates about the modernization project in China involves the idea of movement or mobility, as expressed in the Chinese term *liu*. An important dimension of modern behavior, especially among the young, involves the preference for being constantly on the move, or literally "mobile" (as expressed in the widespread use of mobile phones of young people all round the world). In a theoretically insightful argument, Bakken (2000) has argued that there appears to be a connection between movement, deviance, and danger in the modern city. One indication of this connection is the emergence of a new breed of urban anti-heroes in urban China in the 1990s: the *liumang* (roughly translated as "hooligans").[7] As Bakken points out, the Chinese character *liu* – meaning "to flow" – carries with it the negative connotation of a person who is not tied down spatially, and whose character is brought into question as a result of his (actual and assumed) lack of stability. The character *mang* has been used to describe people who come from "outside the boundaries," "from the wilderness," or, more likely these days, "from the countryside." Combining these two characters into *liumang*, Bakken suggests, creates a word that can effectively describe a person who appears to be unreliable in some way, and not to be trusted: which is certainly the way many residents think about many of the new migrants: the "floating" populations in contemporary Chinese cities (Buruma 2001). In behavioral terms one of the major outcomes of such forces is the emergence of a new culture of "marginality": with more people than ever before involved in pursuits that only a few years earlier had been sharply proscribed. There are now opportunities to buy or to gain access to virtually anything: lurid pictures are for sale in magazines, as posters and calendars, as well as in books and on the Internet. New lines of fashionable clothing, cosmetics, and personal care products intended to accentuate (male and female) sexuality are now ubiquitous. Sex-oriented shops in most of the big cities are now selling birth control supplies, condoms, herbal aphrodisiacs, skimpy leather clothing, and all manner of sex toys and aids. Chinese films now routinely feature explicit sex scenes, and, although visually much tamer, Chinese television advertising is now thoroughly suffused with sexual imagery designed to sell products of all types.

Another important variable is the artificially high sex ratios that are to be found in China, as well as other Asian societies, especially India. This has produced an unprecedented number of surplus adult males, often referred to in the Chinese vernacular as "guang gun-er" or "bare branches." These are young men who will never have families because they can not find spouses (Hudson and den Boer 2004). The surplus men in "bare branches" societies are likely to share a number of characteristics: they belong primarily to the lowest socio-economic classes; they are more likely to be unemployed and underemployed; they usually do not have access to land or other resources; they are typically transients, with few ties to their localities; and they tend to live among and socialize exclusively with other "bare branches" males like themselves, and as a result, are often treated as social outcasts by the other members of the community. Hudson and den Boer also suggest (2004, p. 192) that "bare branches" males are more likely than other males to come into contact with crime, violence, and vice: "Unable to achieve satisfaction in socially approved ways, they spend their meager wages on gambling, alcohol, drugs, and prostitutes in short (but intense) sprees. The disorder created during these sprees predictably has a negative impact on society."

Urban culture change and the individualization of life in the new China

In spite of the liberalizing effects of the opening-up and reform era, the new trends have not been uniformly heralded as beneficial for China. Some cultural critics have suggested that the material advancements notwithstanding, the contemporary era has involved a transition from the "noble" vision of utopian socialism, to one that has only the "baseness of the self" at its core (Zhang 1997). One of the significant "costs" associated with life in contemporary China is "the loss of frameworks in which to situate oneself" resulting in a loss of orientation, associated with "a modern sense of meaninglessness, lack of purpose, and emptiness" (E.Y.J. Wang, 1996, pp. 39–40).

It is from this perspective that we can begin to interpret some of the currently widespread phenomenon of cynicism, alienation, and thoroughly self-serving behavior that is being reported as almost omnipresent in contemporary China (Zha 1995, 1997). This sentiment, originally expressed in literary and artistic circles (J. Wang 1996), is now shared by many China scholars in the social sciences, both Chinese and foreign. Blecher (1997, p. 213), for example, has observed that the "profound uprooting of cultural moorings provided by shared values, commitments and identity" makes it appropriate to speak about "a cultural crisis in contemporary China." We do not have to go far in present-day China to find a cluster of trends that might be considered symptomatic of these new self-serving tendencies, including: rising crime rates; government corruption; marketed pornography; the trafficking of women and girls; drug and alcohol abuse; and commercial sex.

During the first years of the STD/HIV epidemic in China, many officials argued that the vast majority of the threat to public health was coming from beyond China's borders, an idea that was fuelled by popular prejudice, as well as publicly stated (official) pronouncements. Foreign residents and Chinese citizens who traveled abroad or who had regular contact with foreigners, including employees of joint-venture firms and hotels, were (and still are) tested for HIV. The official language used in media pronouncements (particularly in the editorials of state-controlled outlets) about HIV/STDs identified a clear notion of foreign nationals as a dreaded "other" group (Evans 1997). More recently, it has become apparent that at least as much attention should be paid to the domestic dimensions of STDs.

Like China, Indonesia has experienced rapid urbanization in recent decades. Although it is difficult to establish the exact rate of urbanization in either country, the percentage of Indonesians living in urban areas has increased from less than 20 percent in 1971 to more than 40 percent in 2000. As is the case in China, this rapid rate of urbanization has been associated with substantial social and economic change. However, there have been some significant differences between the two nations, in the nature and progress of this change. While the Suharto regime (1966–98) greatly restricted the rights of

Indonesians, at no time was the state able to enforce the same degree of social and spatial control that was experienced in China. There was little effective control over the rate of internal migration, for example, and urbanization was more rapid in Indonesia than China during the 1970s and 1980s. Since 1998, Indonesia has been experiencing a substantial political change involving democratization and an attempt to devolve power and investment within the country. Along with this has been an increase in personal freedoms, although the relative amount of change has not been as dramatic as in China. While China and Indonesia have similar levels of average income, the trajectory has differed in recent years. While prosperity has increased exponentially in China, in Indonesia the Asian economic crisis of 1997 saw a dramatic fall in national wealth, which has still not fully been recovered.

One important demographic feature that China and Indonesia share is that both have witnessed huge rates of growth among young adults (Hugo 2005).[8] Not only are they a large part of the national population, they are also the first generation to grow-up in an era of full education, combined with the widespread penetration of mass media, and subject to the forces of economic and cultural globalization. There is a new youth culture in a new era of consumption, and they disproportionately live in urban areas where these forces are likely to be at their strongest, as a result of the high rate of youth selectivity involved in the mass exodus to cities.

One of the features of the demographic behavior of contemporary Asian youth which has been identified as potentially serious in its implications (Gubhaju 2002; Qiao et al. 2001; Westley and Choe 2002; Xenos and Kabamalan 1998) is the increasing level of risky behavior, compared to previous generations. One important area is in sexual risk-taking (Westley and Choe 2002, p. 61), and although Asia still falls far behind Sub-Saharan Africa in the incidence of premarital sex, Gubhaju (2002, pp. 104–5) has shown that pre-marital sex is clearly on the rise in Asia, despite strong traditional norms opposing it in many nations. Moreover, the risky sexual behaviors of adolescents are compounded by a widespread sexual double standard, which accepts, or even encourages, promiscuity among men but strictly restricts women's sexual behavior (Micollier 2004).

It is apparent that pre-marital sex is of significance in Indonesia and elsewhere in Asia; and that levels of sexual risk-taking are on the increase (Westley and Choe 2002, p. 61). In Indonesia, surveys indicate increasing levels of sexual intercourse among youth, and although the peak incidence of HIV infection is in the 20s age group, there are also significant numbers in the 15–19 age group. In addition to these risk factors, it is evident that condom usage rates are low among sexually active youth in Indonesia (Hugo 2001).

In recent times the increasing rate of intravenous drug use (IDU) has been identified as a major factor in the spread of HIV/AIDS, in Indonesia as well as in China. It has been reported (Vatikiotis and Crispin 2004, p. 36) that in

2002, for example, there were around 160,000 intravenous drug users in Jakarta alone; 55 percent of whom were believed to be sharing needles, and half of them infected by HIV. The rapidly increasing use of drugs in Indonesian cities is an established fact, one that is clearly associated with rapid social and economic changes similar to those occurring in China.

In Indonesia the evolving HIV epidemic must be seen to some extent as a continuation of long standing trends in the spread of STDs, involving the growth of the commercial sex industry in urban areas, combined with high population mobility. However, it is also evident that profound social and economic changes have been sweeping the nation, and that such trends might also be linked to the erosion of traditional values, a decline in the influence of the family, and shifting attitudes toward sexual relationships. This is overlain with a well-established commercial sex sector and the existence of double standards that even tolerate a significant use of CSWs by Indonesian married men (see also Micollier 2004). Women are often in an unequal relationship with men in respect to sexual activity in Indonesia, and there is a strong antipathy among men to the use of condoms. In other words, the contemporary crisis in Indonesia is being driven by a mix of long-standing cultural, gender, and historical forces, combined in an explosive fashion with the impact of rapid social and economic change.

Conclusions and Implications for Future Research

This chapter has looked for contrasts and comparisons between China and Indonesia in an attempt to identify the circumstances under which China's previously unconnected outbreaks of HIV/AIDS might become connected, with the disease spreading to parts of the country not already significantly afflicted. One of the most obvious answers to this question, based in part on the work that has already been reported from Indonesia and other countries in the region, involves the role of migration in the transmission and spread of HIV/AIDS. One of the unintended consequences of China's reform project was the creation – beginning in the mid-1980s – of an enormous opportunity divide between the rural and the urban sectors, which has produced one of the largest population movements in history.

Migration helps to spread the HIV/AIDS epidemic. In addition to the increased exposure to HIV at their places of destination, many of China's migrants travel back and forth to their homes, often several times a year, then return, often to different places, thereby running the risk of contracting new diseases and spreading them to their partners. Some empirical support is offered to support this hypothesis, but pointing a finger at the migrants may be a too-simple conclusion that is inherently discriminatory toward them. There are good reasons, in other words, to conclude here that demographic change does not tell us the whole story about the HIV/AIDS situation in contemporary China.

At the same time that this mass migration was happening, new structural forces were emerging in China, any or all of which might have influenced human behaviors, and not only among migrants. It is the links between these new forces and the rising prevalence of risky behaviors associated with the transmission of STDs and HIV, which we now need to be exploring if any headway is to be made in the direction of preventing their spread. Based on the evidence that this spread may be most rapid among the middle-class sectors of Chinese society, there are good reasons to look beyond the migrant groups for answers, and this chapter has explored some of the reasons why it is appropriate to look at China's cities as the potential breeding grounds for rising rates of prevalence of STDs and HIV/AIDS.

We have tried to situate China's HIV/AIDS dilemma in the context of the sociogeographic transformations that have accompanied the reform project and the expanding polarization of Chinese society. At one end of the spectrum we find the new urban poor, and the increasing marginality of some major sections of urban society – including a larger than ever number of "surplus" or "bare branches" males. This has been accompanied by the simultaneous emergence of a sizeable middle-class, made up of individuals who have prospered during the reforms. Among this group, which also includes many of the wealthy "overseas Chinese" investors from Hong Kong, Taiwan, and Singapore, surplus capital is now available to be pumped into a new street- and club-level world of entertainment, consisting of establishments and activities catering to all of the new "appetites" that are being generated in the post-socialist consumption milieu in China's exploding cities (Farquhar 2002). The workforce for such establishments comes primarily from the ranks of the rural and urban poor at the bottom end of the social ladder; while the nouveau riche at the other end are acting as both the customers and the investors. The stage has been set for a wholly new array of behaviors; some of which are no doubt harmless, but others might involve epidemiologically high-risk activities associated with alcohol, drugs, and sex. In these newly dangerous times, this could prove to be a deadly combination for a small but growing number of participants. Migration has played and will continue to play a part in this process, in the sense that rural women on the move are the sex workers in many of the new establishments; while rural men now living in the cities are among the purveyors of sex and drugs on the streets. However, migration is only one of many variables that have interwoven to produce totally new sets of potential risk factors.

In a series of very complex ways, all of which still need to be theorized and analyzed carefully, the changes associated with China's urban transition could be contributing to a situation in which there is more risk-taking behavior occurring in the cities. Migration has a role to play in this process, but many other forces are also operating, and it is prudent, therefore, to focus research attention in other directions, especially as many of the migrants appear to be settling into their new lives in the cities, and are not likely to return permanently to the countryside anytime soon.

NOTES

1 Skeldon (2004, p. 267) refers to this syndrome of countries not being politically committed to recognition of HIV/AIDS as a major policy issue as the "Three Ds" – Denial, Delay, and Do Nothing.

2 We recognize that cultural variables are of critical importance in understanding the spread of HIV/AIDS in China and Indonesia. As Micollier (2004) has observed, it appears that in many parts of Asia – in sharp distinction with what prevails in most Western countries – no significant stigma is associated with men who choose to have sex with commercial sex workers; in fact, in some instances it is seen as enhancing one's "maleness." In both China and Indonesia this may prove to be of great significance for the future spread of the disease, especially as heterosexual encounters become one of the primary means of transmitting HIV.

3 Not surprisingly, the geographic distribution of such infections shows the highest rates in the region adjacent to the so-called Golden Triangle (bordering Laos, Myanmar, and Thailand), see: www.usembassy-china.org.cn/sandt/CDCAssessment.htm. See also the report prepared by the (US) National Intelligence Council (2002): *The Next Wave of HIV/AIDS: Nigeria, Ethiopia, Russia, India, and China.*

4 Many outsiders argue that the potential for a further and even more rapid spread of HIV in China is increased by the overall lack of AIDS awareness and education, at least until quite recently (Holtzman 2003). The government counters these claims by pointing out that it has launched a number of major national, regional, and local campaigns against AIDS, as well as efforts to improve AIDS education and raise AIDS awareness in China; see: www.china.org.cn/e-news/news823.htm.

5 The situation is different for many of the (mainly) young women who are working in factories – all over China – that have regular social control and surveillance routines. In these situations, where the workers are literally "tied" contractually to the factories, movement away from the workplace is difficult.

6 At the start of the 1980s, the average household in China spent less than 5 percent of its total budget on housing, utilities, transport, education, and medical care combined; in the US the equivalent figure would have been closer to 50 percent or 60 percent.

7 *Liumang* is the word used to describe the perennially popular "hooligan" literature associated with Wang Shuo and other writers, where layabouts and minor hoodlums are often the (anti-) heroes of the day (J. Wang 1996).

8 They are the last generation of births born under a high fertility regime.

REFERENCES

Bakken, B. (2000) *The Exemplary Society: Human Improvement, Social Control, and the Dangers of Modernity*. London: Oxford University Press.

BBC News Online. (2003) "China Facing AIDS 'Time Bomb.'" (Thursday, June 27). Online at: http://news.bbc.co.uk/2/hi/asia-pacific/2069943.stm

Blecher, M. (1997) *China Against the Tides: Restructuring Through Revolution, Radicalism and Reform*. London: Pinter.

Brockett, L. (1996) 'Thai Sex Workers in Sydney.' Unpublished MA thesis, Department of Geography, University of Sydney.

Buruma, I. (2001) *Bad Elements: Chinese Rebels from Los Angeles to Beijing*. New York: Random House.

Chan, A. (2001) *China's Workers Under Assault: The Exploitation of Labor in a Globalizing Economy*. Armonk, NY: ME Sharpe.

Chan, J. (2001) "HIV/AIDS Epidemic in Rural China", *World Socialist Website*, 6 August. Online at: www.wsws.org.

Chan, J. (2003) "Social Discontent Escalates in China" *World Socialist Website*, 12 February. Online at: www.wsws.org.

Chantavanich, S. (2000) *Mobility and HIV/AIDS in the Greater Mekong Subregion*. Asian Research Center for Migration, Institute of Asian Studies, Chulalongkorn University, Bangkok, Thailand. Online at: http://www.adb.org/Documents/Books/HIV_AIDS/Mobility/prelim.pdf

Conachy, J. (2001) "Chinese Think-Tank Warns of Growing Unrest Over Social Inequality." *World Socialist Website*, 15 June. Online at: www.wsws.org.

Eckholm, E. (2001) "When Lies Kill: In China the Right to Truth Meets Life and Death." *New York Times* (Sunday) "Week in Review" Section, pp. 1, 4.

Evans, H. (1997) *Women and Sexuality: Dominant Discourses of Female Sexuality and Gender Since 1949*. Cambridge, UK: Polity Press.

Farquhar, J. (2002) *Appetites: Food and Sex in Post-Socialist China*. Durham, NC: Duke University Press.

Farrer, J. (2002) *Opening Up: Youth Sex Culture and Market Reform in Shanghai*. Chicago, IL: University of Chicago Press.

Forney, M. (2003) "China's Prosperous Surface Masks a Rising Sea of Joblessness That Could Threaten the Country's Stability." *Time*. Online at: http://www.time.com/time/asia/covers/1101020617/cover.html

Goodman, P.S. (2003) "Localities Exploiting a Growing Business." *Washington Post Foreign Service* Saturday, January 4: p. A01. Online at: http://www.aidscience.org/Science/2339.html

Gubhaju, B.B. (2002) "Adolescent Reproductive Health in Asia." *Asia-Pacific Population Journal* 17, 4: pp. 97–119.

Holtzman, D. (2003) "Current HIV/AIDS-Related Knowledge, Attitudes, and Practices Among the General Population in China: Implications for Action." *AIDScience* 3, 1 (January). Online at: http://aidscience.org/articles/aidscience028.asp

Hudson, V.M., and A.M. den Boer. (2004) *Bare Branches: Security Implications of Asia's Surplus Male Population*. Cambridge, MA: The MIT Press.

Hugo, G.J. (1975) "Population Mobility in West Java, Indonesia." Unpublished PhD Thesis, Department of Demography, Australian National University, Canberra.

Hugo, G.J. (1978) *Population Mobility in West Java*. Yogyakarta: Gadjah Mada University Press.

Hugo, G.J. (1982) "Circular Migration in Indonesia." *Population and Development Review* 8, 1: pp. 59–84.

Hugo, G.J. (2001) "Population Mobility and HIV/AIDS in Indonesia." Report prepared for the UNDP South East Asia HIV and Development Program, UNAIDS and ILO, Indonesia. Online at: http://www.hiv-development.org/publications/Indonesia.html

Hugo, G.J. (2003) "Labour Migration Growth and Development in Asia: Past Trends and Future Directions." Paper for ILO Regional Tripartite Meeting on Challenges to Labor Migration Policy and Management in Asia, Bangkok, 30 June–2 July.

Hugo, G.J. (2005) "A Demographic View of Changing Youth in Asia." In F. Gayle (ed.), *Youth in Transition: The Challenges of Generational Change in Asia.* Bankok: UNESCO.

Hyde, S.T. (2001) "Sex Tourism Practices on the Periphery: Eroticizing Ethnicity and Pathologizing Sex on the Lancang." In N.N. Chen et al. (eds.), *China Urban: Ethnographies of Contemporary Culture.* Durham, NC: Duke University Press: pp. 143–62.

Lin, G.C.S. (2002) "The Growth and Structural Change of Chinese Cities: A Contextual and Geographical Analysis." *Cities* 19, 5: pp. 299–316.

McBeth, J. (2004) "Cusp of a Crisis." *Far Eastern Economic Review*, 15 July: p. 46.

Micollier, E. (ed.). (2004) "The Social Significance of Commercial Sex Work in China: Implicitly Shaping a Sexual Culture?" In E. Micollier (ed.), *Sexual Cultures in East Asia: Social Construction of Sexuality and Sexual Risk in a Time of AIDS.* London: Routledge/Curzon Press.

Ministry of Health, Republic of Indonesia. (2002) *Increasing Evidence Threat of HIV/AIDS in Indonesia, Calls for More Concrete Prevention Efforts*, Ministry of Health, Republic of Indonesia, Jakarta.

National Center for HIV, STD, and TB Prevention. (2001) *Report of an HIV/AIDS Assessment in China.* Centers for Disease Control and Prevention, US Department of Health and Human Services (July 30–August 10, *and* August 28–30). Online at: http://www.usembassy-china.org.cn/sandt/CDCAssessment.htm

National Intelligence Council. (2002) *The Next Wave of HIV/AIDS: Nigeria, Ethiopa, Russia, India, and China.* ICA (Intelligence Community Assessment) Document 2002-04D (September). Online at: http://www.odci.gov/nic

Pan, S.M. (1999) *Red Light District.* Beijing: Qunyan Publishing House. Online at a United Nations website: http://www.unchina.org/unaids/enews5.html

Parida, S.K. (2000) "Sexually-Transmitted Infections and Risk Exposure Among HIV Positive Migrant Workers in Brunei Darussalam." In *Population Mobility in Asia: Implications for HIV/AIDS Action Programmes*, Bangkok: UNDP, South East Asia HIV and Development Project: pp. 26–32.

Parish, W.L., and W.M. Chen. (1996) "Urbanization in China: Reassessing an Evolving Model." In J. Gugler (ed.), *The Urban Transformation of the Developing World.* Oxford: Oxford University Press.

Pei, M.X. (2001) "Cracked China." *Foreign Policy Online* (Sept/Oct). On;line at: http://www.foreignpolicy.com/issue_SeptOct_2001/IOWSeptOct2001.html

Qiao, X., X. Guo, L. Zhang, A. Mao, F. Wang, Z. Liang, S. Xia, and J. Li. (2001) "Shanxisheng HIV Ganranzhe zhong Liudong Renyun Xianzhuang Diaocha Fenxi [A Profile of Temporary Migrants among HIV Infected in Shanxi Province]." *Diyijia Zhongguo Aizibing Xingbing Fangzhi Dahui Lunwenji* (Zhongguo Xingbing Aizibing Fangzhi, Zengkan): 35.

Rosenthal, E. (2002) "China Raises HIV Count in New Report." *New York Times* (12 April). Online at the Human Rights Monitor website: http://www.humanrightsmonitor.org/article479.html?POSTNUKESID=51e6aabad40d4f4bddb6e79a6a55db

Schauble, J. (2002) "China Forecasts Grim Future for Rising Jobless." *The Age*. Tuesday April 30. Online at: http://www.theage.com.au/articles/2002/04/29/1019441344886.html

Skeldon, R. (2004) "The Global Challenge of HIV/AIDS." In R. Black and H. White (eds.), *Targetting Development: Critical Perspectives on the Millennium Development Goals*. London: Routledge, Routledge Studies in Development Economies: pp. 256–72.

Smith, C.J. (2005) "Social Geography of Sexually Transmitted Disease in China: Exploring the Role of Migration and Urbanization." *Asia Pacific Viewpoint* 46, 1 (April): 65–80.

Smith, C.J., and X.S. Yang. (2005) "Examining the Connection between Temporary Migration and the Spread of HIV/AIDS in China." *The China Review* 5. 1 (Spring).

Solinger, D.J. (2000) "The Potential for Urban Unrest." In D. Shambaugh (ed.), *Is China Unstable?* Armonk, NY: M.E. Sharpe: pp. 79–94.

Solinger, D.J. (2002) "Labour Market Reform and the Plight of the Laid-off Proletariat." *The China Quarterly* 170 (June): 304–26.

Solinger, D., and K.W. Chan. (2002) "The China Difference: City Studies Under Socialism and Beyond." In J. Eade and C. Mele (eds.), *Understanding the City*. Oxford: Blackwell: pp. 204–21.

Stephenson, J. (2001) "HIV/AIDS in China." *Journal of the American Medical Association* 286, 14, October 10.

Thompson, D. (2003) "HIV/AIDS Epidemic in China Spreads into the General Population." *Population Reference Bureau Online*, April. Online at: http://www.prb.org/Template.cfm?Section=PRB&template=/ContentManagement/ContentDisplay.cfm&ContentID=8501au

UNAIDS. (2000) *Country Profile: China*. Online at: http://www.unchina.org/unaids/ekey2.html.

UNAIDS. (2001) "Immediate Action Essential to Prevent Epidemic Taking Hold in General Population, Says UNAIDS Executive Director." Press Release, November 13.

UNAIDS. (2002a) *Epidemiological Fact Sheets on HIV/AIDS and Sexually Transmitted Infections (China; 2002 Update)*. Geneva: UNAIDS.

UNAIDS. (2002b) *China's Titanic Peril – 2001 Update of the AIDS Situation and Needs Assessment Report*. Geneva: UNAIDS, June. Online at: http://www.unaids.org/whatsnew/newadds/AIDSchina2001update.pdf

UNAIDS (Jakarta). (2001) *HIV/AIDS Report 2000 Indonesia*. Jakarta, Mimeo.

UNAIDS Secretariat. (2004) "HIV/AIDS Activities in Indonesia: An Overview by Funding Source, Location and Coverage. Initial Draft for Limited Circulation." Prepared for the National AIDS Commission, UNAIDS Secretariat.

United States Embassy. (2000) *AIDS in China: From Drugs to Blood to Sex*. Online at: http://www.usembassy-china.org.cn/english/sandt/webads1a.htm

Vatikiotis, M., and S.W. Crispin. (2004) "Asia's Wasted Lives." *Far Eastern Economic Review* 15 July: 34–8.

Wang, E.Y.J. (1996) "Samsara: Self and the Crisis of Visual Narrative." In W. Dissanayake (ed.), *Narratives of Agency: Self-Making in China, India, and Japan*. Minneapolis, MN: University of Minnesota Press: pp. 35–55.

Wang, J. (1996) *High Culture Fever: Politics, Aesthetics, and Ideology in Deng's China*. Berkeley, CA: University of California Press.

Westley, B., and M.K. Choe. (2002) "Asia's Changing Youth Population." In East-West Center, *The Future of Population in Asia*. Honolulu: East-West Center: pp. 57–67.

Whyte, M.K., and W.L. Parish. (1984) *Urban Life in Contemporary China*. Chicago: The University of Chicago Press.

Wu, Z.Y. (2003) "National Center for AIDS/STD Prevention and Control," "Chinese Center for Disease Control and Prevention." Presentations made in Washington, DC, in October 2002 and Beijing, in January 2003.

Xenos, P., and M. Kabamalan. (1998) *The Changing Demographic and Social Profile of Youth in Asia*. Asia-Pacific Research Report, no. 12, Honolulu: East-West Center Program on Population.

Yang, X.S. (2002a) "Temporary Migration and the Spread of STDs/HIV: Is There a Link?" Paper presented at the Association of American Geographers National Meeting, Los Angeles, March.

Yang, X.S. (2002b) "Migration, Socioeconomic Milieu, Migrants' HIV Risk-Taking Behaviour: A Theoretical Framework." Paper presented at the Population Association National Meetings, Atlanta, Georgia, May.

Yang, X.S. (2004) "Temporary Migration and the Spread of STDs/HIV in China: Is There a Link?" *International Migration Review* 38, 1: 212–35.

Zha, J.Y. (1997) "China's Popular Culture in the 1990s." In W.A. Joseph (ed.), *China Briefing: The Contradictions of Change*. Armonk, NY: M.E. Sharpe: 109–50.

Zha, J.Y. (1995) *China Pop: How Soap Operas, Tabloids, and Bestsellers are Transforming a Culture*. New York: The New Press.

Zhang, X.D. (1997) *Chinese Modernism in the Era of Reforms: Cultural Fever, Avant-Garde Fiction, and the New Chinese Cinema*. Durham, NC: Duke University Press.

14

The State's Evolving Relationship with Urban Society: China's Neighborhood Organizations in Comparative Perspective

Benjamin L. Read and Chun-Ming Chen

Introduction[1]

Throughout its nearly 60 years of existence, the People's Republic of China has made mighty efforts to *organize* urban society. It has sought to co-opt, manage, or dominate major forms of organization within the cities, including those based on workplace, trade or profession, religion, and place of residence. Today, the world of organizations within urban Chinese society can no longer be thought of as a monopoly by the Party-state, given the prevalence of private firms and the emergence of NGOs, migrant communities, homeowners' groups, and other phenomena. The official organizational apparatus nonetheless remains highly salient, though poorly understood.

This chapter considers a major portion of this apparatus: the dense and pervasive network of roughly 90,000 neighborhood organizations known as Residents' Committees (RCs).[2] The committee staff are paid by the state to serve as its designated liaisons, in most cases operating out of a permanent local office. The members act as the grassroots contacts and informants of the police and the government, carrying out a number of administrative tasks, from monitoring family-planning compliance to maintaining the household-registry rolls. At the same time, they also provide a range of services to their constituents, listen to and act on their suggestions and complaints, and organize social and public-benefit activities for them to take part in if they choose. (Space constraints preclude a full bibliography on RCs here, but see Benewick,

Tong, and Howell (2004), Derleth and Koldyk (2004), and Read (2000, 2003).)

The RCs and organizations like them in other countries confound a number of social-science categories. Hired by the state yet part of the community in which they serve, the RC staff occupy an unfamiliar type of intermediary position between state and society. They are not accountable to their neighborhoods through anything like a genuine electoral mechanism, yet they often work hard to build an image of responsiveness. They combine service functions with routine monitoring and complicity in occasional episodes of repression; their work involves a mixture of gentleness and domination.

In terms of the various perspectives put forward by Logan and Fainstein in this volume's introduction, modernization theorists would nod their heads at the challenges these grassroots organizations face in staying connected to an increasingly mobile, affluent, and privacy-conscious society. Scholars of state socialism would quickly identify analogous institutions in places like Cuba, Vietnam, and the former Soviet Union. Yet both would be puzzled by the persistence of state-linked neighborhood groups not just in ostensibly post-socialist China but also in other Asian countries that are even more modern and less socialist. While China fits only problematically into the mold of nearby "developmental states," aspects of this body of theory may be most fruitful in considering the RCs and their counterparts abroad.

Theoretical Context

Studies of socialist cities around the world have provided accounts of neighborhood-level groups that resemble those in China (Colomer 2000; Fagen 1969; Friedgut 1979; Koh 2006; Roeder 1989). On the face of it, such general accounts of state socialism would seem to provide the theoretical framework we need. Communist Parties everywhere propagated "mass organizations" through which to incorporate and monitor society. True, these kinds of organizations collapsed in the regime transitions in Eastern Europe and the Soviet Union. Yet this in itself hardly makes the RCs anomalous; while China is "post-socialist" in the sense of having enthusiastically embraced market practices in many parts of the economic realm, the governing apparatus still has much in common with other socialist cases.

We argue, however, that to see the RCs merely within a "comparative communism" context is inadequate for several reasons. Far from allowing them to atrophy, the Chinese government has taken active steps to develop and revitalize these organizations in the past two decades. While their counterparts elsewhere in the socialist bloc are often seen as having left a wholly negative legacy of distrust (Howard 2003), the RCs are perceived in a much more variegated and often positive way. Most tellingly, in many respects they closely resemble counterpart organizations in East and Southeast Asian states that never were socialist and indeed were strongly anticommunist.

These local groups are best understood as manifestations of a broader phenomenon, *grassroots administrative engagement*, in which states create, sponsor, and manage networks of organizations at the most local of levels that facilitate governance and policing by building personal relationships with members of society. Crucially, grassroots engagement is not confined to one type of political system; variants can be found under democracies, semi-democracies, and authoritarian as well as state-socialist regimes. But the countries that develop these institutions do share a set of pre-requisites. Only *strong* states have the bureaucratic coherence and organizational capacity to establish this form of grassroots presence. They share a desire, initially rooted in *security concerns*, to monitor and influence society at the neighborhood or household level, even though security is only one of many purposes served by such institutions. Finally, they are relatively *illiberal*, lacking deep commitment to principled prohibitions on government organization of society.[3]

An example such as Peru, along with the Cuban and Soviet cases mentioned above, demonstrates that these kinds of institutions can be found in many parts of the world (Stepan 1978, ch. 5). Yet some of the most elaborate and persistent examples are found in East and Southeast Asia. Indonesian society, for example, is finely subdivided into small units comprising clusters of households and groups of such clusters (known as *rukun tetanga* and *rukun warga*, respectively) that work closely with the state (Guinness 1986; Sullivan 1992). In Singapore, too, the government's People's Association runs an elaborate network of grassroots groups, such as the Resident's Committees and Neighbourhood Committees (Meow 1987).

Why are these parts of Asia such fertile soil for these institutions? Here it is worth turning to accounts of Japan, South Korea, Taiwan, and Singapore, which identified several features shared by these "developmental states." These countries possessed uncommonly strong states featuring coherent and capable bureaucracies. Far from governing in a hands-off way, administrators actively shaped their constituencies through corporatist links. Far from allowing society and the economy to develop haphazardly, they pursued interventionist strategies of partnering with, guiding and disciplining the firms and sectors that they oversaw (Amsden 1989; Evans 1995; Haggard 1990; Johnson 1982; Wade 1992).

The similarities among these cases have been attributed to several factors. "Late" industrialization (relative to the West) impelled the state to take an especially active role in guiding development. The Cold War created a strong security imperative, and the US provided military protection, open markets, aid funds and policy guidance. Land reforms instilled social leveling and spurred productivity. Finally, some scholars emphasized historical legacies stemming in part from Japanese colonization in the first half of the twentieth century; some also found common roots deeper in the past, such as traditions of recruiting bureaucrats through exams and prizing education as a route to advancement.[4]

Yet the developmental state literature is hardly tailor-made for explaining neighborhood administration. These theories were created in order to explain growth, hence they focus most sharply on *economic* institutions and policies, ignoring other important aspects of these governing systems. A brief look at specific historical links among these cases (Taiwan and South Korea, in addition to China) helps to clarify the common heritage that in part underpins similar approaches to urban governance.

The pre-requisites mentioned above – strong states, security concerns, and collectivist rather than liberal traditions – figure prominently in all three of these cases during the post-World War II era. In addition, rulers in each case had particularly robust historical templates to look to in constructing the machinery of ultra-local governance. The *bao-jia* established by China's Qing (1644–1911) and earlier dynasties grouped households into clusters so that roughly ten formed a *jia* and ten *jia* formed a *bao* (Hsiao 1967). The state designated headmen for each group and assigned them such tasks as keeping lists of the residents and reporting suspicious activity. During their colonization of Taiwan in 1895, Japanese administrators adapted and tightened up the *bao-jia* system, and freighted it with duties relating to public health, road construction, and taxation, in addition to its surveillance and registration functions (Chen 1975; Ts'ai 1990; Yao 2002).

As Japan occupied and colonized other parts of Asia in the twentieth century, it established either *bao-jia* or similar bodies (Chen 1984: 225–6). Korea, whose dynastic heritage also included state-imposed systems of household registration and collective responsibility,[5] was organized into small groups called *ban* in 1917. In 1940, these were re-dubbed "patriotic ban" [*aegukban*] (Seo 2002). "Each association consisted of ten households, and this became the basic unit for a variety of government programs for collection of contributions, imposition of labor service, maintenance of local security, and for rationing" (Eckert et al. 1990, p. 321). Japan introduced similar units, called *tonarigumi*, to Indonesia during its wartime occupation (Sullivan 1992, p. 136).

One might imagine that anything associated with previous rulers (especially the Japanese conquerors) would be rejected by the post-war governments of South Korea, Taiwan, and mainland China. But the ROK, ROC and PRC each reestablished systems of ultra-local administration both in the countryside and the cities. In Korea, the Syngman Rhee government kept in place the basic *ban* structure, once again renamed as "citizen's ban" [*kungminban*].[6] While its heyday of mass participation came in the 1970s under Park Chung Hee's "revitalization," the basic structure still exists. The Kuomintang (KMT), which itself had deployed *bao-jia* during its struggle with the Chinese communists, brought to Taiwan the administrative setup it had employed on the mainland, including urban neighborhood units known as *li*. In China, meanwhile, the new regime purged the former *bao*-heads as it cleansed the cities of KMT legacies, but soon formally mandated the establishment of neighborhood-based Residents' Committees in a 1954 law. Today Japan itself features a dense network of neighborhood associations that cooperate closely with local government (Bestor 1989, Pekkanen 2006).

In each case the new government hardly wished to acknowledge a debt to prior systems of social control. And the new institutions embodied substantial departures from their antecedents, far from merely aping them. The RCs in particular, in keeping with the CCP's brand of mass mobilization and guided participation, took on a great variety of educational and other service tasks, such as literacy classes and newspaper reading groups. Nonetheless, the core functional echo of the past – recruitment of local deputies to facilitate policing and governance through their ties with neighbors – could hardly be mistaken.

This micro-level organization of society persists to this day. In Taiwan, city districts are divided into neighborhoods, known as *li*. Each *li* is directed by a *lizhang*, who is paid no formal salary but receives a monthly stipend of around US$1,500, ostensibly for work-related expenses. Once appointed by the formerly dominant KMT party, the island's approximately 4,800 *lizhang* are now elected to four-year terms by their constituents.[7] Working hand-in-hand with an unelected civil servant known as a *liganshi*, the *lizhang* handle a range of official duties under the supervision of the district authorities. South Korea's roughly 58,000 urban *tongjang* perform a similar set of functions, though they are appointed rather than elected and their stipends are much smaller.[8] Just as the *lizhang* draw on the help of a set of volunteers known as *linzhang*, each of whom takes responsibility for a small section of the neighborhood, so too the *tongjang* are aided by *banjang*. Like China's RCs, these institutions operate in close proximity to residents. In Taipei, there are approximately 2,000 households per *li* and 100 per *lin*, while in Seoul, there are about 250 households per *tong* and 34 per *ban*.[9]

State-sponsored institutions of grassroots engagement have received relatively little study within the social sciences. Yet any thorough analysis of the way cities work in China or elsewhere in East Asia must take them into account. In this chapter, we pursue two goals. The first is to provide an empirical overview of the RC system, focused on Beijing. This not only makes clear how the system works and what it does, but illustrates urban governments' efforts to bolster and revamp this organizational network in recent years in response to several sets of changing circumstances. The second is to lay out in a more analytic fashion the multiple ways in which individuals within the urban population respond to and interact with this system. How much contact do city residents have with these committees? How do the RCs help or harm them? Do people perceive this system in a positive or a negative light?

The chapter draws upon several different types of data. The first author spent 14 months in 1999 and 2000 carrying out research in Beijing, centering on 10 neighborhood sites around the city. He made short trips to six other cities (Qingdao, Shijiazhuang, Shanghai, Guangzhou, Hengyang, and Benxi), in order to visit neighborhoods there, and conducted follow-up research in July 2003, December 2003, and July 2004. A series of private interviews with Beijing residents afforded perspectives on their attitudes toward and participation in neighborhood organizations. These topics were also the focus of a battery of questions on the 2001 Beijing Law and Community Survey (BLCS). He also carried out fieldwork on Taiwan's *lizhang* (December 2003; March–April 2006; January 2007) and South Korea's *tong/ban* system (July 2004). The second author has conducted an opinion survey and other research on Taiwan's *lizhang*.

Residents' Committees in Beijing and Elsewhere

As mentioned earlier, the RCs form an immense network covering a large majority of China's urban neighborhoods, with the exception of some newly built housing developments and recently urbanized areas. Details of the organizational setup vary from place to place, but it follows a single basic template. Table 14.1 shows the place of the RCs in the urban administrative hierarchy. Large Chinese cities have three levels of government: the city government, district governments, and Street Offices. The latter directly manage the RCs, maintaining a specialized staff to liaise with the committees and frequently calling their members in for meetings and instructions. As of 2003, Beijing's RCs averaged around 3,700 constituents, or 1,350 households, each. These numbers indicate that the RCs' jurisdictions are still small enough that its staff can be personally familiar with a substantial proportion of the residents. Through this system, the state reaches deep into the fabric of neighborhood life.

The RCs facilitate a range of state policies and programs, including:

- assisting the police by sharing information with officers from the local substation (*paichusuo*);
- helping to maintaining household registry (*hukou*) records;
- helping to implement the one-child-per-family birth control policy;
- facilitating welfare programs by keeping tabs on disadvantaged individuals; and
- conveying information from the state concerning laws, campaigns, programs, and policies.

Some of these functions, such as assisting with the one-child policy, are not found in Taiwan and South Korea because comparable family planning

Table 14.1 Levels of administration in Beijing and other Chinese cities

Level	*Chinese term*	*Number in Beijing*	*Number nationwide*
City Government	*shi zhengfu*	1	660
District Government	*qu zhengfu*	(in city core) 8	830
Street Office	*jiedao banshichu*	105	5,576
Residents' Committee	*jumin weivuanhui*	1,822	91,893

The national figure for City Governments includes prefecture-level [*diji*] cities, county-level [*xianji*] cities, and the four provincial-level cities. All Beijing figures do not include the outlying suburbs, including nine districts and two counties, that are part of the greater Beijing administrative region.
Source: National figures for Cities, Districts, and Street Offices: *China Statistical Yearbook 2003*, Table 1–1 (2002 data); national figures for RCs: *China Civil Affairs Statistical Yearbook 2002* (2001 data); Beijing Street Offices: *Beijing Statistical Yearbook 2002* (2001 data); Beijing RCs: unpublished figures for the year 2003 from the Beijing city government, on file with Read.

programs do not exist there. Yet the *lizhang* and the *tongjang* are also sometimes asked to help police gather information concerning crime cases. The *lizhang* are only peripherally involved in household registry records, but do help determine welfare eligibility. The *tongjang* help keep the household registry up-to-date, organize civil defense preparation, and together with the *banjang* form part of an elaborate system to convey information from the national, city, and district governments to neighborhood dwellers. Indeed, the *banjang* circulate official publications and convene meetings of *ban* residents at which the primary agenda item generally is the presentation of a set of messages and announcements from higher state authorities – though these meetings, called *bansanghoe*, are in many places held irregularly rather than at the monthly intervals that were once enforced.

Returning to the Chinese case, all of the above programs *could* be carried out by police or social workers without the permanent, neighborhood-based apparatus of the RC. Why then go to the trouble and expense of maintaining grassroots engagement bodies? What these functions all have in common is that they are all made much easier and less disruptive when conducted by individuals who are deeply familiar with the locality and the people in it. Many involve gathering "local knowledge" that bureaucrats in state offices far removed from the neighborhood would otherwise have no access to.

It is important to note that the actual character of these state-assigned functions varies and most of them are ambiguous in their relationship to residents' interests. In each case, the RC could be doing something that intrudes, inconveniences, or otherwise strikes people as noxious. Yet it could also be doing something residents perceive as beneficial. The RC's policing-related tasks certainly do involve helping the authorities sniff out political dissent – which some residents might object to. Yet the great majority of their work with the police revolves around efforts to prevent burglary, talking to former offenders in the neighborhood, and such – all of which is quite likely to be seen in a favorable light.

The previous paragraphs have focused specifically on what could be termed the "hard" functions of the RC, the things it does that have a major payoff to the state in terms of policing and policy implementation. But to understand the breadth and character of contact between RCs and their constituents, one also has to look at the many roles that the committee plays *other* than that of government information-collector. These functions include:

- listening to input from residents, serving as sounding boards for all manner of complaints and suggestions from their constituents;
- mediating disputes between neighbors and within families;
- providing a range of small goods and services, usually free of charge, such as measuring blood pressure;
- leading charity collection drives; and
- coordinating collective action in response to local problems such as theft, trash disposal, and vermin.

These many tasks have both a "face-value" function as well as an underlying purpose. They aim to bring the RC into contact with as many constituents as possible in a benign context. They also strive to build interpersonal familiarity between RC staff and residents, and develop a feeling that the RC is looking out for the interests of the public.

In addition to cultivating this relatively shallow, occasional interaction with many constituents, the RC also provides a welcoming venue for those residents who wish to participate in social or civic activities in a more active and sustained way. Some Beijing committees lead choral groups, others dance classes. Quite prevalent are exercise groups practicing both the traditional *taiji* (taichi) and the newly popular *jianshenqiu*, a rubber ball attached to a bungee cord used to whack oneself therapeutically on the back in a series of synchronized movements. One RC chair in a relatively affluent neighborhood organized her circle of middle-aged-and-older associates into a fashion show team, which would get together to show off their latest apparel. RCs would also lead outings to parks and other attractions, and some maintain activity centers where residents can read books or play low-stakes *majiang* (mah-jongg).

In addition to these recreational endeavors, the RCs organize the most receptive of residents to participate in various forms of service. These individuals, often designated as "small-group heads," "building heads," "courtyard heads," or the like, act as the committee's contact people within a portion of the RC's jurisdiction. Any given neighborhood may have 20 to 50 such people, generically termed "activists." They help the RC with many aspects of its work, from disseminating announcements to keeping in contact with welfare cases. But activists don't just work on behalf of the RC; they may also do chores like going door-to-door to collect fees for sanitation, gas, or electricity; help deliver mail or newspapers; or sweep the hallway floors. Some residents also choose to take a shift in the RC-sponsored security patrols, analogous to Neighborhood Watch groups, which can be seen keeping an eye on local comings and goings and proudly sporting their distinctive red arm-bands.

In a few of the Beijing neighborhoods, the security patrollers were given very small cash payments on a monthly basis in compensation for the hours of service they put in. Those partial exceptions aside, activists serve on a volunteer basis. Their motivation for taking part in this type of voluntary work seems to be the pleasures of sociability, a sense of empowerment, and the fulfillment from playing a valued part, however minor, in the ranks of the city's administrative apparatus.

A Changing Institution

Understanding the evolution of the RCs requires thinking about three aspects of change. The first consists of the structures of the state and the administrative programs they carry out. The second is the evolving physical and spatial setting of residential neighborhoods. The third consists of the needs and preferences of residents relative to the RCs.

Ever since its founding, the state has carried out administrative tasks and exercised political control in urban areas largely through the workplace or "work unit" (*danwei*; Perry and Lü 1997). Nonetheless, city governments have always maintained the RCs as their alley-level liaisons, and their importance to the state has arguably increased over time, at least in certain respects. Even at its peak, the *danwei* only incorporated a portion of the urban population, and in recent years this has shrunk as state firms lay off workers and migrants stream in from the countryside. Although work units in politically important sectors remain closely overseen by the state, the trend is to build up *government* capacity instead.

As the primary purpose of the RCs is to facilitate governance, the rising challenges facing cities led them to devote increasing attention and resources to them over the past two decades. Extolled at the highest levels of the Chinese Communist Party and coordinated by the national Ministry of Civil Affairs, this effort proceeds under the banner of "community building" (*shequ jianshe*). While this project borrows from an imported, public-relations-friendly discourse of community development, it is driven by a highly statist vision centered around government-dominated organizations, particularly the RCs (now often relabeled "Community Residents' Committees.") As with so many CCP projects, the redoubled focus on the RCs has both a state-security motivation as well as a service-delivery motivation. It aims to maintain oversight but also to establish a better administrative "platform" from which to facilitate the provision of more sophisticated forms of welfare.

The physical, built environment of residential neighborhoods has been undergoing far-reaching transformation in the past 15 years. Older neighborhoods of often dilapidated one- or two-story homes have been demolished and rebuilt at great speed in Beijing, Shanghai and other cities. At the same time, residential construction has shifted to the creation of new estates of privately owned condominiums run by professional property management companies (Davis 2003; Wang and Murie 1999; Zhou and Logan 1996).

All this has affected the RCs. The development of the organizational "software" of the committees lags considerably behind the pace of the "hardware" of housing construction. It can take several years for city and district governments to build new RCs (and in some cases on city peripheries, whole new Street Offices to oversee them.) Even after staff are hired, newly built organizations generally must start from scratch in trying to recreate the subtle web of interpersonal ties that underpin their operations in the older neighborhoods. The staff at least initially are strangers to their constituents. Moreover, in modern neighborhoods residents have fewer reasons to seek them out. Maintenance, security, and most fees are handled by the property management company. The RCs have less to offer by way of small favors and conveniences, as more and more needs are met by market-based providers.

If property management companies take away some of the RC's function as the neighborhood's all-purpose go-to office, a new type of organization has the potential to undercut its status as "representative" body. Authorized under national policies issued in 1994 and 2003, homeowners' organizations (*yezhu weiyuanhui*, YWH) are forming in a growing number of new, privately owned housing developments. Perhaps 6,000 to 7,000 of them existed as of 2004.[10] These groups vary widely, from those that are dominated by developers to a minority that are democratically elected by owners themselves. The national guidelines on YWH assert the primacy of the RC over the homeowners' groups. In practice, relations between the two bodies range from cooperative in some cases to conflictual in others. It is as yet unclear just what degree of autonomy the government will ultimately permit for the YWH and how compatible these "two centers" of power in the neighborhood will be. At a minimum, in those instances where they genuinely represent the residents, they complicate the neighborhood's administrative milieu (Read 2003b, 2007).

Given these three sets of evolving circumstances, how are the RCs themselves changing? One reform that has *not* yet taken place is democratization – at least, not in any substantial way as of the 2003 elections. Though according to law and rhetoric the RC members are to be elected by their neighborhood constituents, in fact they are still essentially hand-picked by the Street Offices. In the large majority of cases, elections follow a pre-rehearsed script in which votes are cast only by a group of perhaps 30 to 50 "residents' representatives" who are chosen by the Street Office and the Residents' Committee themselves on the basis of their supportiveness. Candidates are for the most part determined in advance by the Street Office, sometimes in consultation with the RC incumbents. Generally there are no more candidates than there are positions to be filled, although sometimes there is a measure of competition for one or two of the RC positions.

A number of cities have experimented with electoral procedures by relaxing one or more of the above constraints in some cases, although not so far as to render the elections free and fair by democratic standards. In Beijing's 2003 round of RC elections, only a small fraction of neighborhoods (less than 10 percent) held votes where balloting was not limited just to "residents' representatives." In only 11 neighborhoods were all adult residents eligible to vote. These elections involved very little competition: only 9.5 percent of candidates failed to be elected.[11]

Instead of democratization, RC reforms have focused on streamlining the system and recruiting better-credentialed and younger staff. Between 2000 and 2003 alone, the total number of committees in the eight districts at the heart of Beijing was reduced by more than half, from 3,885 to 1,822. This entailed merging many smaller RCs and sending home some of the older staff members. The changes were part of an ongoing process of reform stretching back at least to the early 1990s.

This state-led revitalization of the RCs has made the positions more remunerative and the hiring process increasingly competitive. Once staffed largely by homemakers who were paid little or nothing, the committees now employ middle-school and high-school graduates who often have substantial organizational experience from previous jobs. Typically these are people who have been laid off or taken early retirement from ailing state-sector enterprises. Stipends of 300–500 *yuan* per month are common in Beijing and other large cities, and some better-educated staff recruited under special programs make more than 1,000 *yuan* – hardly enough to get rich on, but among low-skilled middle-aged urbanites in China's unemployment-stricken economy, this represents a stable income that is better than many have.

RC staff are, now as always, far more likely to be female than male.[12] The average employee in Beijing is a woman with a high school education – though significant numbers of people with tertiary degrees now serve. Unlike the stereotypical bound-footed old lady of yesteryear, the average staff member is in her mid forties. Nearly 48 percent of RC staff are Communist Party members, an *increase* from 40 percent in 2000. This reflects diligent efforts by the authorities to maintain the Party committees associated with each RC, to ensure that at least one of the leaders of each RC is a Party member, and generally to keep this institution under Party as well as Street Office supervision.[13]

The nature of the committees' work has evolved over a full half-century of their existence. RCs were established in the early 1950s, shortly after the Communists took control of the cities. They were part of the way the revolutionary leadership consolidated its grip on urban society and tried to transform it. The RCs helped the state to identify and monitor people who were considered threats to the new order. In the Cultural Revolution, they led rallies against people in the neighborhood who had been designated class enemies. They also helped the housing authorities ascertain whose homes were big enough that they should be shared with other households. In the late 1960s and 1970s, when urban families were required to send many of their children out into the countryside to learn from the peasants, the RCs had to check that they were not sneaking back to their parents' homes. Their more politicized and intrusive practices shaped popular perceptions of them throughout the Mao era.

As the Communist Party has carried out reform policies and sharply curtailed its ambitions for class transformation along socialist lines, the RCs have quietly realigned their position with respect to urban society. They still play an important role in policing and policy implementation, which is the biggest reason why city governments have labored to revitalize them. But as discussed below, these monitoring duties have become less antagonistic to most urbanites, while their role as providers of services and sponsors of social and civic activities has been greatly augmented.

RCs and Their Constituents

What is the basic nature of constituents' relations with their RCs? Do not residents live in fear of this organization which, after all, represents the eyes and ears of the state? As we will see, residents' attitudes toward the RCs vary widely. To begin with, it should be pointed out that, for most people, the RCs hardly inspire fear. While in certain respects they are backed by the formidable coercive resources of the state, including the police, the system operates in such a way as to minimize the use of coercion or pressure and reserve it for exceptional circumstances. The RC in fact has relatively few favors and sanctions that it can apply in its dealings with constituents.[14] Unlike the work unit, the RC has no control over people's salaries or their access to housing. Residents can opt not to cooperate with it. Indeed, most of the time they can ignore it entirely if they choose.

As noted earlier, in order to carry out their duties efficiently and smoothly, RCs try to cultivate personal contact and familiarity with their constituents. The Beijing Law and Community Survey provides figures indicating what proportion of Beijing residents have ever approached the staff of their RC for various purposes, like paying fees (26 percent), taking care of documents (29 percent), and receiving low-cost goods or services (48 percent). Nearly a quarter of the sample also reported having visited the RC to express an opinion on issues of local concern, such as neighborhood security. Close to two-thirds indicated that they had sought out the RC for at least one of these reasons. Modest though some of these services are, the neighborhood committees are institutions that considerable numbers of Beijing residents at least occasionally turn to for assistance of one kind or another.

Table 14.2 provides figures on participation in the RCs' social and voluntary activities. Relatively undemanding forms of participation, such as contributing occasionally to RC charity drives, are found among a wide swath of the population (more than 35 percent of our sample), and involve a variety of residents. More time-consuming or committed forms of participation attract around 10 percent of the sample, disproportionally women and retirees. Neighborhoods vary in terms of the amount of RC-related volunteer activity they generate, as one would expect. In new neighborhoods where committees are just being developed, there is relatively little such activity, while RCs are more active in this respect in older neighborhoods.

The more casual forms of interaction that constituents have with their RCs, together with the various activities of a highly-involved minority of residents, add up to a substantial amount of face-to-face contact between RCs and those

Table 14.2 Participation in RC-sponsored activities

	All respondents (%)	*Women (%)*	*Retirees (%)*	*Party members (%)*
Percent in category of respondent who:				
Contribute regularly to RC charity drives	35.4	36.9	49.7	37.5
Participate in RC social events	10.7	11.9	18.7	18.4
Participate in RC patrols	6.8	9.5	17.5	8.8
Hold a neighborhood volunteer post	4.9	7.0	12.9	5.7
Of those who:				
Contribute regularly to RC charity drives		54.1	45.6	25.9
Participate in RC social events		57.9	57.0	42.1
Participate in RC patrols		72.6	83.6	31.5
Hold a neighborhood volunteer post		75.0	86.5	28.8

Source: Beijing Law and Community Survey, 2001.

they administer. Figure 14.1 shows that more than 75 percent of our Beijing sample reported some degree of contact with the RC in the past two years. To be sure, for many residents this is sporadic and infrequent. But more than 30 percent of our sample reported contact with an RC staff member at least every other month or so. Though quite a few citizens may not even know where their RC's office is located, generally speaking the RC is in close enough touch with enough of its constituents to have a good sense of the goings-on within its domain.

What do ordinary citizens think of the RCs? Figure 14.2 offers several ways of answering this. More than 60 percent of the respondents indicated they were at least "fairly satisfied" with their RCs; another 16 percent were neutral. But would people prefer not to have the committees at all? This was asked indirectly through a question measuring perceptions of the RC's dispensability. Over 70 percent of respondents answered that not having an RC would entail either "big trouble" or "a certain amount of trouble" for residents, as opposed to "a small amount of trouble" or "no trouble." This suggests that Beijing residents tend to perceive it as a useful institution, at least on balance. It is worth noting that in Taiwan, as well, most urban residents approve of their counterpart to the RCs; 74 percent said in a 2004 survey that they would like the *lizhang* system kept, while only 19 percent wished it to be abolished.[15] At the same time, this broad-based feeling that the RC is generally satisfactory and useful should not be taken as implying deep trust in this institution. Only

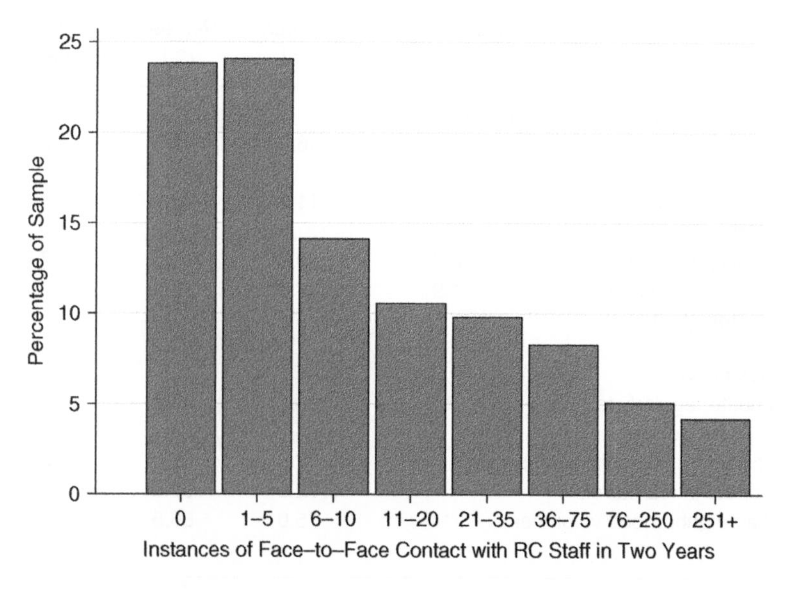

Figure 14.1 Frequency of Beijing citizens' contact with their RC staff

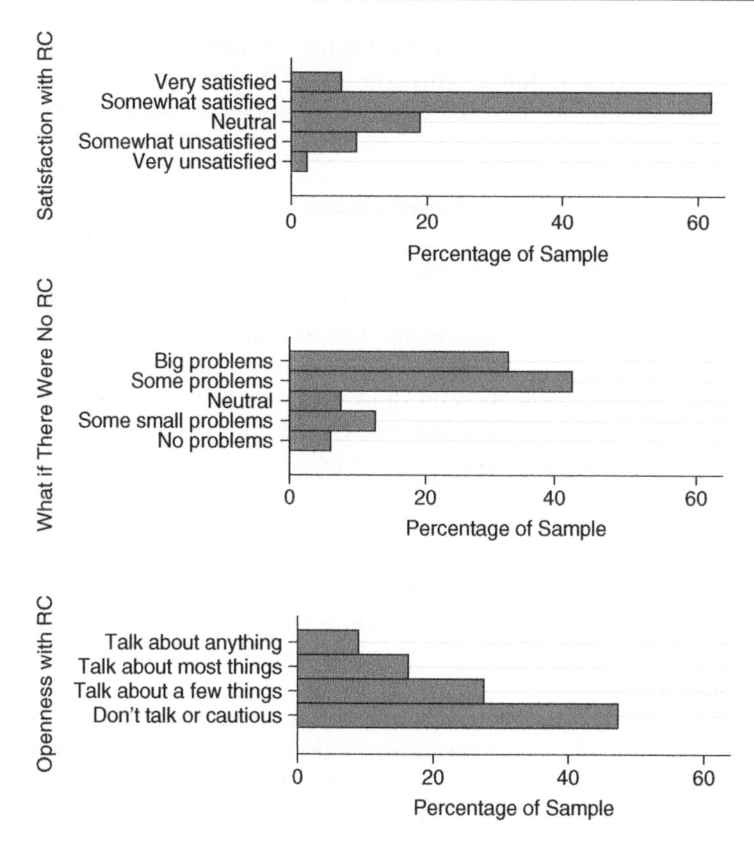

Figure 14.2 Three measures of Beijing citizens' views on the RCs

about a quarter of our respondents selected one of two "trusting" answer categories in our question designed to measure how comfortable people are talking to the members of their RC.

Survey results, taken by themselves, would not be entirely convincing. We might wonder, for instance, whether respondents felt pressure to give positive answers (though the responses to the trust question imply that this was not the case.) Interviews conducted in private also uncovered widespread approval for RCs, albeit with great variation from person to person. Some interviewees complained that the RCs were of no help in solving problems that really mattered to them, like finding a job. Others were angry about specific decisions that RCs had made. But many others looked favorably on their RCs, for a variety of reasons. Some liked knowing that the RC is keeping an eye on things while they're away at work. Others appreciated knowing that they can call someone when the sewer backs up. Some find it useful to contact the RC when they need to get certain kinds of documents officially certified. One factor underlying residents' attitudes seems to be fear of the burgeoning rural migrant population, whom urbanites tend to view with suspicion (Solinger 1999).

Conclusions

This chapter has considered China's Residents' Committees not simply as a Chinese or a state-socialist phenomenon, but as a case of what we refer to as grassroots administrative engagement. The RCs are fundamentally different from autonomous organizations of civil society. Their state-managed nature is off-putting to some residents, who dislike the committee's surveillance and limited ability to represent their interests. Others appreciate the groups' keeping an eye on the neighborhood and its being available to provide various services and hear a wide range of complaints and requests. Still others enjoy participating as volunteers in RC social or service activities. The committees' links to the government, then, cut in more than one way. Many urbanites ignore this institution entirely; most have occasional interactions with it; and a relatively small fraction of the population is closely engaged. Public opinion toward the RCs is mixed, with a perhaps surprisingly large proportion expressing approval.

We have seen that grassroots engagement plays an important role in policing, governance, and policy implementation; indeed, from the state's perspective this is its principal raison d'être. The outcomes that this contributes to can be unequivocally benign or deeply oppressive, and are commonly somewhere in between. For example, the role that local informants play in assisting the police aids in providing security for all their constituents, while also making it easier for states to repress deviants who are branded as threatening. Grassroots engagement can help keep authoritarian regimes in power just as it helps bolster public health and social welfare. Regardless of our normative feelings about the appropriateness of these undertakings, the results they contribute to matter.

Returning to the broad theoretical paradigms under consideration in this volume, we argued at the outset that other state socialist systems (whether current or former) do not provide the most useful frame of reference. What about modernization theory? The evidence presented above is mixed. The kinds of networks that these local organizations involve are not easy to categorize as either "traditional" or "modern." As discussed, these institutions in some ways hark back to premodern systems of social control. Yet in the first half of the 20th century, whether in Taiwan, South Korea, or mainland China, similar techniques were employed by modernizing states to coopt and incorporate preexisting forms of community. As we have seen, residents' contacts with, and participation in, the RCs are largely voluntary in nature. Yet there are also traditional elements in that neighborhood leaders in China sometimes draw on people's willingness to give respect to their elders. This is so even if, in other ways, a laid-off female Communist Party member fits no one's image of a "traditional elite." (Taiwan's *lizhang*, who are overwhelmingly male, older than Beijing's RC staff on average, and sometimes have the status of local landowners, clan leaders, or temple managers, admit to even more traditional characteristics.) Thus, perhaps applying a tradition/modernity lens to this institution is to generate a "misplaced polarity" (Gusfield 1967).

In fairness, however, it must be said that residential neighborhoods in cities like Beijing are undergoing change at a breathtaking pace, and this form of modernization is indeed relevant to the RC system. City residents themselves commonly speak of the coldness and anonymity of neighborly relations in the new apartment complexes, compared to older neighborhoods they remember, perhaps with a bit of nostalgic imagination. Modernity, among other things, implies functional specialization, a broader scope of market-based relationships, and greater freedom of choice. Urbanites today are much more able to choose where they live than in the past. In the new apartment complexes, professional management companies play roles that once were the province of the RC or other government agencies. Private homeowners' groups (where they exist) provide an alternative channel through which residents raise demands and pursue their interests. These trends diminish some of the bases on which RCs maintain contact with their communities, even as they strive creatively to maintain other means of establishing rapport and develop still others.

This chapter has suggested that looking comparatively at ultra-local organizations in cities elsewhere in East Asia is perhaps the most useful way to put this aspect of urban China in context. Differences notwithstanding, China's governing apparatus shares certain generic features as well as historical roots with those of Taiwan and South Korea. Indeed, research on the "developmental state" would do well to go beyond its economic dimensions toward a more holistic understanding that would include urban governance. While growth was what called the world's attention to these superstar performers, a closer inspection reveals that the ability to conjure economic development was but one aspect of a constellation of features both of the state and its relation to society.

Parallel institutions elsewhere, like Taiwan's urban *lizhang* network and Korea's *tong/ban* system, remind us that state-fostered local institutions do not necessarily wither away even in the context of the kind of dramatic regime transition that took place in those countries in the late 1980s and 1990s. The *lizhang* (though not the *tongjang* and *banjang*) also show that democratically elected (as opposed to government-appointed) local representatives can fit snugly into a system of close state-society ties. In turn, the Chinese RCs suggest that even in the absence of electoral accountability, a significant portion of urban constituents may be positively disposed toward this type of state institution, even as others treat it coldly. In short, while there is no one "Asian model" of grassroots administrative engagement, we may well expect the organizational structure of China's cities to continue to develop in ways that parallel other metropolises within the region rather than mimicking their North American or European counterparts.

NOTES

1 Fieldwork in China, South Korea, and Taiwan upon which this chapter draws was sponsored by a Fulbright-Hays Doctoral Dissertation Research Abroad Fellowship; a grant from the Committee on Scholarly Communication with China of the American Council of Learned Societies, a Graduate Student Small Grant from the Urban China Research Network, the Chiang Ching-kuo Foundation, and support from International Programs at the University of Iowa. None of these organizations is responsible for the findings.
2 In Chinese these organizations are known formally as *jumin weiyuanhui* or *shequ jumin weiyuanhui*, or *juweihui* for short.
3 These pre-conditions do not suffice to explain grassroots administrative engagement, but they help to show how some states are unable or unwilling to develop such institutions. Thus cases are rare or attenuated in the poorest and weakest of developing-world states, and in countries with a longstanding liberal tradition.
4 Specifically on these historical roots, see Johnson 1982, p. 23; Kohli 1994, 2004 (c.f. Haggard, Kang, and Moon 1997); Amsden 1989, pp. 37, 63–4; Evans 1995, pp. 51, 55; and Wade 1990, pp. 228–31, 337–42.
5 See Lee 1984, p. 184, for a brief discussion of Choson dynasty (1392–1910) systems that organized households into units of five, and required subjects to carry identification tags.
6 Seo 2002, p. 2.
7 Figures on the *lizhang* are available from the Statistical Yearbook of the Ministry of the Interior, online at http://www.moi.gov.tw/stat/
8 Officials of the South Korean Ministry of Government Administration and Home Affairs, in an interview on July 12, 2004, provided the figure of 57,993 *tong* in cities nationwide as of December 31, 2003.
9 The Taipei figures are derived from the city's Civil Affairs bureau website at http://www.ca.taipei.gov.tw/civil/page.htm. The Seoul figures stem from a chart provided by Suh Young Kwon of the city's Local Autonomy Assistance Division in an interview on July 13, 2004.
10 Shanghai is far ahead of other cities in terms of the number of officially recognized homeowners' groups. By the end of 2003, 4,756 homeowners committees had been established in Shanghai, at least an order of magnitude more than the figures for other cities. Shanghai is also distinct in allowing significant numbers of *yeweihui* in formerly state-owned neighborhoods (*fanggaifang*) in addition to commercial housing (*shangpinfang*). The numbers come from an interview with Shanghai housing official Xin Yiming, June 30, 2004.
11 Unpublished figures from the Beijing city government, on file with the author.
12 Unpublished figures from the Beijing city government, on file with the author. One reason for the disproportionately large number of women among RC staff

is that women often face layoffs or mandatory early retirement at younger ages than men. Apart from this, however, the highly gendered way in which RC service is conceived – it is widely seen as "women's work" – makes a difference as well. Popular perceptions were shaped in the 1950s and 1960s, when RC programs were particularly aimed at integrating into the new socialist order women who were not affiliated with work units. Although precise figures are unavailable, a large majority of Seoul's *banjang* also seem to be women. In contrast, about 90 percent of the *lizhang* elected in Taipei in 2003 were men.

13 This is termed "community Party building." The principal function of the Party committees that belong to the RCs is to maintain contact with Party members who, due to age and/or retirement, are no longer affiliated with the Party organization at their work unit.

14 This is not equally true for all residents, of course. Those who depend on state welfare support, for instance, may require the RC's favor. Also, in a few cases within any typical community, migrants or long-term residents will need the RC's imprimatur (granted in exchange for rent or fees) in order to run a shop or other small business in the neighborhood.

15 Data are from a Ministry of the Interior survey conducted by the second author.

REFERENCES

Amsden, A.H. (1989) *Asia's Next Giant: South Korea and Late Industrialization.* New York: Oxford University Press.

Benewick, R., I. Tong, and J. Howell. (2004) "Self-Governance and Community: A Preliminary Comparison between Villagers' Committees and Urban Community Councils." *China Information* 18: 11–28.

Bestor, T.C. (1989) *Neighborhood Tokyo.* Stanford: Stanford University Press.

Chen, C.-C. (1975) "The Japanese Adaptation of the *pao-chia* System in Taiwan, 1895–1945." *Journal of Asian Studies* 34: 391–416.

Chen, C.-C. (1984) "Police and Community Control Systems in the Empire." In R.H. Myers and M.R. Peattie (eds.), *The Japanese Colonial Empire, 1895–1945.* Princeton: Princeton University Press: pp. 213–39.

Colomer, J.M. (2000) "Watching Neighbors: The Cuban Model of Social Control." *Cuban Studies* 31: 118–38.

Davis, D.S. (2003) "From Welfare Benefit to Capitalized Asset: The Re-Commodification of Residential Space in Urban China." In R. Forrest and J. Lee (eds.), *Housing and Social Change: East-West Perspectives.* London: Routledge.

Derleth, J., and D. Koldyk. (2004) "The Shequ Experiment: Grassroots Political Reform in Urban China." *Journal of Contemporary China* 13: 747–77.

Eckert, C.J. et al. (1990) *Korea Old and New: A History.* Seoul: Ilchokak Publishers.

Evans, P. (1995) *Embedded Autonomy: States and Industrial Transformation.* Princeton: Princeton University Press.

Fagen, R.R. (1969) *The Transformation of Political Culture in Cuba.* Stanford: Stanford University Press.

Friedgut, T. (1979) *Political Participation in the USSR.* Princeton: Princeton University Press.

Guinness, P. (1986) *Harmony and Hierarchy in a Javanese Kampung*. Singapore: Oxford University Press.
Gusfield, J.R. (1967) "Tradition and Modernity: Misplaced Polarities in the Study of Social Change." *American Journal of Sociology* 72: 351–62.
Haggard, S. (1990) *Pathways From the Periphery: The Politics of Growth in the Newly Industrializing Countries*. Ithaca, NY: Cornell University Press.
Haggard, S., D. Kang, and C.-I. Moon. (1997) "Japanese Colonialism and Korean Development: A Critique. *World Development* 25: 867–81.
Howard, M.M. (2003) *The Weakness of Civil Society in Post-Communist Europe*. Cambridge, UK and New York: Cambridge University Press.
Hsiao, K.-C. (1967) *Rural China: Imperial Control in the Nineteenth Century*. Seattle: University of Washington.
Johnson, C. (1982) *MITI and the Japanese Miracle: The Growth of Industrial Policy, 1925–1975*. Stanford: Stanford University Press.
Koh, D.W.H. (2006) *Wards of Hanoi*. Singapore: Institute of Southeast Asian Studies.
Kohli, A. (1994) "Where Do High Growth Political Economies Come From? The Japanese Lineage of Korea's 'Developmental State.'" *World Development* 22: 1269–93.
Kohli, A. (2004) *State-Directed Development: Political Power and Industrialization on the Global Periphery*. Cambridge: Cambridge University Press.
Lee, K.-B. (1984) *A New History of Korea*. Cambridge, MA: Harvard University Press.
Meow, S.C. (1987) "Parapolitical Institutions." In J.S.T. Quah, C.H. Chee, and S.C. Meow (eds.), *Government and Politics of Singapore*, revised edn. Singapore: Oxford University Press: pp. 173–94.
Pekkanen, R. (2006) *Japan's Dual Civil Society: Members Without Advocates*. Stanford: Stanford University Press.
Perry, E.J., and X. Lü. (1997) *Danwei: The Changing Chinese Workplace in Historical and Comparative Perspective*. Armonk, NY: M.E. Sharpe.
Read, B.L. (2000) "Revitalizing the State's Urban 'Nerve Tips.'" *China Quarterly* 163: 806–20.
Read, B.L. (2003) "State, Social Networks, and Citizens in China's Urban Neighborhoods." Dissertation, Harvard University.
Read, B.L. (2003b) "Democratizing the Neighborhood? New Private Housing and Homeowner Self-Organization in Urban China." *The China Journal* 49: 31–59.
Read, B.L. (2007) "Inadvertent Political Reform via Private Associations: Assessing Homeowners' Groups in New Neighborhoods." In E. Perry and M. Goldman (eds.), *Grassroots Political Reform in Contemporary China*. Cambridge, MA: Harvard University Press: pp. 149–173.
Roeder, P.G. (1989) "Modernization and Participation in the Leninist Developmental Strategy." *American Political Science Review* 83: 859–84.
Seo, J. (2002) "Bansanghoe (1976–present) and the Politics of Symbolism in Korea." Paper Delivered at the 2002 Annual Meeting of the American Political Science Association.
Solinger, D.J. (1999) *Contesting Citizenship in Urban China: Peasant Migrants, the State, and the Logic of the Market*. Berkeley: University of California Press.
Stepan, A. (1978) *The State and Society: Peru in Comparative Perspective*. Princeton: Princeton University Press.

Sullivan, J. (1992) *Local Government and Community in Java: An Urban Case-Study*. Singapore: Oxford University Press.

Ts'ai, H.-Y.C. (1990) "One Kind of Control: The 'Hoko' System in Taiwan under Japanese Rule, 1895–1945." Dissertation, Columbia University.

Wade, R. (1992) *Governing the Market: Economic Theory and the Role of Government in East Asian Industrialization*. Princeton: Princeton University Press.

Wang, Y.P., and A. Murie. (1999) *Housing Policy and Practice in China*. New York: St. Martin's Press.

Yao, J. (2002) "Governing the Colonised: Governmentality in the Japanese Colonisation of Taiwan, 1895–1945." Dissertation, University of Essex.

Zhou, M., and J.R. Logan. (1996) "Market Transition and the Commodification of Housing in Urban China." *International Journal of Urban and Regional Research* 20: 400–21.

Subject index

Page numbers in *italics* refer to figures and tables: those with suffix 'n' refer to Notes.

Author index

Note: page numbers in *italics* refer to figures and tables; those with suffix 'n' refer to Notes.

Printed and bound by CPI Group (UK) Ltd, Croydon, CR0 4YY

27/10/2024

14580367-0002